D0903468

DATE DU

THRESHOLDS IN GEOMORPHOLOGY

EDITED BY: **DONALD R. COATES**
State University of New York
at Binghamton

JOHN D. VITEK
Oklahoma State University

Produced by Dowden & Culver Inc.
Stroudsburg, Pennsylvania, USA

for

GEORGE ALLEN & UNWIN
London • Boston • Sydney

To our wives Jeanne and Peggy

First published in 1980

GEORGE ALLEN & UNWIN LTD
40 Museum Street, London WC1A 1LU

© Dowden & Culver, Inc., 1980

British Library Cataloguing in Publication Data

Thresholds in geomorphology.
  1. Geomorphology—Congresses
  I. Coates, Donald R   II. Vitek, John D
  551.4   GB400.2

ISBN 0-04-551033-4

Manufactured in the United States of America.

The geomorphology symposium series conducted by the Department of Geological Sciences, State University of New York at Binghamton, was inaugurated in October 1970 and meetings have been held each fall since then. Each symposium has resulted in a published proceedings volume, and the topics of the nine symposia cover a wide range of subjects as noted in their titles: (1) Environmental Geomorphology, (2) Quantitative Geomorphology, (3) Coastal Geomorphology, (4) Fluvial Geomorphology, (5) Glacial Geomorphology, (6) Theories of Landform Development, (7) Geomorphology and Engineering, (8) Geomorphology of Arid Regions, and (9) Geomorphology Thresholds. Thus, four themes have emphasized geomorphic processes (3, 4, 5, and 6), two have dealt with practical applications of geomorphology (1 and 7), and there has been one volume on methodology (2) and one volume on specific climatic effects (8). It is fitting, therefore, to add to this topical range a theme that is now in the vanguard of new conceptual ideas. We firmly believe that the marshalling of research around the central core of thresholds can provide new and important advances in geomorphology. The establishment of limits and boundary conditions for geomorphic processes and landforms is vital to the growth of the discipline. The ever-changing earth's surface is constantly in a state of flux between equilibrium and disequilibrium forces. Knowledge and prediction of the way these systems operate may well be the new frontier of geomorphology.

Organization for the symposium that gave birth to this book began in September 1976. At that time, Jack Vitek and Ray Frederking proposed the subject of thresholds in geomorphology as a topic for one of the future symposiums in the Binghamton series. This proved to be an intriguing idea and after discussions with many people involved with such concepts and research it was decided to use the topic "Geomorphology Thresholds" as the central theme for the Ninth Annual Geomorphology Symposium. Accordingly, invitations to participate in the meeting were sent to those investigators known to be active in such an area of specialization. In addition, after the symposium title had been announced, many scientists volunteered their services. From this pool of international talent, the symposium speakers were selected. This book, therefore, represents not only those manuscripts presented at the meeting on October 19–21, 1978, and accepted for publication, but several additional papers that the editors and the publisher felt were necessary to provide a timely and comprehensive appraisal of thresholds in geomorphology. This collection of papers contains what we consider to be the cornerstones of a subject that is becoming increasingly important in geomorphology.

Examination of geomorphology texts quickly reveals the lack of universal paradigms. The usual scheme treats geomorphic factors such as climate, weathering, and lithology before the various geomorphic processes are examined as discrete entities. With the exception of the Law of Uniformitarianism, there are no universal geomorphic themes that have been used to organize the discipline into a cohesive pattern. In contrast, the science of chemistry utilizes, for example, many laws that apply to all chemical processes. We believe the idea and conceptual base that thresholds offer can serve as a common denominator to link and unite geomorphic processes and landforms.

The strength and stature of any proceedings volume rests on the expertise of the contributors. We are fortunate to have assembled such an able and widespread group of scientists. Recognition is given not only to our United States colleagues but also to our brethren in Canada, England, and New Zealand. This international flavor provides an important dimension to the book. We were also deliberate in choosing as authors a group of scientists who represent a wide range of interests, and come from many diverse universities and government agencies. This volume, therefore, contains a wealth of ideas from internationally recognized authorities who have shared their talents and made possible this first compilation of significant papers that focus on the status, methodologies, and bright future of thresholds in geomorphology.

The twenty-two chapters written for this book can be conveniently divided into five parts. Such an organization of the material provides a consistent framework that reveals the scope and fabric of geomorphic thresholds.

PART 1. Historical Background. The three chapters that comprise this part set the stage for the volume and provide insight into the importance of recognizing threshold concepts as an integral theme in geomorphology. In the opening chapter, by Coates and Vitek, a generalized overview is given of the status of threshold concepts in geomorphology and other geology disciplines. Although thresholds have played important roles in decision making as a result of such disasters as the St. Francis Dam and Mississippi River floods, increased precision in forecasting potential changes and hazards should be a goal of geomorphology. In Chapter 2, Peltier expands the historical base for the development of geomorphology as a science. He shows that public interest, scientific theory, and technological capability went hand in hand in the progress of the discipline. Finally, several examples of geomorphic thresholds are used by Fairbridge in Chapter 3 to illustrate a connective summation and provide basic threshold definition. Energy transfer systems provide the necessary mechanics for their understanding.

PART 2. Fluvial Landforms. The four chapters in this part evaluate the role that thresholds play in landform development. Discussion starts with erosion at the stream head and progresses through the formation of gullies and valleys, and then to deposition in deltaic environments. In Chapter 4 Kirkby shows that one of the crucial thresholds in the landscape is that associated with stream heads and drainage basins. The stream head represents a changeover in dominance from one process to another, and criteria are developed to test its character. Bradford and Piest analyze gully development in Chapter 5. Thresholds of instability within the processes of gully bank mass wasting are defined from consolidated, drained direct shear tests. The relative strengths of stratigraphic units determine both gully erosion rate and geometry. In Chapter 6 Salisbury evaluates valley morphology and tests statistically the downvalley shapes of stream junctions. Surprisingly, the merging of two streams does not always produce a predictable change in valley form. The concluding chapter, by Roberts, Suhayda, and Coleman, contains a study of the Mississippi River delta and reveals the unusual importance that mass movement processes have in this sedimentary environment. Here slumping and submarine landslide activity are dominant threshold-exceeding events and have contributed immensely to slope configuration throughout the region. Recognition of such processes is vital in offshore oil production activities.

PART 3. Hydrogeologic Regimes. These five chapters provide mathematical frameworks that quantify the threshold concept in surface hydrologic systems. McKerchar in Chapter 8 uses both stochastic and deterministic methods to model catchment response to precipitation. This is continued in Chapter 9, where Rao shows that significant changes in levels of hydrologic time series can define thresholds for this aspect of the water cycle. Karcz emphasizes in Chapter 10 that success of the nonequilibrium theory in analysis of hydro-dynamic instabilities can provide important description and interpretation for the evolution of bedforms and stream patterns. Chapter 11 by Howard then goes on to define the three basic segments of natural streams—bedrock-floored channels, fine-bed alluvial channels, and coarse-bed alluvial channels—and shows that few stream reaches of intermediate characteristics exist. The changes between channel types are attributed to thresholds in the behavior of sediment transport and deposition. Bull in Chapter 12 concludes that a threshold may be regarded as a balance between opposing tendencies and shows that a system of ratios can define this relationship. The numerator contains those variables that operate to produce change, and the denominator includes variables that resist change. When this ratio is 1.0, the system is in a threshold condition.

PART 4. Thresholds in Other Geomorphic Processes. This section of the volume includes analyses of threshold conditions in mass movements, glaciation, river ice, groundwater, and coastal systems. Chapter 13 opens this section with Gardner discussing how rockfalls and rockslides are rapid adjustments to disequilibrium which signal that a threshold intrinsic or extrinsic to the geomorphic system has been exceeded. King provides many examples of thresholds in glacial systems in Chapter 14, and emphasizes the wide range of controls and feedback that operate. Typical thresholds include the growth and decay of glaciers, jökulhlaups, tunnel valleys vs. eskers, drumlins vs. Rogen moraines, and fjords vs. submarine topography. Chapter 15 continues to catalog landform changes in cold climates, as Smith focuses on the four processes of ice drives, ice jams, anchor ice, and channel icings. These river ice forces can be dominant in determination of channel properties in cold regions. There are five threshold limit effects discussed for karst terrains by Ford in Chapter 16: solution processes and mineral solubility; rock control; textural-kinetic control; porosity, permeability, and groundwater circulation; and climatic morphogenesis. For solution processes to dominate a landscape the karst processes must compete with others so as to produce a more effective threshold-exceeding force. In Chapter 17, Hayden, Dolan, and Ross analyze barrier island migration rates of the Atlantic coast Assateague and Core Banks islands and conclude that equilibrium morphologies exist which reflect adjustments to prevailing wave energies, sediment size, and sediment supply. Although the character of the two islands differs significantly, the islands are responding to sea-level rise with nearly identical migration rates.

PART 5. Thresholds and Man. This section shows the importance of thresholds to man, because not only can man's activities provide the stimuli to exceed natural threshold conditions on the earth's surface, but his understanding of thresholds, both natural and man-induced, is vital for proper environmental management. DeGraff and Romesburg in Chapter 18 assess the need for providing landslide-susceptibility evaluation in large areas for wildlands management, and they develop a program for this assessment by the use of the matrix approach. This technique is designed as a planning tool so that priorities in land use can be resolved. An unusual and timely case history is provided by Leighton in Chapter 19. The Bluebird Canyon Landslide, at Laguna Beach, California, on October 2, 1978, was a perfect vehicle for the presentation which was dramatically presented at the Symposium. This vivid portrayal was the ideal example of thresholds and of how man, although previously warned, had not taken advantage of the prior prediction by scientists of the hazardous terrain. The theme that man is creating accelerated erosion in California's deserts is discussed by Wilshire in Chapter 20. Here human impacts are contributing significantly to terrain destabilization by increasing the area directly vulnerable to wind erosion, and increasing the amount of sediment brought by water to the areas of natural wind erosion. Parizek shows in Chapter 21 the wide range of diffuse pollutants that enter hydrologic systems. Their introduction into the system is vastly different within and outside of hydrologically active areas, hence land management practices and structural measures for their control should differ for the two types or regions. It is fitting that Schumm, who originated threshold terminology in geomorphology, should provide the concluding chapter in this volume. In Chapter 22 he shows the types of applications that can result from evaluation of geomorphic thresholds. For instance, they can be used to explain anomalous erosion features and to predict future erosion and deposition changes that will occur as a result of either man's activities or as a normal part of the evolution of landscape.

Perhaps George P. Marsh (in his classic of 1864, *Man and Nature, Physical Geography as Modified by Human Action*) has summed up what we feel is the purpose of this volume:

> In these pages it is my aim to stimulate, not to satisfy curiosity, and it is no part of my object to save my readers the labor of observation or of thought.

Finally, we profusely thank and are indebted to all contributors to the symposium and this volume. This was possible because colleagues were willing to share their ideas with others for the advancement of science. We gratefully acknowledge the travel grant from the National Science Foundation, which aided in attracting speakers from distant places. Although this book represents the end of one phase of the symposium, it can also provide a formal beginning for a new and fundamental concept. Adoption or rejection of the threshold as a methodological strategem rests with all of us. If the symposium and this volume lead to greater understanding of our geomorphic system, such knowledge can be translated into more effective teaching and research, as well as application to man's utilization of planet Earth.

DONALD R. COATES AND JOHN D. VITEK

CONTENTS

CONTRIBUTORS

Joe M. Bradford
U.S. Department of Agriculture, University of Missouri, Columbia (present address: U.S. Department of Agriculture, Purdue University, West Lafayette, Indiana)

William B. Bull
Department of Geosciences, University of Arizona, Tucson

Donald R. Coates
Department of Geological Sciences and Environmental Studies, State University of New York at Binghamton

J. M. Coleman
Coastal Studies Institute, Louisiana State University, Baton Rouge

Jerome V. DeGraff
USDA Forest Service, Fishlake National Forest, Richfield, Utah

Robert Dolan
Department of Environmental Sciences, University of Virginia, Charlottesville

Rhodes W. Fairbridge
Department of Geology, Columbia University, New York

D. C. Ford
Department of Geography, McMaster University, Hamilton, Ontario, Canada

James S. Gardner
Department of Geography, Faculty of Environmental Studies, University of Waterloo, Ontario, Canada

Bruce P. Hayden
Department of Environmental Sciences, University of Virginia, Charlottesville

Alan D. Howard
Department of Environmental Sciences, University of Virginia, Charlottesville

Iaakov Karcz
Department of Geological Sciences and Environmental Studies, State University of New York at Binghamton, and Israel Geological Survey, Jerusalem

Cuchlaine A. M. King
Department of Geography, University of Nottingham, Nottingham, England

M. J. Kirkby
School of Geography, University of Leeds, England

F. Beach Leighton
Leighton & Associates, Inc., Irvine, California

A. I. McKerchar
Ministry of Works and Development, Water and Soil Division, Christchurch, New Zealand

Richard R. Parizek
Department of Geosciences, The Pennsylvania State University, University Park

Louis C. Peltier
Office of Environmental Planning, Montgomery County, Rockville, Maryland

Robert F. Piest
U.S. Department of Agriculture, University of Missouri, Columbia

A. Ramachandra Rao
School of Civil Engineering, Purdue University, West Lafayette, Indiana

H. H. Roberts
Coastal Studies Institute, Louisiana State University, Baton Rouge

H. Charles Romesburg
Department of Forestry and Outdoor Recreation, Utah State University, Logan

Phyllis Ross
Department of Environmental Sciences, University of Virginia, Charlottesville

Neil E. Salisbury
Department of Geography, University of Oklahoma, Norman

S. A. Schumm
Department of Geology, Colorado State University, Fort Collins

Derald G. Smith
Department of Geography, The University of Calgary, Calgary, Alberta, Canada

J. N. Suhayda
Coastal Studies Institute, Louisiana State University, Baton Rouge

John D. Vitek
Department of Geography, Oklahoma State University, Stillwater

Howard G. Wilshire
U.S. Geological Survey, Menlo Park, California

THRESHOLDS IN GEOMORPHOLOGY

I

HISTORICAL BACKGROUND

PERSPECTIVES ON GEOMORPHIC THRESHOLDS

Donald R. Coates and John D. Vitek

INTRODUCTION

Although the idea of critical limits, boundary conditions, and yield points form an important part of other disciplines, similar terminology constitutes only a small part of the literature in geomorphology. In this chapter we will refer to such limiting conditions and tolerance levels as "thresholds." Failure modes are well established in metallurgy, Mohr envelopes of stability and instability are vital components in rock mechanics and structural geology, and melting and freezing points are crucial in a full understanding of igneous petrology. Not only does a nomenclature gap exist in geomorphology, but research on topics that concentrate on imbalance in landscape systems are very sparse. There are several reasons that contribute to this oversight, which we address in this chapter. Moreover, we will identify the place of threshold topics in the mainstream of geomorphology. It is imperative that research generate the conceptual framework and the prerequisite data base to aid in the formation of predictive and planning strategies for proper environmental management.

THREE COMPETING GEOLOGIC DOCTRINES

Two doctrines that have been instrumental in the development of geology have been the antagonistic tenets of catastrophism and uniformitarianism. In our chapter, and throughout this volume, arguments are made that thresholds constitute a fundamental base in geomorphology, and thus form a third doctrine that might serve as an arbiter between the two extreme polar positions. This difference might be epitomized by contrasting a catastrophist who would explain upon viewing the Grand Canyon, "Look how the earth suddenly was rent by some great convulsion and water is now flowing at the bottom of the abyss." However, the uniformist would counter by saying, "No, instead, gradually through the millions of years the Colorado River occupies the very valley it created by erosion of rock formerly in the gorge." Although agreeing that erosion had carved the canyon, the thresholdist would point out that fluvial processes and gravity movements have changed drastically through time. Periods of vast change would be followed by quiescent times of little erosion, and such alterations would mark positions of geomorphic thresholds.

From biblical times and into the 1800s the prevailing doctrine for the explanation of most earth events and differences in animals was to resort to catastrophism and divine intervention. Noah's Flood was a favorite event that explained everything in terms of the Deluge. The products of erosion, hills and mountains, were attributed to the violent turbulence of the waters and storms that accompanied the Flood. (Furthermore, the Flood provided a convenient method to implant fossil remains, and to displace huge ice masses and scatter them on unusual places.) Another favorite "proof" of catstruphism was calculation of the youthfulness of the earth's creation. One method for doing this was analysis of Bible genealogies. Such studies led Archbishop James Ussher in 1650 and Dr. John Lightfoot, vice-chancellor of Cambridge University in 1654, to conclude that the Earth was created at 9:00 A.M. on October 26, 4004 B.C.

With such biblical pronouncements, interpretations of a universal flood, a creation process that was accomplished in six days, and a total time of only a few thousand years to refashion the landscape, it was logical that natural philosophers would believe that catastrophism offered the best explanation for the forces of change. Thus, earth features were not formed gradually, but suddenly. This seemed in accord with observations of the ancients, who lived in a part of the world which was repeatedly subjected to violent earth forces. The Mediterranean region of the Greeks and Romans was constantly afflicted with volcanic eruptions, earthquakes, floods, and landslides. Tales of Atlantis and the destruction of entire cities also helped create the impression that not only was mankind affected by cataclysms but that the earth itself had been sculptured by massive and devastating catastrophes.

Although the erosion of valleys by rivers had been stated in the literature before the eighteenth century, it remained for Sir James Hutton in 1788 and 1795, and his interpreter Sir John Playfair in 1802, to place fluvial erosion into a comprehensive picture. Their works formed the focus for an alternative doctrine of landscape sculpture that was to bear the name of "uniformitarianism." Hutton wrote:

> The natural operations of this globe, by which the size and shape of our land are changed, are so slow as to be altogether imperceptible to men who are employed in pursuing the various occupations of life and literature (1795, II, p. 563).

Such ideas of gradualism and a challenge of prevailing views received many responses, but the severest rebuttal came from the highly respected Baron Georges Cuvier in 1815. Cuvier applied the catastrophism doctrine to both the form and materials of the earth as well as the various animal groups and extinctions.

> Thus the various catastrophes of our planet have not only caused the different parts of our continent to rise by degrees from the basin of the sea, but it has also frequently happened, that lands which have been laid dry have been again covered by the water . . . . These repeated irruptions and retreats of the sea have neither been slow nor gradual: most of the castrophes which have occasioned them have been sudden . . . . Life upon the earth in those times was often overtaken by these frightful occurrences. Living things without number were swept out of existence by catastrophes. Those inhabiting the dry lands were engulfed by deluges, others whose home was in the waters perished when the sea bottom suddenly became dry land: whole races were extinguished leaving mere traces of their existence . . . (Cuvier, 1815, p. 14, 15).

Now that the battle between catastrophists and uniformists was in full bloom, it remained for Sir Charles Lyell to refocus a newly orchestrated doctrine of uniformitarianism. Although Lyell felt that his contribution to the argument provided the principal basis for uniformitarianism, he clearly drew on the works of Hutton, Playfair, and George Poulet Scrope. For example, Scrope had shown in 1825 and 1826 the rather continuous nature of volcanic deposits and that when given sufficient time all lava formations could be accounted for by igneous forces no greater than were operating in his day. In addition, similar observations were said to be true for other aspects of the geological changes on the earth's surface. Lyell wrote in a letter that "I was taught by Buckland the catastrophical or paroxysmal theory . . . ." but not until 1827 did he start to rebel against such ideas which he summarized in his 1830 textbook. He stated

> that all former changes of the organic and inorganic are referrable to one unin-terrupted succession of physical events, governed by the laws now in opera-tion . . . .

Such an all-encompassing uniformity formed the basis for strenuous opposititon not only by contemporary geologists but also in more recent times (Gould, 1965). The inter-pretation that was applied to Lyell's work attacked the methodological limitation that the past should be studied only by analogy with natural agencies operating at the present. Indeed, this was overdrawn, but even Lyell fell into the trap. For example, his strict adherence to such a philosophy blinded his ideas about glaciation, and (unaccustomed to the work of glaciers) to acknowledge that glacial landforms and deposits had been created by glacial processes. He remained steadfast in opposition until the 1850s.

During the 1800s one by one the tenets of the original catastrophe theory were un-dermined: the 6-day earth creation, the 40-day Noah Flood, the 5,800-year age of the earth, the origin of the earth in a primeval diluvium, and divinely ordained cataclysms. When it became clear that moderation was needed, other approaches were used. (Chorley et al., 1964, describe the neocatastrophe diehards.) Schindewolf (1962), however, championed a regeneration of catastrophe ideas. More recently, Dury (1975, 1978) has called attention to the necessity for revitalization of some type of neocatastrophism and a new journal, *Catastrophic Geology*, is being published in Brazil. In addition, some of the neocatastrophism is based, in part, on mathematics and history. Mathematicians, using topology theory, have developed models concerned with discontinuous change caused by the operation of continuous changes in the controlling variables (Wagstaff, 1976). These models and the evidence that the rare event has played an important role in geologic history (Gretetner, 1967) could initiate a reassessment of catastrophism, or the reinter-pretation of such threshold events. Although Gretetner (1967) and Gould (1965) believe that uniformitarianism should be dropped from the vocabulary because rates of geologic activity are not uniform throughout geologic time, they have not proposed catastrophism or thresholds as alternative nomenclature.

We feel it is time to stop the pendulum from swinging crazily from one side of the arc (catastrophism) to the other side (uniformitarianism). The doctrine that can fulfill this moderation is that of thresholds. Within the natural system boundary conditions exist that when exceeded can cause sudden and vast changes. This concept is applicable not only to volcanic activity, earthquakes, landslides, and flooding but to other realms in the sculpturing of the earth's crust. The new geology should embrace the best parts of both ends of the spectrum.

HISTORICAL BACKGROUND

The discipline of geomorphology was largely an outgrowth of physical geography and its derivative of physiography. Cataloging and describing landforms was an early function of geomorphology. Fortunately, a few researchers also became interested in the origin of landforms, which led to the study of the dynamics of processes that were responsible for sculpturing the landscape. William Morris Davis is representative of those pursuing the descriptive direction, whereas Grove Karl Gilbert personified the smaller group dedicated to understanding the mechanics of the physical processes. John Wesley Powell and others occupied the middle ground of those searching for the genesis of features but who did not carry their work into the dynamics of the system. After 1890, and until the 1950s, the Davisian school dominated the thrust of geomorphology with the classification of landscapes. Such emphasis left little room for contributions to the understanding of dynamic processes. Moreover, the entire Davisian approach was rooted with deductions and generalizations, derived largely from intuition. Several weaknesses were implicit in such work:

1. It did not allow for significant experimentation.
2. It was not sufficiently predictable for the short range, and thus was unacceptable for environmental planning or engineering design.
3. It was inbred and unidisciplinary, incapable of growth in terms of thermodynamic principles.

A second idea closely linked with the Davisian evolutionary model of a "geographic cycle" of landscape development, and indeed necessary for its cornerstone, was that of uniformitarianism. All landforms were viewed as gradually changing through time, so that at any given moment only imperceptible modifications were postulated. Such a system was like a calculus of slow, infinitely small events producing miniscule alterations of the earth's surface. This school of thought championed modern adherents, exemplified by Wolman and Miller (1960) and Leopold, Wolman, and Miller (1964). Unfortunately such ideas were extrapolated to being relevant throughout the entire scope of geomorphology, whereas in actuality they treat only one aspect of the hydrologic system. The pendulum representing uniformity has swung too far and its wholesale adoption without qualifications is unwarranted.

A third inhibitant to research on thresholds has been the preoccupation of many geomorphologists with the search for the holy grail of equilibrium systems. Gilbert (1877), in his classic Henry Mountain tome, was the first to espouse dynamic equilibrium ideas which were quantified in his later works (1914). Concepts of balanced systems were expanded and carried in slightly diverse directions, but the common thread uniting such work was the framework of the harmony and equilibria of the land–water ecosystem. Typical of this work was that of Strahler (1950) and Hack (1960) on equilibrium ideas, and Kesseli (1941) and Mackin (1948) on the concept of grade. In the 1960s some new wrinkles were introduced that emphasized extension of these concepts into the realm of steady-state systems, entropy, and quasi-equilibrium (e.g., Chorley, 1962; Leopold and Langbein, 1962; Langbein and Leopold, 1964). Thus the common denominator of such work has been recognition of factors related to the stability of systems. The goals of such studies were the establishment of criteria, arguments, and equations that identified a balanced and harmonious landscape.

The element that has been missing from these paradigms is the failure to analyze the other side of the coin—how to identify, describe, and measure an unbalanced system or one that is on the verge of becoming disharmonious. Such conditions seem to have been assumed or taken for granted, but their scope was not clearly enunciated or investigated. Although some of the rudiments that underlay the concept of what is now called "thresholds" were available in sporadic publications, no concerted or comprehensive analysis of its meaning or appropriate terminology was developed. Schumm (1973) set the stage for the incorporation of thresholds into the mainstream of geomorphology. Subsequently, there has been a rapid growth of interest in the topic, which has culminated in this volume and the symposium that led to its publication.

Why has it taken geomorphologists longer than other scientists to examine thresholds? Observation of geomorphic processes, by necessity, must be completed in the field. Although some processes can be simulated in a laboratory, the complex association among variables affecting processes and resultant forms restricts our powers of observation. As technology improved, our ability to measure processes has evolved to the point that we are capable of quantifying "thresholds," that is, the point or points at which change occurs. As a result, we have a better understanding of (1) the variables involved in a process, (2) relations between variables, and (3) the changes that will occur. Hence, the present is the time for geomorphologists to examine threshold concepts and applications.

Furthermore, as we begin to utilize thresholds, it seems that we are one step from the ultimate geomorphic analysis—examination of energy input, output, and resultant change with respect to energy dissipation in geomorphic processes. Documenting the complex expenditure of energy in any geomorphic process, given the complex interaction of variables, requires powers of observation greater than presently available. Assuming a continuous advancement of technology, such analyses should represent a future goal for geomorphologists. For the present, observing, quantifying, and applying threshold information to environmental problems should provide a useful focus for geomorphologists for years to come.

THRESHOLD TYPES AND DEFINITIONS

The term "threshold" in prior usage related to a piece of stone or timber at the bottom or under a door. It became synonymous with an entranceway or a beginning point. Later usage of the term expanded into medicine, such as a threshold of pain, and into psychology, wherein it denoted a response that had been triggered by a stimulus. Thresholds are described in other disciplines, such as geography (Brunet, 1968), and have been used to denote some landform elements, as "fjord threshold," which is the sill at the distal part of a fjord, separating the glaciated trough from the normal marine subaqueous profile. Schumm (1973) was the first to specifically apply the concept of thresholds to geomorphic systems and to develop a nomenclature. The occurrence of floods (Fig. 1) and landslides (Fig. 2), probably the best known geomorphic thresholds, indicate that a threshold has been exceeded.

Although geomorphic ideas related to thresholds are becoming more common in the literature, as shown in this volume, there are several different variations on the theme and the constituted conceptual base. A unifying thread shared by these ideas is that the forces involved have finally produced a significant change that is manifested within the system.

The variety of different threshold types can be classified into several groupings:

1. *External and internal thresholds.* This division of thresholds is dependent upon whether the acting force(s) operates within the system (intrinsic) or outside the system (extrinsic) (Schumm, 1973). The determining factor is the relation of dependent and in-

(a)

(b)

8

(c)

FIGURE 1 *These photographs are typical of the Mississippi River basin flood of 1973. It reached a flood of record between St. Louis and Cairo and was the greatest flood since 1937 south of Cairo. For 90 days the Mississippi was out of its banks at Vicksburg. Total damages exceeded $1 billion, with 69,000 people made homeless, and 6.7 million hectares of land were inundated. (a) and (b) show flooding of farmlands. The residence time of floodwaters on the land prevented many farmers from planting crops, thus increasing the flood damages. Courtesy Robert Lockridge, J., Corps of Engineers. (c) Severe bank erosion by flood waters of the Red River, endangering the man-made levee. (Courtesy Frederic Chatry, Corps of Engineers.)*

dependent variables. In a watershed, basin size and the lithologic and structural characteristics of the bedrock are intrinsic factors. Even basin relief at any given time can be considered as internal to the system. The external forces that change the landscape elements are linked to climate and elevation change, the latter induced by diastrophic forces (orogeny and epeirogeny). The topographic slopes of the basin are complex geometric forms that result from the interplay of rock type and structure, soil, vegetation types and amount, as well as the drainage density of streams. It is possible for a change in any of these components to initiate change in the other elements of the system. Thus, thresholds can be a function of either external or internal factors.

2. *Process and/or landform thresholds.* A threshold can occur in a process, or as a part of the landform. Moreover, these factors can be juxtaposed whereby the process change causes a landform discontinuity, or the landform shape may induce a process

(a)

(b)

FIGURE 2 *(a) The Gros Ventre Slide, Wyoming, occurred on June 23, 1925. This landslide is classed as a rock avalanche with a slide speed in excess of 160 km/hr that displaced 38,200,000 m³ of rocks. Conditions that contributed to the landslide include strata dipping valleyward, deep melting snows and heavy rains that had saturated the rocks and regolith, and a slight earthquake tremor. (Courtesy Wyoming Travel Commission.) (b) Hurricane Camille did extensive damage to inland eastern United States in 1969. In Virginia extensive flooding and debris avalanches combined to cause damages in excess of $100 million. Along one stream, Davis Creek, 50 lives were lost. This view shows property damage in a small tributary of Davis Creek, where the debris avalanches and flooding combined to devastate the area. (Courtesy Virginia Division of Mineral Resources.)*

change. The magnitude and velocity of water during a flood may produce an entire new topography, such as formation of the Salton Sea, with flooding of the Colorado River. The tendency of many hillslopes to develop a rather predictable hierarchy of thresholds— waxing slope, free face, debris slope, and waning slope—shows the intricate combination of process and landform. Prime examples of landforms causing drastic flow regime differences in water are waterfalls and knickpoints.

3. *Scale of the threshold.* This can be measured in a variety of ways, and indeed the judgment criteria that are adopted may determine whether to label an event or a landform change as a true "threshold." Thus, categorizations are important because even thresholds have limits. For example, how large a change and over how long a period must an event exhibit to be classified as a bonafide threshold? Thresholds can be big or small, with long or short duration. Their impacts may also be lasting or temporal. Thus, what are the degrees that separate a threshold from a nonthreshold event or feature? One determinant is the environmental setting. A stress impingent upon a receptor under one range of circumstances may produce a notable change, whereas the same stress in a different setting may yield only inconsequential reactions. One way out of this dilemma is to use the method of ratios (see Bull, Chapter 12), whereby the forces involved in the system are quantified. A threshold ratio is a value that expresses the relation of the forces acting to create change compared with those forces that operate to resist change. When the derived value exceeds 1.0, a threshold has been reached. This approach is employed in construction projects to determine the safety factor of slope designs. Moreover, this technique is common practice in design and control of material prone to landsliding.

4. *Form of the threshold.* The physical characteristics of the threshold phenomenon can also be used as descriptors for comparison. For example, changes in material behavior are evaluated in terms of elasticity, plasticity, and viscosity. Hydraulic changes include the changes of flow from laminar, to turbulent, to jet, and shooting. When time, landform shape, or gradient is one of the variables, the threshold may be described as a step, ramp, or transient type (see Rao, Chapter 9). A step or jump threshold contains an abrupt change from one level to another, with retention of levels for significant time periods. A ramp threshold contains a change in slope in which the inflection point indicates a new trend and set of slope conditions. A transient threshold occurs when there is a sharp slope departure to a new condition, but an equally rapid return to prethreshold conditions.

5. *Man-induced thresholds.* Although man is part of the environment, it can be of vital concern to determine his role in creating significant change. It is important to separate natural from man-made changes for the simple reason that a planner may be in a better position to control the forces of mankind than the forces of nature. In some instances, as will be discussed later, man may be a major cause of large-scale changes such as the desertification process (Fig. 3).

The types of thresholds could be further divided regarding the special manner in which they apply to each of the geomorphic processes. For example, Bagnold (1941) has quantified many of the relations for movement and entrainment of materials by eolian activity. There is a large engineering literature on stability of slopes under a variety of conditions, and Van Burkalow (1945) collected much experimental data for the angle of repose of granular materials. Indeed, gravity movement and landsliding are especially amenable to treatment with threshold concepts.

Study of the behavior of solid materials is usually the realm of such disciplines as rock mechanics and structural geology. The nomenclature adopted to describe change in

FIGURE 3   *Salt accumulation from improper land use in Western Australia. Such processes can be one step in the desertification of regions. (Photo by Charles Finkl, Jr.)*

solid material includes such terms as yield point, rupture, ultimate strength, and elastic limit. It should be emphasized that changes produced, as when elastic behavior is replaced by plastic behavior, are accomplished by the application of a gradually increasing stress. Such modifications are not dependent upon sudden bursts of new energy. Similarly, when boundary conditions are exceeded, geomorphic processes and landforms undergo changes. These altered conditions may be dramatic or subtle, may have reverberation throughout the entire system, but may have been initiated by only small increments of force. Thus, the triggering point need not be of unusual character. Constructional river terraces have usually been explained by regional uplift or climatic change, but Schumm (1973) has shown that they can be created by intrinsic events associated with full extension of fingertip tributaries to the watershed boundary. Such change, fed back into the system, ushers in a new set of erosion-deposition relations. This reversal in energy conditions, which might be called "topographic fatigue," was caused by the cumulation of many small events. A point is reached in which there is one straw that breaks the camel's back.

RECOGNITION AND PREDICTION

One of the important benefits that emerges from an understanding of thresholds is the practical use of such knowledge for environmental planning. If the advent of a threshold condition can be anticipated and predicted, an important service can be rendered to the land manager who must carefully invest limited funds at locales that are in greatest jeopardy. In such cases an ounce of prevention can be worth a pound of cure, because treating a malady with the possibility of irreversible trends before it has gained full strength can yield high societal benefits. Thus, recognition of likely threshold events and places should be of utmost concern to earth scientists, especially environmental geomorphologists. The determination of threshold phenomena includes the following principles.

1. *Knowledge of the historical record for the area.* There is no substitute for the compilation of long and accurate records regarding historical developments of the phenomenon in question. The precision for predicting is enhanced when there is great regularity and many cases of former threshold events. For example, in coastal planning it is vital to know what points along the shoreline are most fragile and ready for a damaging change (Fig. 4). Towards such a goal G. Giese was asked to provide planning information

FIGURE 4 *New inlets formed in barrier beaches on the south shore of Long Island, New York, from the September 1938 hurricane. This was the first major hurricane to slam into northeastern United States in more than 20 years. (Fairchild photo.)*

for the proper management of eastern Cape Cod beaches by the Chatham Conservation Council. In his report Giese (1978) assembled a series of maps that covered the period from 1772 to the present, from which he prepared drawings that depict coastal configuration at 20-year time intervals. From this presentation he hypothesized that the beach undergoes changes that are cyclic every 150 years. Such a data set was then extrapolated to predict that between 1985 and 1995 the barrier beach at a point known as North Beach opposite Minister's Point would be breached. This coastal incision would widen and become an inlet to the mainland. Of course, this projection, being based on only one previous threshold occurrence, is an overzealous prediction with a large element of conjecture. If, on the other hand, the same locality had repeatedly changed in similar fashion, the exactitude for such a prediction would be enhanced. Thus, it is of paramount importance in threshold prediction to demonstrate that a particular stimulus has repeatedly produced the same response. However, dealing with time-series recurrence intervals for events becomes a matter of statistics and probability (see Rao, Chapter 9).

2. *The use of threshold analogs.* In many areas being developed, the norm is insufficient information about the environment. Often, the only available expedient is to import data developed at another locality and attempt to properly translate it into the new setting. Of course, the fidelity for predicting similar behavior can only be assured if indeed the two areas respond in the same way to geomorphic forces. The following case history (Eckholm, 1976) illustrates the caution that is necessary when performing such extrapolations. In the Congo (now Zaire) Belgian agronomists attempted to utilize European methods of farming by deep plowing, row planting, and crop rotation. They were shocked by the abysmal failure that such methods produced. Instead of the abundant yields predicted, as was the situation in Europe, the same methods caused nearly complete ruination of the African tropical soils. A threshold had been exceeded with great loss in soil nutrients and fertility. Thus, similar methodologies can create entirely different results if the chosen comparison is inappropriate. The similitude for a more perfect analog must be contained in all elements of the system.

3. *The development of an accurate data base.* An observational and experimental amalgam of information is necessary to properly evaluate the degree of stability of an area and its proclivity to undergo a threshold-type change. For example, when dealing with soil productivity, such factors as tolerance levels of erosion, soil salinity (Fig. 3), cropping method, and water use must be known if sustained crop yields are to be maintained throughout the years. Another part of the data bank should deal with the types of symptoms or precursors that might signal the approach of a threshold, in the same manner that changes in dilatancy are used in earthquake prediction. The Dust Bowl years in the 1930s throughout the Great Plains produced thresholds in erosional devastation as well as human suffering and financial loss. The contributing factors involved in these damages were carefully analyzed and certain remedies initiated to prevent such a recurrence. Somewhat similar drought conditions were encountered in the area in the 1950s and mid-1970s. However, the damages were greatly reduced because of the prior planning and protective measures on the lands, and the early recognition of the forces so that additional preventative management could be instituted. Thus, an adequate data base, greatly aided by experimental work and conservation methods, defused what could have been a new set of environmentally destructive thresholds.

THRESHOLD EXAMPLES

Although the literature of geology and geomorphology includes many examples of thresholds, they usually have not been labeled as such. A classic example is the Hjulstrom set of curves, which depict the limiting conditions that determine erosion, transportation, and deposition of sediment as a function of stream velocity and particle size. Diagrams that show the ranges of planimetric shapes of the three main stream types, straight, braided, and meandering, can also be considered as threshold diagrams. The history of the North Platte River illustrates that variations in channel slope or in discharge can determine stream shape. The North Platte River was formerly a braided stream with a mean annual discharge of 2300 cfs (cubic feet per second), but diversions by man have reduced annual flow to 560 cfs and the river is now classified as the meandering type.

Examples that illustrate wherein critical limits are a main component of the concept are the numbers associated with such mathematical formulations as the Reynolds number

FIGURE 5 *The coulees, dry waterfalls, and scalloped terrain are vestiges of what were some of the largest geomorphic threshold events in earth history. Within a matter of hours many cubic kilometers of water roared through this region, carving a bizarre landscape. This process happened several times during the ice ages, when glacial dams were breached by water imponded in lakes 300 m deep and hundreds of kilometers in area. (Photo by Victor Baker.)*

and the Froude number. In the Reynolds number (R) the velocity (V), and depth (d) are related to the viscosity (v) and density (n) so that $R = Vd/(v/n)$. This results in a dimensionless expression, but a generalized meaning is that for numbers less than 500, laminar flow in streams will occur, whereas turbulent flow occurs at higher numbers. The Froude number (F) represents a landform expression of a dimensionless character where bottom channel forms in stream beds with sand deposits are a function of $F = V/\sqrt{gD}$ where V is velocity, g the gravitational force, and D the water depth. Water flow is tranquil when $F < 1$ and a plane bed is produced. However, ripples form when F is 1 or slightly more. With F much greater than 1, dunes form and may ultimately pass into antidunes. Thus, a series of thresholds develop, and when passed produce variations in the channel bottom topography.

In addition to hydrologic thresholds, there are many examples from other areas of geomorphology. King (Chapter 14) has described several thresholds in glacial processes such as glacial surges and jökulhalaups. Landform thresholds are present in the erosional moraine form reversal. Baker has examined the channeled scablands, the most spectacular short-term landscape threshold on the planet (Fig. 5). In Washington and Oregon an area 10,000 km^2 was affected by flow volumes of 40 km^3/hr when an ice dam was breached and drained a 600-m-deep lake that covered about 7500 km^2. In even more grandiose fashion one might argue that the initiation of glacial cycles, and glacial ages, constitutes threshold-creating events.

A glacial age may be considered as a catastrophic event in the context of earth history. The rock record extends back nearly 3.9 billion years, but in all this time only seven or eight major glaciations have occurred, each lasting a few million years. Thus, glaciations are transient thresholds. Various theories have been used to explain the advances and retreats during a single glacial age, and most of these are linked to some type of threshold condition. Ewing and Donn (1956) have theorized thresholds of process with diversion of the Gulf Stream, and of landform, as the Newfoundland sill, which were important triggers in northern hemisphere glacial cycles. Tanner (1965) favored thresholds in the crust and isostatic disruption as causal factors. Wilson (1964) argued that massive changes in the pressure melting point of basal ice caused by the thickness of Antarctic ice were capable of creating extraordinary ice movement on a continental scale. Such theories hold in common that to explain an unusual phenomenon requires there must be highly unusual circumstances.

Topographic thresholds are present in many terrains: waterfalls, cliffs, mountain fronts, etc. Landform vestiges such as wind gaps may represent that a threshold was exceeded and stream diversion occurred. In arid and semiarid regions, above-normal amounts of rain, although they may be cyclic, can be considered as threshold-producing events. The weathered debris that has remained on slopes, and in some channels, for months or even years is suddenly eroded and transported to a new environment, all within a time span that lasted perhaps 1×10^5 as long as its accumulation period. Patton and Baker (1977) point out that the bedrock terrane of the Edwards Plateau, Texas, contains stream channels whose configuration is controlled by catastrophic floods with a recurrence interval of about 400 years—a 1×10^6 event.

Thresholds are, of course, associated with the other geomorphic processes: gravity movement and landslides, groundwater action and sinkholes, coastal processes and overwash, and inlet formation (Fig. 4). Man is the newest geomorphic force capable of producing thresholds, even on a grandiose scale. Although the origin of arroyo denudation in the West is still in dispute, many observers agree that man at least played some role in producing the accelerated erosion in the region. Other contributing causes include changes in storm frequency and intensity, in themselves natural threshold events which may have been influenced by man. There is now an increasing body of knowledge (Glantz, 1977) which indicates that man is a leading contender in the formation of some deserts and in the more recent desertification process (Fig. 3). Bryson and Baerreis (1957) provide strong arguments that the Harrapan civilization in western India and Pakistan was influential in causing increased aridity in deserts of the region. The recent desertification in the Sahel during 1968–73 was enhanced by overgrazing, overcultivation, and deforestation, which produced dramatic changes in the troposphere. The changed albedo, caused by lightening of the earth's surface, and increased presence of atmospheric dust prevented the formation of convective overturn of upper air masses, which was necessary for the development of rain. Although the air contained more moisture during the drought years than in the pre-drought period, man's impact resulted in a tropospheric lid that prevented rainfall.

Man's activities also may provide the trigger to initiate thresholds. Construction of dams with their impounded reservoirs produce an entire array of feedback mechanisms that can be threshold-creating forces. For example, a new erosion cycle is inaugerated downstream because siltation in the reservoir has increased the water power downstream. The eroded debris from the immediate downstream channel is transported to a new locale, where it is deposited and can lead to flooding problems, such as that experienced at Needles, California. Even a new coastal erosion regime can commence. The coastal waters, deprived of the usual land sediments, are now free to expend more energy in beach erosion, such as is happening in California beaches where streams have been dammed. The weight of water in reservoirs can even trigger earthquakes, as evidence indicates at more than 40 places throughout the world.

Man's withdrawal of materials and fluids from the surface and subsurface can also lead to threshold events. Although the gradual subsidence of land by excessive ground-water depletion, as in California and the Houston–Galveston area, Texas, might not be considered appropriate dramatic thresholds, certainly the drastic collapse and sinkhole development caused by dewatering for mining activities in South Africa qualifies as a threshold phenomenon. In the Witwatersrand District man-induced sinkhole activity has caused death and property losses.

In many instances man inadvertently produces stimuli that exceed terrain thresholds (Fig. 6 and 7). The Coastal Plain of Israel was the site for the creation of man-made bad-

FIGURE 6 *Accelerated erosion on Westhampton beach, Long Island, New York. The primary duneline has been completely destroyed along this stretch of beach. The principal cause of erosion here, which is five times the norm, is attributed to a field of groins which occur updrift (and in far background of photo) from this property.*

land topography by two different influences. In one case improper land cultivation by Bedouins was the cause, and in the other construction of the Ottoman Railroad produced new erosion cycles along its route (Aghassy, 1973). The Willow River valley in Iowa offers another example of severe terrain modification. Channelization of the river resulted in a new cycle of stream downcutting, affected tributary drainages, and led to incisement of the main river that exceeded 13 m (Daniels, 1960). A new threshold development is generated with the extensive use of ORVs (off-road vehicles). The largest scale changes

FIGURE 7 *Extensive terrain degradation and landsliding caused by clearcutting of forests, Drift Creek basin, Oregon. Here man-induced activities have been responsible for initiating the disequilibrium in the natural system. (Photo by Fred Swanson.)*

FIGURE 8 *An intensely used hillside by off-road vehicles (ORVs) southwest of Red Rock Canyon, California. The heavy rains of September 1976 have caused extensive changes on this man-devegetated slope, with gullying and deposition of sand at the base of the slope. (Photo by John Nakata.)*

are made in the arid lands of California, where motorcycle use and races affect huge land areas (Fig. 8). In less than two decades, ORVs have destroyed soils in California which equal one-half all lands damaged by mining in the United States during the past 150 years (Howard Wilshire, 1978, pers. comm.).

Geomorphologists can often provide vital information on projects where man plans to tamper with natural processes (Coates, 1976, has termed this field of endeavor "geomorphic engineering"). Kolb (1976) derived the information that is necessary to prevent the formation of sand boils, a major hazard along man-made levees of the Mississippi River (Fig. 9). Appropriate geomorphic studies, therefore, prior to construction activities may

FIGURE 9 *Sand boils and seepage water on a mainline Mississippi River levee near Greenville, Louisiana. The river is in the lower left of the photograph, and seven sand boils have formed as result of high waters from the 1973 flood, exerting pressure on subsurface materials that were in a tenuous state of equilibrium. (Photo by Charles Kolb.)*

reveal potentially critical sites where thresholds are likely to occur. For example, the Corps of Engineers was knowledgeable about possible damage that can result from liquefaction of fine-grained sediments. Since such materials were present in the Cowanesque Valley, Pennsylvania, the site of a proposed dam and reservoir, the area was mapped (Coates, 1966) so that an appropriate engineering design could be made and damage avoided in construction of the dam and adjacent highway. In such cases the knowledge of thresholds was used to prevent a critical event from destroying an anthropogene landscape.

SOCIAL IMPACTS OF THRESHOLDS

Thresholds have played a significant role in the decision-making process. In this context a threshold event has often created a human catastrophe which has prompted governmental action and new directions in environmental management.

1. *Failure of the St. Francis Dam, California.* In 1928, the dam was destroyed and the floodwaters killed 500 people and caused $10 million in property damages. Aroused public officials helped usher in a new era of geological investigations for all future dams in the state, and to expand the importance and involvement of engineering geologists on other projects.

2. *The Dust Bowl period of the 1930s.* A convergence of events led to great havoc in the Great Plains. These factors included the depression, drought, more windstorms than normal, and man's lack of land stewardship. The staggering soil and property losses, and human suffering, catalyzed a new trend in governmental planning and involvement in terrain conservation. The Soil Conservation Service was a direct legacy of such threshold events.

3. *The disastrous 1935–36 floods.* During two adjacent years the Mississippi Basin was exceptionally hard-hit by massive flooding. These events led to passage of the Flood Control Act of 1936, which vested tremendous authority with the Corps of Engineers for the development of projects throughout the Basin. This act was responsible for providing the Corps with an entire new pulpit for action and growth.

4. *The 1952 damaging storms in southern California.* Property losses ran into the millions of dollars from flooding and landsliding caused by exceptional rains, especially in the Los Angeles area. The discovery that much of the loss was from areas that had been improperly developed led to creation of the Grading Ordinances. These were the first laws that specifically mandated that geologists must be involved in decisions for new developments. Thus, the geologic profession received important public exposure and was recognized for the significant contributions it could make in societal affairs.

On a much larger scale, throughout the millennia, mankind has been involved with massive threshold-type activities, often termed revolutions. Although geomorphology has not been involved in all of these, each has ramifications throughout science, and in most there are influences that do affect the land-water ecosystem. Such revolutions completely change the direction of civilization and produce modification in life-styles and settlement patterns. Typical of these are the agricultural revolution starting as early as 4000 B.C. in the Mideast. Others include the industrial revolution in the early 1700s in Europe, and in rather quick succession in the twentieth century the petroleum revolution, urban revolution, communications revolution, and nuclear revolution. All have left, and continue to impress, vital impacts on the density of mankind and the quality of life.

CONCLUSION

In 1973, Schumm suggested that the concepts of geomorphic thresholds and complex response of geomorphic systems are of equal importance to concepts of cyclical dynamic equilibrium in understanding landform genesis over space and time. We agree but go one step further in proposing paradigm status for the concept of geomorphic thresholds. Geomorphology lacks a unifying concept, that is, methodology pertinent to the entire discipline, not simply process specific concepts. Thresholds are present in all geomorphic processes, although knowledge about them in each process varies considerably. Numerous examples in this book of the human interaction with geomorphic thresholds illustrates the importance of the concept. Ignoring thresholds invites tragedies in every aspect of human utilization of the surface.

At the 1975 Binghamton Symposium, Higgins stated that geomorphic theories seem to come in their own cycles of 25 years—Davis (1899), Penck (1953), King (1953)—the time seems right for renewed interest in the synthesis of geomorphic thought. Although important paradigms do not conform to Higgins 25-years cycle—Hack (1960), Chorley (1962), and Wolman and Miller (1960)—we believe that the time is right for the conceptualization of thresholds as a new geomorphic paradigm. Tuttle (1975) stated that "a geomorphic theory is, in a sense, like a garment. We use the one that best fits our data or best serves a particular need or purpose in a given circumstance." Let us don the "garment" of geomorphic thresholds. This requires an evaluation of references with related ideas in order to form a solid foundation for the new paradigm. The 1978 symposium and this book represent the first attempt to provide the foundation for a widely applicable paradigm—geomorphic thresholds.

REFERENCES CITED

Aghassy, J., 1973, Man-induced badlands topography: *in* Coates, D. R., ed., Environmental geomorphology and landscape conservation, vol. III, nonurban regions: Stroudsburg, Pa., Dowden, Hutchinson & Ross, p. 124–136.

Bagnold, R. A., 1941, The physics of blown sand and desert dunes: London, Methuen, 265 p.

Brunet, R., 1968, Les phénomènes de discontinuité en géographie: Centre de Recherches et Documentation Cartographique et Géographique, v. 7, 117 p.

Bryson, R. A., and Baerreis, D. A., 1967, Possibilities of major climatic modifications and their implications: Northwest India, a case for study: Bull. Amer. Met. Soc., v. 48, p. 136–142.

Chorley, R. J., 1962, Geomorphology and general systems theory: U.S. Geol. Survey Prof. Paper 500-B, 10 p.

——, Dunn, A. J., and Beckinsale, R. P., 1964, The history of the study of landforms: volume one: Geomorphology before Davis: London, Methuen, 678 p.

Coates, D. R., 1966, Report on the geomorphology of the Cowanesque Basin, Pennsylvania: U.S. Army Corps of Engineers Cowanesque Reservoir Study, Baltimore, 27+ p.

——, 1976, Geomorphic engineering: *in* Coates, D. R., ed., Geomorphology and engineering: Stroudsburg, Dowden, Hutchinson & Ross, Inc., p. 3–21.

Cuvier, G., 1815, Essay on the theory of the earth (trans. by R. Jameson): 2nd ed., Edinburgh, 322 p.

Daniels, R. B., 1960, Entrenchment of the Willow drainage ditch, Harrison County, Iowa: Amer. J. Sci., v. 258, p. 161–176.

Davis, W. M., 1899, The geographical cycle: Geographical J., v. 14, p. 481–504.

Dury, G. H., 1975, Neocatastrophism?: Anais da Acad. Brasil de Ciencias, v. 47 (Suplmento), p. 135–151.

——, 1978, Neocatastrophism—A Further Look: Program and Abstracts, Ninth Annual Geomorphology Symposium, Binghamton, N.Y., p. 8.

Eckholm, E. P., 1976, Losing ground: New York, W. W. Norton, 223 p.

Ewing, M., and Donn, W. L., 1956, A theory of ice ages: Science, v. 123, p. 1061–1066.

Giese, G., 1978, Report on North Beach, Cape Cod, Massachusetts: The Cape Codder, v. 33, n. 31, 4 p.

Gilbert, G. K., 1877, Report on the geology of the Henry Mountains: 2nd ed., Washington, D.C., Govt. Printing Office, 1880, 170 p.

——, 1914, The transportation of debris by running water: U.S. Geol. Survey Prof. Paper 86, 263 p.

Glantz, M. H., ed., 1977, Desertification: Boulder, Colo., Westview Press, 346 p.

Gould, S. J., 1965, Is uniformitarianism necessary?: Amer. J. Sci., v. 263, p. 223–228.

Gretetner, P. E., 1967, Significance of the rare event in geology: Amer. Assoc. Petrol. Geol. Bull., v. 51, p. 2197–2206.

Hack, J. T., 1960, Interpretation of erosional topography in humid temperate regions: Amer. J. Sci., v. 258-A, p. 80–97.

Hutton, J., 1788, Theory of the earth: Roy. Soc. Edin. Trans., v. 1, p. 209–304.

——, 1795, Theory of the earth: 2 vols., Edinburgh.

Kesseli, J. E., 1941, The concept of the graded river: J. Geol., v. 49, p. 561-588.

King, L. C., 1953, Canons of landscape evolution: Geol. Soc. Amer. Bull, v. 64, p. 721–753.

Kolb, C. C., 1976, Geologic control of sand boils along Mississippi River levees: *in* Coates, D. R., ed., Geomorphology and engineering: Stroudsburg, Pa., Dowden, Hutchinson & Ross, p. 99–113.

Langbein, W. B., and Leopold, L. B., 1964, Quasi-equilibrium states in channel morphology: Amer. J. Sci., v. 262, p. 782–794.

Leopold, L. B., and Langbein, W. B., 1962, The concept of entropy in landscape evolution: U.S. Geol. Survey Prof. Paper 500-A, 20 p.

——, Wolman, M. G., and Miller, J. P., 1964, Fluvial processes in geomorphology: San Francisco, W. H. Freeman, 522 p.

Lyell, C., 1830, Principles of geology: 1st ed., vol. 1, Edinburgh.

Mackin, J. H., 1948, Concept of the graded river: Geol. Soc. Amer. Bull., v. 59, p. 463–512.

Patton, P. C., and Baker, V. R., 1977, Geomorphic response of central Texas stream channels to catastrophic rainfall and runoff: *in* Doehring, D. O., ed., Geomorphology in arid regions: Binghamton, N. Y., State Univ. of New York, Publications in Geomorphology, p. 189–217.

Penck, W., 1953, Morphological analysis of land forms: translated and edited by Czech, H., and Boswell, K. C., London, Macmillan, 429 p.

Schindewolf, O. H., 1962, Neokatastrophismus?: Deutsch Geol. Ges. Zeit., v. 114, p. 430–445.

Schumm, S. A., 1973, Geomorphic thresholds and complex response of drainage systems: *in* Morisawa, M., ed., Fluvial geomorphology: Binghamton, N. Y., State Univ. of New York, Publications in Geomorphology, p. 299–310.

Strahler, A. N., 1950, Equilibrium theory of erosional slopes approached by frequency distribution analysis: Amer. J. Sci., v. 248, p. 673–696, 800–814.

Tanner, W. F., 1965, Cause and development of an ice age: J. Geol., v. 73, p. 413–430.

Tuttle, S. D., 1975, How many peneplains can sit on top of a mountain?: *in* Melhorn, W. N., and Flemal, R. C., Theories of landform development: Binghamton, N. Y., State Univ. of New York, Publications in Geomorphology, p. 299–305.

Van Burkalow, A., 1945, Angle of repose and angle of sliding friction: an experimental study: Geol. Soc. Amer. Bull., v. 56, p. 669–708.

Wagstaff, J. M., 1976, Some thoughts about geography and catastrophe theory: Area, v. 8, p. 316–320.

Wilson, A. T., 1964, Origin of ice ages: an ice shelf theory of Pleistocene glaciation: Nature, v. 20, n. 4915, p. 147–149.

Wolman, M. G., and Miller, J. P., 1960, Magnitude and frequency of geomorphic processes: J. Geol., v. 68, p. 54–74.

2

EVENTS IN THE DEVELOPMENT OF GEOMORPHOLOGY

Louis C. Peltier

INTRODUCTION

This chapter is concerned with land classification, rivers, beaches, soil erosion, slopes, landslides, and subsidence. These components of geomorphology are selected as illustrative of the complex evolution of the science. It is proposed to show that public interest, scientific theory, and technological capability went hand in hand in the progress of the science. Modern geomorphology is the composite product of legislative action, applied mechanics, and the ideas of natural history. Because each of these components is dynamic and subject to change, geomorphology is also dynamic and changing in its directions and emphasis.

Public interest is generally expressed in the executive orders and legislative acts of governments. These actions reflect the dominant interest of the time; they must be interpreted in terms of the political importance of the situation, the policy and objectives of the state at that time, and the degree to which corrective or ameliorative action is then feasible. These actions commonly direct engineering works and supporting research. They serve to stimulate progress and lead to a convergence of interest between geomorphology and engineering. Catastrophic events in populated and technologically advanced areas can lead to legislative action, but these are not the only causes of such an expression of public interest.

Faced with a requirement to design and construct protective and useful works, the engineering profession has found it necessary to explore the physics of geomorphic process. Their findings have been essential in the interpretation of geomorphic features. Of particular importance has been the observation that short-term events of major intensity such as storms, floods, and heavy rains are most significant in creating lasting effects upon the landscape. In the development of reliable designs, it became necessary to recognize that events could occur in different degrees of intensity or magnitude, that events of any class of intensity tend to occur at unique recurrence intervals, and, further, that events of different magnitudes could create quite different effects.

Geomorphology, as a science, has predominantly followed the lines of natural history and has been concerned with the interpretation of landforms in terms of the sequence of events which produced them. Initially, this interpretation was strongly influenced by the facts and concerns of tectonic history. More recently, developments in paleoclimatology and geochronology have provided a more detailed division of time and encouraged analyses to be made with greater precision and detail (Costa, 1978). As one result, there has been somewhat greater attention paid to the effects of "micro-catastrophes."

A common sequence of events in the progress of geomorphology, particularly applied geomorphology, appears to have followed seven steps:

1. Observation of an unusual landform or process.

2. Institution of experimental or temporary engineering measures to correct the situation.

3. Proposal of a theory to explain the situation and process.

4. Development of engineering capability to correct and control the situation.

5. A catastrophic event in an important place accompanied by a demand for amelioration and protection.

6. Legislative action.

7. Accelerated research and refinement of theory.

In the discussion that follows, it will be seen that some catastrophic events led to increased interest and activity, but that others did not. It will also be seen that different places at different times were clearly most active in the development of science, and it is suggested that an overriding requirement for progress is an environment which favors research and publication.

LAND CLASSIFICATION

The dominant concern in the United States over the past 200 years, insofar as geomorphology is concerned, has been with the development, value, and use of land. Its history illustrates a changing emphasis with changing needs, but a concern for land development, including its preservation, has been a dominant factor in public policy.

The initial concern, illustrated by the Land Disposal Act of 1785, was for settlement. This concern persisted through the nineteenth century. A theoretical basis for land development did not appear until later, with the writings of Ricardo (1817) and von Thünen (1826). However, it became recognized that all lands were not equally suited for settlement and that corrective measures should be encouraged. In a series of laws (Swamp and Overflowed Land Act of 1850, Desert Land Act of 1877, Timber Lands Act of 1891, Irrigation Act of 1902, and the Grazing Land Act of 1934) the Congress, by implication, created a land classification system.

In 1892, Hahn (1892) proposed a general system of land classification. This was, presumably, the first general system, although van Aartsen (1944) describes an early system of agricultural land classification proposed by Albrecht Thaer in 1813.

The United States continued to be a predominantly agricultural nation until about 1930. However, as shown by Baker (1921), changes in agricultural practices led physical factors, including geomorphology, to be of increasing importance.

At about the same time, the zoning movement began to become important in the United States through the formulation of a "Standard State Zoning Enabling Act" (1925) and the U.S. Supreme Court support of zoning in the famous case of "Village of Euclid v. Ambler Realty Co." (1926). Land was increasingly to be classified in terms of its desirability for urban development. Land-use planning grew in importance, and the findings of such early studies as those by Garnett (1935), Beard (1948), and MacGregor (1957) were refined into systems of land suitability such as described by Kiefer (1967). The effect of this change has been to bring geomorphology and engineering geology closer together, for the question is no longer merely "What is the landform and how and when was it formed?" but "What is the landform and how can it best be used?" The interpretation of landforms came to be placed in a broader context of social and economic needs.

The National Flood Insurance Act of 1968 and the National Environmental Protection Act of 1969 have imposed new and challenging requirements on geomorphology.

Land must be classified in terms of flood risk. Land must be evaluated in terms of its aesthetic and recreational value as well as of its suitability for construction. One might even look to the future and anticipate that we may be required to evaluate land in terms of exposure to wind damage and air pollution. As a result of recent developments, geomorphology is being required to look much more closely at landforms and their expanding implications.

RIVERS

Rivers and stream floods have long attracted attention. However, the scientific investigation of rivers appears to begin with Torricelli (1643) and to have expanded through the work of Newton (1713), Hutton (1785), and Playfair (1802) in Britain and Coulomb (1800) in France.

In the United States, development of the western territories during the early nineteenth century was dependent upon traffic on the rivers. Concurrently, this was a period of active canal building. Lyell (1837) quoted Hall's "Travels in North America" to describe the driftwood snags in the Mississippi River. The Navigation Act of 1824 and the Harbors and Rivers Act of 1852 addressed the need to remove these snags. The Appropriation Act of 1850 led to the notable report by Humphreys and Abbott (1861), which marks the beginning of the scientific investigations of rivers in the United States.

During the period 1870–87, the major theoretical development took place in Europe, notably by Cotterill (1876), Boussinesq (1877), Thomson (1877), and Reynolds (1887). In the United States this period was marked by the establishment of the Mississippi River Commission (1879) and the Missouri River Commission (1884) and by the Henry Mountain report of Gilbert (1877). The significant Act of 1879 was primarily motivated by the end of the post-Civil War reconstruction and the desire to improve the economy of the South. The objectives were essentially economic and political.

The development of agriculture in California following the gold rush of 1849 ultimately led to a conflict between mining and agriculture (Kelley, 1959). Problems of alluviation of riverine lands arose and the California Debris Commission was established in 1893. Engineering works, described by Wadsworth (1911), were built and research was undertaken which culminated in the reports by Gilbert (1914, 1917).

In about 1905 the Colorado River was diverted into the Salton Sink with the development of the Salton Sea, as described by Grunsky (1907) and Cory (1913). This event had a significant local impact, but the population affected at this time was small.

Congressional interest in the Mississippi River, until early in the twentieth century, was primarily directed toward the support of navigation even though flood control was mentioned in the legislation of this period. It was not until after the flood of 1882 that the Congress appropriated any funds for flood control and levee construction. The floods of 1913 (Brown, 1913), 1927 (Faris, 1927), and 1936, shown in Table 1 occurred in populated areas and attracted particular attention. The assurance given to Congress after the 1927 flood, that such floods could be contained, led to the Mississippi River Flood Control Act of 1927, which included provision for the establishment of the Waterways Experiment Station at Vicksburg, Mississippi. Research using models was undertaken leading to such results as those reported by Friedkin (1945). The Act of 1927 was followed by the Flood Control Act of 1936.

TABLE 1 *Major Natural Stream Floods in the United States, 1913–76*

Mississippi River Basin	Rocky Mountain streams (flash floods)
1913	1972 (Rapid City, S.D.)
1927	1976 (Loveland, Colo.)
1936	Atlantic Coast streams
Pacific Coast streams	1927
1862	1936
1955	1955 (Hurricane Diane)
1964	1972 (Hurricane Agnes)
1969	

Beginning with the experiences of the 1936 flood and continuing over succeeding years, it became increasingly apparent that it was not feasible to control enough streams to prevent all significant floods. The floodplains were becoming increasingly occupied by residences and economic activities. As a result, the concept of floodplain zoning and flood insurance arose, impelled by the floods in New England in 1954. One product of this concern was the National Flood Insurance Act of 1968.

Concurrent with these developments of the early twentieth century were scientific advances in the understanding of the behavior of rivers at various stages of flow, particularly in their ability to scour and transport sediment. These advances are illustrated by the papers of Hill (1905), Jarvis (1924), Leighly (1934), Hjulstrom (1935), Friedkin (1945), Rubey (1952), and Bagnold (1977). In this development most attention has been given to a very few rivers, such as the Mississippi River (Humphreys and Abbott (1861), Colorado River (Kelly, 1924), and the Yellow River (Todd and Eliassen, 1938).

BEACHES

As in the case of stream floods and landslides, coastal inundations have been spectacular and have attracted attention over a long period of time, particularly in the North Sea area. Brooks (1949) suggests that the "Cimbrian flood," which occurred in the North Sea region about 120 B.C. to 114 B.C., led to the migrations of the Cimbri and Teuton peoples. Salmon (1757), Morse (1819), Lyell (1837), and Brooks (1949) mention 20 years in which coastal flooding took place in the North Sea between 1176 and 1825. These are given in Table 2.

Protective sea walls, dikes, and drains were developed at an early time, but it was not until the nineteenth century that theory began to be developed to enable a rational analysis of coastal processes and the resulting morphology. Noteworthy among these were the theoretical works of Poisson (1816), Cauchy (1827), Stokes (1847), Rankine (1863), and Russell (1879). These works on waves and wave motion permitted the concept of wave refraction, developed for optics by Snell in 1621, to be applied to water waves.

In the United States, interest in beaches began about the mid-nineteenth century and followed two separate lines of concern: beach protection and the natural history of beaches. Beach erosion problems become important with the growth of seashore recreation, which began in the New Jersey coastal area as early as 1788 (Quinn, 1977). By the time of the American Civil War (1861–65) seashore recreation had become important. Early reports, such as those of Cook (1857) and Haupt (1890), showed that the coastal

TABLE 2 *Major Coastal Floods of the North Sea Area, 1176–1825*

1176	1421	1655
1218–19	1500	1665
1240	1530	1719
1250–51	1568	1729
1277	1570	1736
1282	1634	1824–25
1300	1649	

zone was dynamic and served, essentially, to define the problem. However, it was not until about 1930 that concerted effort at solution was made.

It was recognized at an early date that it is the severe storms which create the most spectacular effects. Along the Atlantic Coast of the United States the prevailing winds are from offshore, but it is the onshore winds which have the dominant effect on these beaches. Along this coast the major storms are hurricanes (Table 3) and detailed beach morphology reflects these storms. Thus, one is led to consider a philosophy of "micro-catastrophism," in which observed features are to be interpreted in terms of dynamic events separated by intervals of little, if any, change.

The early work on beach geomorphology stems primarily from the works of Gilbert (1885, 1890), Gulliver (1899), and Fenneman (1902) which were largely based upon studies of interior lakes. Shaw's (1913) work on mud lump islands, important in terms of the delineation of territorial seas, is unique and not in context with the trends of its time. This early work was summarized and applied by Johnson (1919, 1925). However, in the absence of the controls of Pleistocene chronology, he gave undue attention to the problem of coastal subsidence.

Effort at the development of measures for shoreline protection, which stimulated considerable activity in the study of shore processes, is interpreted to have begun with Dent (1916). His work was followed, in 1922, by the establishment of the Engineering Advisory Board on Coast Erosion of New Jersey, in 1923 by the establishment of the National Research Council Committee on Shoreline Studies and, in 1926, by the organization of the American Shore and Beach Preservation Association. These organizations were instrumental in the passage of the Shoreline Erosion Investigation Act of 1930 and the Beach Improvement and Protection Act of 1936. Concurrently, the important paper on inlets by Brown (1928) appeared. This paper is noteworthy as one of the first analyses of the hydrology of a tidal estuary.

In 1929, the U.S. Army Corps of Engineers had organized a Board on Sand Movement and Beach Erosion. It was replaced by the Beach Erosion Board authorized by the Act of 1930. Under this board both field studies and model studies, such as reported by Hall (1940) and Mason (1941), were undertaken.

World War II (1942–45), with its extensive requirements for amphibious military operations, had a major impact on the progress of coastal geomorphology. The Beach Erosion Board took the lead in the preparation of intelligence reports in the United States beginning with the Cherbourg to Dunkirk Report in 1942. Similar studies were prepared by England and Germany. A notable product of this period was Dryden's (1944) report on coral reefs.

TABLE 3 *Major Atlantic Coastal Storms of the United States, 1900–75*

Gulf Coast, including Florida			Atlantic Coast north of Florida		
Year	Date	Center	Year	Date	Center
1900	Aug. 27–Sept. 15	Galveston, Texas	1909	Dec. 25	Northeast Coast
1909	Sept. 10–21	New Orleans, La.	1913	Sept. 3	Cape Lookout, N.C.
1915	Aug. 5–23	Galveston, Tex.	1924	Aug. 25–26	Florida–Nova Scotia
1915	Sept. 22–Oct. 1	Burrwood, La.	1933	Aug. 23	Cape Hatteras, N.C.
1919	Sept. 2–15	Sand Key, Fla.	1938	Sept. 10–22	Blue Hills, Mass.
1926	Sept. 11–22	Miami, Fla.	1944	Sept. 9–16	Cape Henry, Va.
1928	Sept, 6–20	Lake Okeechobee, Fla.	1954	Aug. 25–31	Block Id., R.I.
			1954	Sept. 2–14	Block Id., R.I.
1935	Aug. 29–Sept. 10	Tampa, Fla.	1954	Oct. 5–18	New York, N.Y.
1944	Oct. 12–23	Dry Tortugas Is., Fla.	1955	Aug. 3–14	Ft. Macon, N.C.
			1955	Aug. 7–21	Wilmington, N.C.
1947	Sept. 4–21	Hillsboro Light, Fla.	1960	Aug. 29–Sept. 13	Block Id., R.I.
1957	June 25–28	Sabine Pass, Tex.			
1961	Sept. 3–15	Port Lavaca, Tex.			
1964	Aug. 20–Sept. 5	Miami, Fla.			
1964	Aug. 28–Sept. 16	St. Augustine, Fla.			
1965	Aug. 27–Sept. 12	Port Sulphur, Fla.			
1967	Sept. 5–22	Brownsville, Tex.			
1969	Aug. 14–22	Boothville, La.			
1970	July 23–Aug. 5	Corpus Christi, Tex.			
1972	June 14–23	Key West, Fla.			
1975	Sept. 13–24	Ozark, Ala.			

Following World War II, the passage of Public Law 166 (1945) formally established the research activity of the Beach Erosion Board and provided for the study of broadly based problems. In 1949, the first contracts were let to universities for research. Concurrently, the U.S. Navy, Office of Naval Research, actively encouraged research on beaches and coastal geomorphology in support of their responsibilities for amphibious operations.

Research activity increased rapidly both in the United States and elsewhere in the world. This activity is illustrated by the rate of publication for a 5-year period (Schou, 1964) summarized in Table 4. As shown here, 24 countries reported the publication of 1847 books, papers, and reports dealing with coastal geomorphology, or an average of 369 per year. In the United States, greatest attention was given to wave mechanics and coastal sediments. A review of Table 4 shows that, during this period (1959–1963), the number of publications, taken as a measure of activity, is proportional to neither the length of coastline nor the frequency or importance of coastal catastrophes. Instead, it seems to be related most closely to a cultural environment which favors research and publication.

This favorable cultural environment has shifted, with the passage of time, from one country to another. Activity in coastal studies seems to have first centered in France in 1816–45, then shifted to England for the period 1846–85. Activity in the United States

TABLE 4 *Publications in Coastal Geomorphology, 1959–63*

Major Publishing Countries

United States	586	Israel	47
West Germany	176	Belgium	44
Soviet Union	176	Poland	24
France	159	Denmark	24
East Germany	142	Norway	20
Japan	108	Canada	19
United Kingdom	85	Eire	18
Netherlands	57	Finland	16
Sweden	51	Pakistan	11
Australia	49	All others	35

Distribution of Effort Within the United States

Topic	Number	Percent
Waves and tides	102	17.4%
Transport and sedimentation	92	15.7
Engineering design	51	8.7
Beaches	33	5.6
Estuaries and inlets	31	5.2
Storms and wind effects	31	5.2
Currents	30	5.1
Shoreline and sea-level changes	28	4.7
Regional coastal studies	25	4.2
Sea bottom geomorphology	24	4.0
All others	139	23.7

appears to have begun about 1885 and become important about 1930. Since 1945 the United States clearly appears to have been the most active.

SOIL EROSION

The fact of soil erosion had long been recognized, but it was not until agricultural practices had reached a high level of development, as described by Baker (1921), that this process began to receive much attention. In 1911, Glenn had described soil erosion in the Appalachian region and the effect of forest clearing upon the flow of streams. At the same time Free (1911) published his report on deflation. The fact and process of soil erosion were well known long before there was concerted effort to control erosion.

A major effort to control soil erosion came in the 1930s in the context of the conservation movement. It was the direct result of the massive erosion associated with the Dust Bowl of that decade, as reported by Allen (1955), Hagner (1958), and Simms (1970). By this time agricultural land had not only become populated, but now spectacular clouds of airborne dust generated by the dust storm of March 20, 1935, spread across the country to become visible to the members of Congress in Washington. It provided critical support to C. R. Kellogg and other proponents of the Soil Conservation and Domestic Allotment Act of 1935 in which the Soil Conservation Service was established. In accordance with this act particular attention was given to accelerated soil erosion, as had been suggested by Lowdermilk (1934). Research activity increased as illustrated by the reports of Ireland, et al. (1939), Happ, et al. (1940) and Musgrave (1947).

At about this same time interest began to develop in establishing a closer connection between geomorphology and hydrology, as illustrated by the work of Little (1940), Horton (1945), and Strahler (1956). The suggestion was made that a corresponding relationship between geomorphology and climatology, earlier discussed by Penck (1905) and Thorbecke (1927), merited further attention in the distinction between natural and man-made accelerated erosion (Peltier, 1950).

SLOPES

The study of slopes appears to have received its initial impetus as a result of the Franco-Prussian War (1870–1871), during which the German army first made extensive use of maps in planning its strategy. One response was the establishment of the French army's geographical service and the subsequent report of de la Noë and de la Margerie (1888). This interest grew concurrently with the early development of military operations research and the growth of the study of geometric probabilities, as illustrated by Czuber (1902). Thus, the initial interest lay not only in what the landform was, but also in what it meant in terms of using it or performing some activity upon it.

The theoretical basis for interest in slopes developed largely during the 30-year period from 1894 to 1924 and was marked by the writings of Albrecht Penck (1894, 1905), Davis (1898), Marr (1901), and Walther Penck (1924). These and other writers showed that slopes could be interpreted in terms of tectonic and environmental history and therefore had an important place in the interpretation of natural history. Their work led to increased attention to slope characteristics, particularly in Germany.

A second episode of theory development arose about 1950 with the interest in relating landforms to hydrology, already mentioned. This interest led to a much more detailed observation of slopes, as illustrated by Fair (1947), and Strahler (1950).

Presumably as a result of activity during World War II in military terrain analysis and interpretation, the post-World War II episode was also marked by increased interest in the interpretation of slopes for practical purposes. In general, this concern was accompanied by a rejection of the theoretical interpretations of the turn of the century in favor of an empirical and pragmatic approach of the kind which had earlier been suggested by Glock (1932) and others. This empirical approach is illustrated by the writings of Belcher (1948), Garrett (1966), and Kiefer (1967). Most recently, a series of papers compiled by Melhorn and Flemal (1975) suggests a trend toward combining the theoretical, hydrologic, quantitative, and empirical interpretations of slopes.

LANDSLIDES

As in the case of fluvial and coastal floods, landslides are spectacular catastrophes and attract attention. Twenty-three of the better-known major landslides are set forth in Table 5. It was not, however, until the analyses of Coulomb (1776), Rankine (1858), and Boussinesq (1885) that a basis for analysis of landslides was set forth. Heim described Alpine landslides as early as 1881 (Heim, 1932), and Holland (1894) had reported on a spectacular slide in India.

American interest in landslides appears to have arisen from the slides which affected the Canadian Pacific Railway (Stanton, 1897) and from the subsequent slides in Alberta (McConnell and Brock, 1904), Quebec (Ells, 1908), the San Juan Mountains (Howe,

TABLE 5   *Major Landslides, 1246–1972*

Year	Place
1248	Mt. Grenier, Savoy (St. Andre)
1618	Mt. Anto, Chiavenna, Switzerland (Pleurs)
1772	Mt. Piz, Treviso, Venice
1806	Rossberg, Switzerland (Goldau)
1831	Elm, Switzerland
1893	Gohna, India
1903	Frank, Alberta
	Lievre River, Quebec
1908–11	Gros Ventre Mts., Wyoming
1911	Murgab River, Russia
1910–15	Culebra Cut, Panama Canal Zone
1922	Durango, Colorado
1925	Kelly, Wyoming
1936	Juneau, Alaska
1955	Nicolet, Quebec
1959	Madison River, Wyoming-Montana (Hegben)
1962	Mt. Huascaran, Peru
1963	Vaiont Canyon, Italy
1964	Anchorage, Alaska
1966	Aberfan, Wales
1970	Mt. Huascaran, Peru (Yongay)
1971	St. Jean-Vianney, Quebec
1972	Big Sur, California

1909), and Wyoming (Blackwelder, 1912; Branson, 1917). The troublesome slides encountered during the construction of the Panama Canal (MacDonald, 1915) led to renewed engineering interest. However, it was the occurrence of slides in the Sodertalje Canal in Sweden which led to the next important advance in the understanding of landslides: the concept of rotational motion (Frontard, 1922; Fellenius, 1927; Terzaghi, 1936).

SUBSIDENCE

Subsidence, although closely related in mechanics to landslides, has been of concern primarily to miners and particularly to coal miners. A very early description and analysis of mine subsidence near Liège, Belgium, was published by Dumont (1871). In 1885 Fayol presented an interpretation of the process and forms which was introduced into England by Galloway (1897). It provided the prevailing theory for half a century until more refined analyses such as that of Rellensmann (1957) appeared.

In the 1940s subsidence of different origin, first noticed in California, attracted attention (Tolman and Poland, 1940; Gilluly and Grant, 1949; Poland and Davis, 1956; Poland, 1958). This was a subsidence caused by the removal of subsurface fluids and the reduction of pore pressure. Similar conditions have been reported from Texas (Deere, 1961), and Venice, Italy (Lewis and Schrefler, 1978).

In this instance there has not been a great popular concern for the problem. The theoretical base does not provide a clear relationship to other geomorphological theories. Except in the case of karst topography it is a kind of man-induced geomorphology, and geomorphologists have given little attention to subsidence. Orchard (1956) is an exception. However, the recent paper by Jones (1971) suggests that piping and associated subsidence may play a part in the development of drainage systems. Thus, interest in subsidence might increase.

CONCLUSION

The advance of geomorphology has depended upon a need to develop an interpretation of a recognized feature, in a region in which there is a favorable environment for research and publication, and supported by an adequate base in theory. Favorable regions have shifted over time beginning in Italy in the seventeenth century and shifting first to France in the eighteenth and early nineteenth centuries, then to England and Germany in the nineteenth and early twentieth centuries, and finally to the United States in the twentieth century. In a simple case, catastrophic events attract attention, their effects upon people call for amelioration, governments direct corrective action, scientists and engineers increase their studies, and progress results. However, more frequently, the social needs are slowly recognized so that in only a few cases can governmental and scientific action be clearly related to a specific event.

Initially, geomorphology developed along two separate paths: natural history, and geophysics—engineering. With improved methods of a chronology and increased precision of observation, the two approaches have tended to converge. This convergence has been supported by improvements in geomorphic theory and advances in geophysics and paleogeography.

National interest, aroused by national needs, wars, and catastrophic events in important places, has been reflected in legislation. These laws have stimulated research. National needs during the period 1785–1977 have passed through evolutionary changes from (1) an initial concern for land development, including land improvement and the support of navigation, to (2) land protection and preservation, and finally, to (3) land evaluation, determination of optimum use, selective development, and the protection of both the environment and the population. These gradual changes have imposed corresponding requirements for change in attitudes and objectives on the part of the scientific and engineering community. With each change there was an increase in research activity and an accompanying advance in the development of geomorphology. In this way such factors, once scarcely considered to be a part of geomorphology, as mobility, trafficability, construction costs, metabolic costs of movement, visibility, aesthetic value, vulnerability to inundation or other hazard, and the popular perception of value have all become concerns of geomorphology.

As a result, the objectives and scope of geomorphology have been expanded by social demand. It is no longer sufficient to postulate a natural history alone. The emphasis is shifting to a concern for the meaning of landforms in a socioeconomic context.

REFERENCES CITED

Allen, S. W., 1955, Conserving natural resources: principles and practice in a democracy: New York, McGraw-Hill, 347 p.

Bagnold, R. A., 1977, Bed load transport by natural rivers: Water Resources Res. v. 13, n. 2, p. 303–312.

Baker, O. E., 1921, The increasing importance of the physical conditions in determining the utilization of land for agricultural and forest production in the United States: Annals Assoc. Amer. Geographers, v. 11, p. 17–46.

Beard, C. N., 1948, Land forms and land use east of Monterey Bay: Econ. Geogr., v. 24, n. 4, p. 286–295.

Belcher, D. J., 1948, The engineering significance of land forms: Highway Res. Board Bull. 13, November, p. 9–29.

Blackwelder, E., 1912, The Gros Ventre slide, an active earth flow: Geol. Soc. Amer., Bull. v. 23, n. 4, p. 487–492.

Boussinesq, J., 1877, Essai sur la théorie des eaux courantes: Mémoires présentées par divers savants à l'Académie des Sciences de l'Institut National de France, v. 23, n. 1, p. 1–680.

——, 1885, Application des potentiels à l'étude de l'équilibre et du mouvement des solides élastiques: Paris, Gauthier-Villard.

Branson, E. B., 1917, Bull Lake Creek rock slide in the Wind River Mountains of Wyoming: Geol. Soc. Amer. Bull., v. 28, v. 2, p. 347–350.

Brooks, C. E. P., 1949, Climate through the ages: 2nd ed., New York, McGraw-Hill, 395 p.

Brown, E. I., 1928, Inlets on sandy coasts: Proc. Amer. Soc. Civil Eng., v. 54, n. 2, p. 505–553.

Brown, R. E., 1913, The Ohio River flood of 1913: Bull. Amer. Geogr. Soc., n. 45, p. 500–509.

Cauchy, A-L., 1827, Théorie de la propagation des ondes à la surface d'un fluide pesant d'une profondeur indéfinie: Mémoires présentées par divers savants à l'Académie Royale des Sciences de l'Institut de France, v. 1, p. 3–312.

Cook, G. H., 1857, On the subsidence of the land on the seacoast of New Jersey and Long Island: Amer. J. Sci., 2nd ser., v. 24, p. 341–355.

Cory, H. T., 1913, Irrigation and river control in the Colorado River Delta: Transactions, Amer. Soc. Civil Engineers, v. 76, p. 1204–1453.

Costa, J. E., 1978, Holocene stratigraphy in flood frequency analysis: Water Resources Res., v. 14, n. 4, p. 626–632.

Cotterill, J. H., 1876, On the distribution of energy in a mass of liquid in a state of steady motion: Phil. Magazine and J. Sci., v. 1, 5th ser., n. 2, p. 108–111.

Coulomb, C. A., 1776, Essai sur une application des règles des maximus et minimus à quelques problèmes de statique relatifs à l'architecture: Mémoires de l'Académie Royale présentées par divers savants, v. 7 (also included in Hayman, J., 1972, Coulomb's memoire on statics: an essay in the history of civil engineering: New York, Cambridge University Press, 1x + 211 p., on p. 1–40).

——, 1800, Expériences destineés à déterminer la cohérence des fluides et les lois de leurs résistances dans les mouvements très lentes: Mémoires de la classe des Sciences Mathématiques et Physiques de l'Institut, Paris, v. 3, p. 246–305.

Czuber, E., 1902, Probabilités et moyennes géométriques (tr. by H. Schuermans): Paris, v + 244 p.

Davis, W. M., 1898, The grading of mountain slopes: Science, n. s., v. 7, p. 81.

Deere, D. U., 1961, Subsidence due to mining: a case history from the Gulf Coast region of Texas: Proc. Fourth Rock Mechanics Symposium, Penn. State Univ.

de la Noë, G., and de la Margerie, E., 1888, Les formes du terrain: Paris, Service Géographique de l'Armée, 205 p.

Dryden, A. L., 1944, Surface features of coral reefs: U. S. Army Corps of Engineers, Beach Erosion Board, Tech. Memo. 4, May.

Dumont, G., 1871, Des affaisements du sol, produits par l'exploitation houillère: Mémoire adressée à l'administration communale de Liège: Liège, Belgium, L. de Thier, xxxviii + 336 p.

Ells, R. W., 1908, Report on the landslide at Notre-Dame de la Salette, Lièvre River, Quebec, Canada: Dept. Mines, Geol. Survey Branch, Ottawa, n. 1030, p. 3–10.

Fair, T. J. D., 1947, Slope form and development in the interior of Natal, South Africa: Trans. and Proc. Geol. Soc. South Africa, v. 50, p. 105–119.

Faris, R. L., 1927, The Mississippi River problem: The Military Engineer, v. 19, n. 104, p. 110–117.

Fellenius, W., 1927, Erdstatische Berechnungen: Berlin, W. Ernst and Sohn.

Fenneman, N. M., 1902, On the lakes of southeastern Wisconsin: Wisconsin Geological and Natural History Survey, Madison, Wis., xv + 178 p.

Free, E. E., 1911, The movement of soil material by the wind, with a bibliography of eolian geology by S. C. Stuntz, and E. E. Free: U. S. Dept. Agriculture, Soils Bull. 68, 272 p.

Friedkin, J. F., coordinator, 1945, A laboratory study of the meandering of alluvial rivers: War Dept., U. S. Army Corps of Engineers, Mississippi River Commission, U. S. Waterways Experiment Station, Vicksburg, Miss. 40 p.

Frontard, M., 1922, Cycloïdes de glissement de terres: Comptes rendus hebdomadaires des séances de l'Académie des Sciences, Paris, séance de 13 mars, v. 174, p. 740–742.

Galloway, W., 1897, Subsidence caused by the workings in mines: Proc. South Wales Inst. Eng., v. 20, n. 5, p. 304–342.
(essentially republished in the Colliery Guardian and Journal of the Coal and Iron Trades, London, v. 74, p. 569–570, Sept. 24, 1897, and p. 617–618, 620, Oct. 1, 1897).

Garnett, A., 1935, Topography and settlement in the Alps: Geographical Rev. v. 25, n. 4, p. 601–614.

Garrett, E. E., 1966, The parametric versus the quantitative-genetic approach to terrain analysis: Proc. Inst. Environmental Sci., p. 73–80.

Gilbert, G. K., 1877, Report on the geology of the Henry Mountains: U.S. Geographical and Geological Survey of the Rocky Mountain Region (Powell), Washington, D.C.; see especially p. 99–150.

——, 1885, Topographic features of lake shores: U.S. Geol. Survey, 5th Annual Report, Washington D.C., p. 69–123.

——, 1890, Lake Bonneville: U.S. Geol. Survey, Monogr. 1, Washington, D.C. 438 p.

——, 1914, The transportation of debris by running water: U.S. Geol. Survey Prof. Paper 86, Washington D.C., 263 p.

——, 1917, Hydraulic-mining debris in the Sierra Nevada: U.S. Geol. Survey Prof. Paper 105, Washington, D.C., 154 p.

Gilluly, J., and Grant, U. S., 1949, Subsidence in the Long Beach harbor area, California: Geol. Soc. Amer. Bull., v. 60, n. 3, p. 461–530.

Glenn, L. C., 1911, Denudation and erosion in the southern Appalachian region and the Monongahela basin: U.S. Geol. Survey Prof. Paper 72, Washington, D.C., 137 p.

Glock, W. S., 1932, Available relief as a factor of control in the profile of a landform: J. Geol. v. 40, n. 1, p. 74–83.

Grunsky, C. E., 1907, The lower Colorado River and the Salton basin: Trans. Amer. Soc. Civil Engineers, v. 59, December, p. 1–51.

Gulliver, F. P., 1899, Shoreline topography: Proceedings, Amer. Acad. Arts Sci., v. 34, p. 149–258.

Hagner, D. C., 1958, Conservation in America: New York, J. B. Lippincott, 240 p.

Hahn, E., 1892, Die Wirtschaftformen der Erde: Petermanns Mitteilungen, v. 38, p. 8–12.

Hall, W. C., 1940, A model study of the effect of submerged breakwaters on wave action: U.S. Army Corps of Engineers, Beach Erosion Board, Tech. Memo. 1, May 15.

Happ, S. C., Rittenhouse, G., and Dobson, C. E., 1940, Some principles of accelerated stream and valley sedimentation: U.S. Dept. Agriculture, Tech. Bull. 695, Washington, D.C., p. 1–134.

Haupt, L. M., 1890, Littoral movements on the New Jersey coast, with remarks on beach protection and jetty reaction: Trans. Amer. Soc. Civil Engineers, v. 23, September, p. 123–141.

Heim, A., 1932, Bergsturz und Menschenleben: Beiblatt zur Vierteljahrschrift der Naturforschenden Gesellschaft in Zurich, Jahrgang 77, n. 20, p. 1–218.

Hill, E. P., 1905, The velocity of water flowing down a steep slope: Minutes of Proc. Inst. Civil Engineers (England), v. 171, p. 345–349.

Hjulstrom, F., 1935, Studies of the morphological activity of rivers as illustrated by the River Fyris: Bull. Geol. Inst. Uppsala, p. 221–527.

Holland, T. H., 1894, Report on the Gohna landslip, Garhwal: Records of the Geological Survey of India, v. 27, pt. 2, p. 55–65. Geol. Survey of India, Calcutta.

Horton, R. E., 1945, Erosional development of streams and their drainage basins: hydrophysical approach to quantitative morphology: Geol. Soc. Am. Bull. v. 56, n. 3, p. 275–370.

Howe, E., 1909, Landslides in the San Juan Mountains, Colorado: U.S. Geol. Survey Prof. Paper 67, 58 p.

Humphreys, A. A., and Abbott, H. L., 1861, Report upon the physics and hydraulics of the Mississippi River: upon the protection of the alluvial region against overflow: upon the deepening of the mouths: Philadelphia, J. B. Lippincott, xxiii + 456 p.

Hutton, J., 1785, System of the Earth: Royal Society of Edinburgh (reprinted in System of the Earth, 1785, Theory of the Earth, 1788, Observations on Granite, 1794, Together with Playfair's Biography of Hutton xviii + 203 p., Darien, Conn., Hagner Pub. Co., 1970).

Ireland, H. A., Sharpe, C. F. S., and Eargle, D. H., 1939, Principles of gully erosion in the Piedmont of South Carolina: U.S. Dept. Agriculture, Tech. Bull. 633, Washington, D.C., 143 p.

Jarvis, C. S., 1924, Flood flow characteristics: Proceedings, Amer. Soc. Civil Engineers, v. 50, n. 10, p. 1545–1581.

Johnson, D. W., 1919, Shore processes and shoreline development: New York, Wiley, xvii + 584 p.

———, 1925, The New England-Acadian shoreline: New York, Wiley, xx + 608 p.

Jones, A., 1971, Soil piping and stream channel initiation: Water Resources Research, v. 7, n. 3, p. 602–610.

Kelley, R. L., 1959, Gold versus grain, the hydraulic mining controversy in California's Sacramento Valley: Glendale, Calif., A. H. Clark Co., 327 p.

Kelly, W., 1924, The Colorado River problem: Proceedings, Amer. Soc. Civil Engineers, v. 50, n. 6, p. 795–836.

Kiefer, R. W., 1967, Terrain analysis for metropolitan fringe area planning: Proceedings, Amer. Soc. Civil Engineers, v. 93, n. UP-4, p. 119–139.

Leighly, J., 1934, Turbulence and the transportation of rock debris by streams: Geogr. Rev., v. 24, n. 3, p. 453–464.

Lewis, R. W., and Schrefler, B., 1978, A fully coupled consolidation model of the subsidence of Venice: Water Resources Res., v. 14, n. 2, p. 223–230.

Little, J. M., 1940, Erosional topography and erosion: San Francisco, A. Carlisle, 104 p.

Lowdermilk, W. C., 1934, Acceleration of erosion above geologic norms: Trans Amer. Geophys. Union, pt. II, p. 505–509.

Lyell, C., 1837, Principles of geology: being an inquiry how far the former changes of the Earth's surface are referable to causes now in operation: 2 vols., Philadelphia, James Kay, Jun. & Brother.

MacDonald, D. F., 1915, Some engineering problems of the Panama Canal in their relation to geology and topography: U.S. Dept. Interior, Bureau of Mines, Bull. 86, Washington, D.C., G.P.O., 88 p.

MacGregor, D. R., 1957, Some observations on the geographical significance of slopes: Geography, v. 42, p. 167–173.

Marr, J. E., 1901, The origin of moels and their subsequent dissection: Geogr. J., v. 17, n. 1, p. 63–69.

Mason, M. A., 1941, A study of progressive oscillatory waves in water: U.S. Beach Erosion Board, Tech. Rep. 1, Washington, D.C., G.P.O., May, vi + 39 p.

McConnell, R. G., and Brock, R. W., 1904, Report on the great landslide at Frank, Alberta, 1903: Canada Geological Survey, Annual Report, pt. 8, for 1903, Ottawa, Government Printing Bureau.

Melhorn, W. N., and Flemal, R. C., eds., 1975, Theories of landform development: Proceedings of the 6th Annual Geomorphology Symposia, State University of New York at Binghamton, Binghamton, N. Y., vii + 306 p.

Morse, J., 1819, The American universal geography: or a view of the present state of all the kingdoms, states, and colonies in the known world: 2 vols., Charlestown, Mass., Lincoln & Edmands.

Musgrave, G. W., 1947, The quantitative evaluation of factors in water erosion - a first approximation: J. Soil Water Conserv., v. 2, p. 133–138.

Newton, I., 1713, Principia: published in Chittenden, N. W., ed., Newton's Principia: the mathematical principles of natural philosophy by Sir Isaac Newton: 1848, New York, D. Adee, vii + 581 p.

Orchard, R. J., 1956, Surface effects of mining - the main factors: Colliery Guardian, v.

193, n. 4980, p. 159–164.

Peltier, L. C., 1950, The geographic cycle in periglacial regions as it is related to climatic geomorphology: Annals, Assoc. Am. Geographers, v. 40, n. 3, p. 214–236.

Penck, A., 1894, Morphologie der Erdoberfläche: 2 vols. Stuttgart.

——, 1905, Climatic features in the land surface: Amer. J. Sci., 4th ser., v. 19, p. 165–174.

Penck, W., 1924, Die Morphologische Analyse, Ein Kapitel der Physikalischen Geologie: Stuttgart, J. Engelhorn, xx + 283 p.

Playfair, J., 1802, Illustrations of the Huttonian theory of the earth: Edinburgh, W. Creech, xx + 528 p.

Poisson, S. D., 1816, Mémoire sur la théorie des ondes: Mémoires de l'Académie Royale des Sciences de l'Institut de France, v. 1, p. 71–186.

Poland, J. F., 1958, Land subsidence due to ground-water development: Proceedings, Amer. Soc. Civil Engineers, v. 84, n. IR-3, pt. 1, p. 106–119.

——, and Davis, G. H., 1956, Subsidence of the land surface in the Tulare–Wasco (Delano) and Los Banos–Kettleman City Area, San Joaquin Valley, California: Trans. Amer. Geophys. Union, v. 37, p. 287–296.

Quinn, M. L., 1977, History of the Beach Erosion Board, U.S. Army Corps of Engineers 1930–1963: U.S. Army Corps of Engineers, Misc. Rep. 77-9, August.

Rankine, W. J. M., 1858, On the stability of loose earth: Phil. Trans. Roy. Soc. London, v. 147, p. 9–27.

——,1863, On the exact form and motion of waves at and near the surface of deep water: Phil. Trans. Roy. Soc. London, v. 152, p. 127–138.

Rellensmann, O., 1957, Rock mechanics in regard to static loading caused by mining excavating: p. 35-44, *in* Behavior of materials in Earth's crust: Second Symposium on Rock Mechanics, Colorado School of Mines Quart., v. 52, n. 3, 306 p.

Reynolds, O., 1887, On certain laws relating to the regime of rivers and estuaries, and on the possibility of experiments on a small scale: British Association Reports, p. 555–562.

Ricardo, D., 1817, Principles of political economy and taxation: 300 p. reprinted E. P. Dutton, New York, 1912.

Rubey, W. W., 1952, Geology and mineral resources of the Hardin and Brussels quadrangles, Illinois: U.S. Geol. Survey Prof. Paper 218, 179 p.

Russell, J. S., 1879, On the true nature of the wave of translation, and the part it plays in removing the water out of the way of a ship with least resistance: Naval Architects Trans. (England), v. 20, p. 59–84.

Salmon, T., 1757, A new geographical and historical grammar: London, William Johnston, xvi & 640 p.

Schou, A., 1964, International Geographical Union, Commission on Coastal Geomorphology, Bibliography 1959–1963: Kobenhavn, Folia Geographica Danica, v. 10, n. 1, 68 p.

Shaw, E. W., 1913, The mud lumps at the mouths of the Mississippi: U.S. Geol. Survey Prof. Paper 85-B, p. 11–27.

Simms, D. H., 1970, The soil conservation service: New York, Praeger, 238 p.

Stanton, R. B., 1897, The great landslides on the Canadian Pacific Railway in British Columbia: Minutes and Proc. Inst. Civil Engineers (England), v. 132, Dec. 14, p. 1–20.

Stokes, G. G., 1847, On the theory of oscillatory waves: Trans. Cambridge Phil. Soc., v. 8, pt. 4, p. 441–455.

Strahler, A. N., 1950, Equilibrium theory of erosional slopes approached by frequency distribution analysis: Amer. J. Sci., v. 248, n. 10, p. 673–696; v. 248, n. 11, p. 800–814.

———, 1956, Quantitative slope analysis: Geol. Soc. Amer. Bull., v. 67, n. 5, p. 571–596.

Terzaghi, K., 1936, Stability of slopes in natural clay: Proceedings, International Conference on Soil Mechanics, v. 1, Cambridge, Mass., p. 161–165.

Thomson, J., 1877, On the origin of windings of rivers in alluvial plains, with remarks on the flow of water round bends in pipes: Proc. Roy. Soc. London, v. 25, p. 5–8.

Thorbecke, F., ed., 1927, Morphologie der Klimazonen: Düsseldorfer Geogr. Vorträge und Erörterungen, 99 p.

Todd, O. J. and Eliassen, S., 1938, The Yellow River problem: Proc. Amer. Soc. Civil Engineers, v. 64, n. 10, p. 1921–1991.

Tolman, C. F., and Poland, J. F., 1940, Ground-water, salt-water infiltration and ground-surface recession in Santa Clara Valley, Santa Clara County, California: Trans. Amer. Geophys. Union, pt. 1, p. 23–35.

Torricelli, E., 1643, De motu gravium projectorum. (This was Torricelli's best known work. He is also reported to have written on the floods in the Chaiam Valley.)

U.S. Army Corps of Engineers, Beach Erosion Board, Landing area report: Cherbourg to Dunkirk: July, 1942.

U.S. Congress, An ordinance for ascertaining the mode of dispersing of lands in the Western Territory: 28 Jour. Continental Congress 375, 1785 (Land Disposal Act of 1785).

U.S. Congress, An act to enable the State of Arkansas and other States to reclaim the "swamp lands" within their limits (Swamp Lands Act of 1850).
Approved Sept. 28, 1850
Statutes at Large and Treaties of the United States of America, v. 9, p. 519–520, Boston, Charles C. Little & James Brown, 1851.

U.S. Congress, An act making appropriations for the civil and diplomatic expenses of government for the year ending the thirtieth of June, eighteen hundred and fifty one, and for other purposes
Approved September 30, 1850
The Statutes at Large and Treaties of the United States of America, v. 9, p. 523–544, Boston, Little, Brown, 1854.

U.S. Congress, An act making appropriations for the improvement of certain harbors and rivers
Approved August 30, 1852
The Statutes at Large and Treaties of the United States of America, v. 10, p. 56–61, Boston, Little, Brown, 1855.

U.S. Congress, An act to provide for the sale of desert lands in certain States and Territories (Desert Land Act)
Approved March 3, 1877
The Statutes at Large of the United States of America, v. 20, p. 88–89, Washington, D.C., G.P.O., 1879.

U.S. Congress, An act authorizing the citizens of Colorado, Nevada and the Territories to fell and remove timber on the public domain for mining and domestic purposes
Approved June 3, 1878

The Statutes at Large of the United States of America, v. 20, p. 88–89, Washington, D.C., G.P.O., 1879.

U.S. Congress, An act to provide for appointment of a Mississippi River Commission for improvement of said river from the Head of Passes near its mouth to its headwaters. Approved June 28, 1879
The Statutes at Large of the United States of America, v. 21, p. 37–38, Washington, D.C., G.P.O., 1881.

U.S. Congress, An act making appropriations for the construction repair, and preservation of certain public works on rivers and harbors, and for other purposes: (Rivers and Harbors Act of 1888). Received by the President July 31, 1888 and became law without his approval.
The Statutes at Large of the United States of America, v. 25, p. 400–433, Washington, D.C., G.P.O., 1889.

U.S. Congress, An act making appropriations for the construction, repair and preservation of certain public works on rivers and harbors, and for other purposes
Approved Sept. 19, 1890
The Statutes at Large of the United States of America, v. 26, p. 426–465, Washington, D.C., G.P.O., 1890.

U.S. Congress, An act to create the California Debris Commission and regulate hydraulic mining in the State of California
(California Debris Commission Act) Approved March 1, 1893
The Statutes at Large of the United States of America, v. 27, p. 507–511, Washington, D.C., G.P.O., 1893.

U.S.Congress, An act appropriating the receipts from the sale and disposal of public lands in certain States and Territories to the construction of irrigation works for the reclamation of arid lands (Irrigation Act of 1902)
Approved June 17, 1902
The Statutes at Large of the United States of America, v. 32, pt. 1, p. 388–390, Washington, D.C., G.P.O., 1903.

U.S. Congress, An act for the control of floods on the Mississippi River and its tributaries and for other purposes (Flood Control Act of 1928)
Approved May 15, 1928
The Statutes at Large of the United States of America, v. 45, pt. 1, p. 534–539, Washington, D.C., G.P.O., 1929.

U.S. Congress, An act authorizing the construction, repair, and preservation of certain public works on rivers and harbors and for other purposes (Rivers and Harbors Act of 1930; Shoreline Erosion Investigation Act)
Approved July 3, 1930
The Statutes at Large of the United States of America, v. 46 pt. 1, p. 918–949, Washington, D.C., G.P.O., 1931.

U.S. Congress, An act to provide for the protection of land resources against soil erosion and for other purposes (Soil Conservation Act)
Approved April 27, 1935
The Statutes at Large of the United States of America, v. 49, pt. 1, p. 163–164, Washington, D.C., G.P.O., 1936.

U.S. Congress, An act for the improvement and protection of the beaches along the shores of the United States

Approved June 26, 1936
The Statutes at Large of the United States of America, v. 49, pt. 1, p. 1982–1983, Washington, D.C., G.P.O., 1936.

U.S. Congress, An act authorizing the construction of certain public works on rivers and harbors for flood control and for other purposes (Flood Control Act of 1936) Approved June 22, 1936
The Statutes at Large of the United States v. 49, pt. 1, p. 1570–1597, Washington, D.C., G.P.O., 1936.

U.S. Congress, An act authorizing Federal participation in the cost of protecting the shores of publicly owned property (Beach Erosion Act)
Approved August 13, 1946
United States Statutes at Large, vol. 60, pt. 1, p. 1056–1057, Washington, D.C., G.P.O., 1947.

U.S. Congress, An act to make certain changes in the functions of the Beach Erosion Board and the Board of Engineers for Rivers and Harbors, and for other purposes
Approved November 7, 1963
United States Statutes at Large, v. 77, p. 304–305, Washington, D.C., G.P.O., 1964.

U.S. Congress, An act to assist in the provision of housing for low and moderate income families and to extend and amend laws relating to housing and urban development (Housing and Urban Development Act of 1968)
Approved August 1, 1968
United States Statutes at Large, v. 82, p. 476–611, Washington, D.C., G.P.O., 1969.

U.S. Congress, An act to establish a national policy for the environment, to provide for the establishment of a council on environmental quality, and for other purposes (National Environmental Policy Act of 1969)
Approved January 1, 1970
United States Statutes at Large, v. 83, p. 852–856, Washington, D.C., G.P.O., 1970.

U.S. Congress, An act to expand the national flood insurance program by substantially increasing limits of coverage and total amounts of insurance authorized to be outstanding and by requiring known flood-prone communities to participate in the program, and for other purposes (Flood Disaster Protection Act of 1973)
Approved January 2, 1974
United States Statutes at Large, v. 87, p. 975–984, Washington, D.C., G.P.O., 1974.

U.S. Department of Commerce, 1925, Thirteenth annual report of the Secretary of Commerce: Washington, D.C., G.P.O., p. 17–18, 123 p.

U.S. Supreme Court, 1926, Village of Euclid et al. v. Ambler Realty Company
United States Reports, v. 272, October, p. 365–397.

Van Aartsen, J. P., 1944, Land classification in relation to its agricultural value: a review of systems applied: International Review of Agriculture, Monthly Crop Report and Agricultural Statistics, year 35, n. 7, 8, p. 79–98, n. 11, 12, p. 139–166.

Von Thünen, J. H., 1826, Der isolierte Staat in Beziehung auf Landwirtschaft und Nationökonomie: Hamburg, Perthes.

Wadsworth, H. H., 1911, The failure of the Yuba River debris barrier, and the efforts made for its maintenance: Transactions, Amer. Soc. Civil Engineers, v. 71, March, p. 217–227.

3

THRESHOLDS AND ENERGY TRANSFER
IN GEOMORPHOLOGY

Rhodes W. Fairbridge

INTRODUCTION

A threshold in geomorphology or physical geography is an upper limit to some cumulative process, beyond which that particular sequence of events is terminated, and a totally new sequence introduced. One might compare it with the events that ensue if the professor falls asleep in his bath while the water is running; as long as the bath continues to fill, the discharge cycle is temporarily interrupted and the only physical effect is the slightly increased load. But now the water level reaches the rim, overflow takes place, floods the downstairs . . . domestic screams—we need not elaborate. Thus, we see that *a threshold is a boundary condition that separates two distinct but interconnected processes, fed by an identical energy source.*

Examples of threshold or crescendo effects may be recognized in different categories according to energy source and site. Those relating to *tectonic phenomena* are controlled by the thermal energy within the earth, interacting with the gravitative energy of the planetary rotation and its modulation due to surface mass transfer (as by creation of ice sheets) or to the mass attraction changes of solar system (celestial) mechanics. Many instances can be found in the earth's *hydrologic systems*: rainfall and sheet flood, lake filling, fluvial sedimentation, estuarine accumulation, and delta progradation. Related systems, but restricted to low-temperature sites, are the *glacial processes* where the viscosity of the ice is subject to a complex interplay of endogenetic and exogenetic heat sources and the load-pressure constraints. Glacial control of the world's water-ice budget then introduces the phenomenon of *glacio-eustasy*; because of its transfers of mass loading this creates a feedback to the tectonic category via glacio-isostasy, hydro-isostasy, and geodetic changes in the earth's shape (the geoid) and its spin rate. Also related to eustasy are the earth's *climate* by the oceanicity-continentality effects, as well as dependent aspects such as soil moisture, vegetation cover, and denudation potential, which feeds back to the hydrologic-fluvial systems. A third area of eustatic threshold control is in *sedimentation*; shifts in sedimentation sites also modify crustal loading parameters and yet another feedback is instituted. *Geochemical* thresholds in earth history have been considered in another study (Fairbridge, 1967).

Throughout all these systems one may readily detect the common thread through cause:effect:feedback. The threshold is a temporal constraint that permits a Davisian stage designation into *youth-maturity-old age* (see recent discussion: Bloom, 1978, p. 300), which is susceptible to quantitative analysis.

TECTONIC SYSTEMS

Two scales may be considered: megatectonic and neotectonic. On the dimensions of plate tectonics, a "nice question" is posed by the problem of what initiates a subduction. Evidently, subduction is a process that does not follow a steady-state condition. Sea-floor spreading is discontinuous and at rates varying through at least one order of magnitude. Subduction is initiated: it proceeds for periods of the order of some 10^7 yr and then ceases. Two threshold boundary conditions are seen in connection with subduction. First, following the taphrogenic continental rupture there will be a long-term "trailing-edge" sedimentary cycle during which a miogeosynclinal wedge will build up on the continental margin. The floor of this wedge will progressively sink, until it reaches the classical orthogeosynclinal form, and there is a critical point (20–30 km depth and 100–200 m. yr. after rupture) when melting of the floor will start: this is the subduction threshold point.

The second threshold is reached when the oceanic part of the plate is consumed and the one drifting continent collides with another. An orogenic cycle is now initiated; some subduction may continue for a relatively brief episode, but essentially a block has now been created and sea-floor spreading in this particular plate complex comes to an end. Mass displacements on the earth's surface cannot go unrecognized in geodesy. They must affect both the earth's axis and its spin rate. The seeds are now sown for the next taphrogenic cycle. This is the feedback: "the King is dead—long live the King"!

On the neotectonic scale (Pavoni and Green, 1975), modern geodetic surveys detect what are known as "Recent Vertical Crustal Movements" (RVCMs); that is, leveling surveys across the country repeated after periods of 20–50 yr disclose that some areas are systematically rising, while others sink. Up to 75% of the uplifts coincide roughly, but not exactly, with ancient uplifts. Recent surveys of the U.S. East Coast show that there is a 6 mm/yr rise in progress on the Cape Fear Arch, an axis that has been rising, on and off, since the Mesozoic (Holdahl and Morrison, 1974; Brown and Oliver, 1976). On the other hand, another important uplift is detected on the coast of Florida, where there is no such earlier record of uplift. And furthermore, if one converts 6 mm/yr to the million year time frame of plate tectonic calculations, it becomes 6 km/10^6 yr. Evidently, this is a quite unrealistic figure in these geotectonic settings. Geomorphic studies on the Russian shield suggest that comparable RVCMs there do not continue in the same mathematic sign for more than about 500 yr at the most; in short, they fluctuate (Mescherikov, in Fairbridge, 1968, pp 223, 768). Mörner (1976a) has recently proposed that the threshold limit is controlled by core-controlled motions of the earth's gravity field. Periodic runaway in mantle convection has been considered by Anderson and Perkins (1974), and by Rice and Fairbridge (1975). The writer's sea-level studies suggest that there are gravitative pulses that effect the geoid in a mathematically uniform train: in 45–360–1080 yr series that show potential correlations with significant periods in planetary motions and in their gravitational effects on the sun (Hillaire-Marcel and Fairbridge, 1978).

HYDROLOGIC SYSTEMS

When a raindrop falls on a dry sandy hillslope, it is immediately absorbed, a process that will continue until saturation is reached. After this threshold boundary is exceeded, the sheet-flood process will be initiated; after another boundary this will be replaced by rill development, and the stream cycle begins.

Consider now the erosion-sedimentation picture. As is now well understood, in semi-arid regimes, a short, sharp thunderstorm will generate a brief downpour, initiating erosion in one sector, but downstream the water is quickly absorbed into the dry creek bed and redeposition takes place. Rapid aggradation is favored by climatic desiccation ("Huntington's Principle": see Fairbridge, 1968, p. 1125).

The writer learned about this phenomenon firsthand on the Nile (Fairbridge, 1962, 1976). Radiocarbon dating of the 40-m-high alluvial terraces in the middle sector disclosed that they were formed at the peak of the last glacial phase. Africa in general, and Ethiopia in particular, at that time underwent a phase of intense aridity, apparently due to the cooling of the world ocean surface. Rains became highly episodic and the mid-Nile terraces grew higher and higher. Discharge to the Mediterranean, judging from the sediments there, ceased altogether for brief periods. Around 13,500 YBP the rains returned. Lake Victoria, which had become an isolated, carbonate regime in the glacial phase, now overflowed for the first time since the mid-Wisconsinan period. Each of the six cataracts of the middle Nile which had become buried in silt were now successively exhumed, and a deep canyon was cut, probably within a few decades (in the soft, unconsolidated silts). A stream of silt suddenly appeared in the eastern Mediterranean and is recorded even in deep-sea cores.

The Nile seems to be a classic example of a hydrologic threshold. When episodic precipitation and discharge is inadequate to traverse the entire thalweg, terrace building ensues. At the moment the discharge volume exceeds a critical point, terrace erosion and dissection is initiated and the site of redeposition is shifted downstream, eventually to the delta and then out to the deep sea.

A somewhat different model can be recognized in the case of the Spokane Flood, in the state of Washington (Baker, 1973a, 1973b), which was almost contemporaneous with the Nile example. In Columbia River valley, sands and gravels had been filling the canyon in what was probably a highly episodic flow regime during the last glacial phase; it would have been hyperarid except during the melt season. About a 300-m-high terrace fill accumulated, wall to wall, and then around 12,000 BP the famous ice dam of Glacial Lake Missoula broke, cutting a deep swarth, down to bedrock in places, within hours; multiple overflow channels created the extraordinary "scablands." This catastrophic event was evidently one of many such floods, each one a crescendo or threshold example of heroic proportions but no different in principle from the Nile cutting or that of many other "normal" river regimes.

GLACIAL SYSTEMS

According to the glaciologists there are two basically distinct types of glacier, cold, dry-based ones, and wet-based ones associated either with heat-flow warmth or with climatically milder situations (Andrews, 1975). A climatic fluctuation of more than decadal length (with or without critical accumulation) can therefore convert a sluggish, dry-based motion into an accelerated wet-based regime. According to the Wilson hypothesis, a critical point could be reached in the warming of Antarctic ice streams that could create a runaway situation, a "surging glacier" that could suddenly introduce a catastrophic addition of fresh water and volume to the world ocean: a climatic-eustatic consequence of a threshold effect. Even on a small scale, a single mountain glacier can surge forward 10–20 km, within a very few years, with potentially disastrous results for those living down-valley.

EUSTASY

Best known of the eustatic processes is glacio-eustasy, due to creation or melting of glacial ice. Slower and more subtle are tectono-eustatic effects related to plate tectonics, and these will be disregarded for the moment, although in long-term geological time, they are probably greater. Glacio-eustatic changes are fast-acting, often more than 1 cm/yr, continuing for centuries and even millennia. As the ice load is removed from the continental areas, the formerly depressed areas rise, but to a lesser extent the former marginal bulges subside; a critical threshold point on the earth's surface that is diachronous and shifts inwards with progressive melting is this inflection between uplift and subsidence (Walcott, 1975a, b; Mörner, 1976b). Initially, it is clear that eustatic sea-level rise was faster than crustal rebound (e.g., creating the Champlain Sea), but a threshold was reached when the bedrock began to rise isostatically faster than the ocean was rising eustatically. This still is precisely marked on the land surface by the "Marine Limit," at over 300 m elevation near the S.E. Hudson Bay coast, where it is dated at around 8300 YBP; above this trim line the till surface is not wave-washed (Hillaire-Marcel and Fairbridge, 1978).

Other affects of glacio-eustasy include, for example, hydro-isostasy: the increasing water load on the ocean's crustal surface, caused a post-6000-YBP fall of sea level in tropical latitudes that has been calculated at up to 2–4 m (Farrell and Clark, 1976). Cooling of the earth's climate in this same period have been responsible for increasing the size of many glaciers, so that in addition some fraction of the recent fall is glacio-eustatic. At the same time the renewed mass transfer of the water-ice load to high latitudes has increased the earth's spin rate, tending to raise equatorial sea level. The observed net effect is a 3-m fall in MSL, but it is a mixed effect: a eustatic-isostatic-spin rate phenomenon. The change in coastal regime from a transgressive (drowning) trend up till 6000 YBP, to an oscillatory, but slightly regressive trend after that date had a profound effect on all coastal morphology as well as on human and general ecology.

CLIMATIC SYSTEMS

Over both Precambrian and Phanerozoic geologic time, there have been ice ages at around 200–250 m. yr. intervals, each lasting for perhaps 10 million yr and modulated by planetary motions into cycles dominated by a 100,000 yr periodicity. Several interesting threshold phenomena and feedback machanisms are recognizable. During a cooling trend, the polar high-pressure cells tend to expand displacing the prevailing westerlies somewhat toward the equator. Key threshold effects occur when on the land the winter snow fails to melt in summer, thus vastly increasing the albedo and global heat loss, and when at sea the pack ice consolidates, creating another high-albedo area, and inhibiting heat transfer from the ocean. Medium-latitude icesheets are now created, but these in turn tend to set up high-pressure regions, thus, according to Lamb (1972), changing the zonal air flow to a blocking regime. Thus, the loess found in western Europe was blown by easterly winds. When the ice reached its maximum extent, southern Illinois in North America, central Germany in Europe, around 17,000 YBP, a threshold was reached. A cold sea surface, a displaced Gulf Stream, and an ice-covered North Atlantic combined to starve the ice sheets of Canada and Scandinavia. The resulting melt of partly stagnant ice sheets was at a catastrophic rate.

A direct result of ice sheet expansion is glacio-eustatic fall, so that an indirect result is then a change in the *oceanicity-continentality ratio* on the land areas. Changes from continental to maritime dominance were threshold effects that were acutely evident in such places as southern Sweden, when the salt water suddenly entered the Baltic; the varve counting specifies the actual year (c. 9965 B.P. sidereal: Mörner, 1977), although some weak links in the chronology still exist. A similar critical invasion brought salt water through the Champlain Sea to Lake Huron (where whale bones are recorded) around 13,500 YBP.

In subtropical latitudes the postglacial warming brought monsoonal rains. A few seasons of summer rains over the extended dune fields (southern Sahara, northern Kalahari) would be enough to initiate a low-albedo, water-conserving, vegetated savanna regime in place of a high-albedo desert. Again, the threshold point was achieved within a few years, essentially a geological "instant." It has proved all too easy in recent years for the same threshold point to be passed in reverse; a vegetated dune landscape, subjected to a desiccating climate regime and triggered by the overgrazing of domestic herds, can revert to a desert of drifting sand in just a few years.

SEDIMENTATION SYSTEMS

The hydrologic system sees the alternate development of erosive and depositional crescendos. The fluvial deposition is essentially ephemeral. A more enduring resting point is found in the *estuarine environment*. Here, freshwater meets the intrusive saltwater wedge, and on coasts of high wave energy, there is a new transport of sediment *into* the estuarine embayment or river mouth. Nevertheless, an estuary as a physical body of water is also an ephemeral creation (Dyer, 1973). Initially, it is usually created by a eustatic rise, modified perhaps by some structural deformation, but dominated by the 6000 YBP stabilization (±3m) around the present mean sea level. Since that threshold was reached in the drowning process, the open estuaries have progressively begun to fill. Some are still open; others, fed by high sediment supply, have already passed their filled status and are prograding deltas, creating their own estuarine embayments as they head toward the shelf margin (Guilcher, 1963; Walker, 1975). Here is the next threshold point: from here on, much of the sediment goes straight into the heads of submarine canyons and then out into the abyssal fans and plains. Radiocarbon dating of the Mississippi delta discloses a record of distributary growth, fluctuating from side to side, at intervals of around 500–1500 yr. Interpreted as an exclusive sedimentologic-compaction phenomenon (e.g., Morgan, 1970), the writer emphasizes a climatic-eustatic control (Fairbridge, 1968, p. 863).

Inasmuch as most major delta sites are inherited from the failed arm rift of plate-margin triple points, they are also the sites of long-term crustal subsidence, consequent upon the lithospheric cooling that follows the high heat-flow phase of continental rifting.

The sediment deposited in the estuarine or deltaic environment has therefore a high potential for becoming semipermanently stored in its original sedimentological form. This status will continue and the deposits will be subjected to no more than compactional and diagenetic modification for periods of the order of 200 m. yr., when the next threshold point is reached: the deep-burial temperature crescendo that will initiate the next sub-duction cycle—as outlined at the beginning of this essay.

CONCLUSIONS

Thresholds are seen as inevitable signposts in the irreversible course of geologic time. As cycles may be visualized as spiralling around the "Arrow of Time," so the thresholds may be seen as nodal points on those cycles. In some cases the turning point is very gradual and drawn-out; in others, it is passed within hours. Each cyclic phase is seen to be susceptible to a Davisian youth-maturity-old age terminology, and that in turn can provide a quantitative basis for future analysis.

To summarize: in any natural system, a threshold is a turning point or boundary condition that separates two distinct phases of interconnected processes, a dynamic system that is powered by the same energy source. Examples of such systems are provided from tectonics (megatectonics and neotectonics), hydrologic processes, glaciology, eustasy, climatology, and sedimentology. Each is shown to generate feedback to the other systems, and eventually, through geologic periods of the order of 10^8 yr., the last example, sedimentation, feeds back into the first, megatectonics—thus the ultimate threshold.

REFERENCES CITED

Anderson, O. L., and Perkins, P. C., 1974, Runaway temperatures in the asthenosphere resulting from viscous heating: J. Geophys. Res., v. 79, p. 2136–2138.

Andrews, J. T., 1975, Glacial systems: North Scituate, Mass.; Duxbury Press (Wadsworth), 191 p.

Baker, V. R., 1973a, Paleohydrology and sedimentology of Lake Missoula flooding in eastern Washington: Geol. Soc. Amer., Sp. Pap. 144, 79 p.

——— , 1973b, Erosional forms and processes for catastrophic Pleistocene Missoula floods in eastern Washington: in Morisawa, M., ed., Fluvial geomorphology: Binghamton, N.Y., SUNY Publ. Geomorph., p. 123–148.

Bloom, A. L., 1978, Geomorphology: a systematic analysis of late Cenozoic landforms: Englewood Cliffs, N.J., Prentice-Hall, 510 p.

Brown, L. D., and Oliver, J. E., 1976, Vertical crustal movements from leveling data and their relation to geologic structure in the eastern United States: Rev. Geophys. Space Phys., v. 14, p. 13–35.

Damuth, J. E., and Fairbridge, R. W., 1970, Equatorial Atlantic deep-sea arkosic sands and ice-age aridity in tropical South America: Geol. Soc. Amer. Bull., v. 81, p. 189–206.

Dyer, K. R., 1973, Estuaries: a physical introduction: London, Wiley, 140 p.

Fairbridge, R. W., 1962, New radiocarbon dates of Nile sediments: Nature, 196, p. 108–110.

———, 1967, Carbonate rocks and paleoclimatology in the biogeochemical history of the planet: *in* Chilingar, G. V. et al., eds, Carbonate rocks: Amsterdam, Elsevier, p. 399–432.

———, ed., 1968, The encyclopedia of geomorphology: New York, Reinhold (reprint by D.H.R./Academic Press), 1295 p.

———, 1976, Effects of Holocene climatic change on some tropical geomorphic processes: Quat. Res., v. 6, p. 529–556.

Farrell, W. E., and Clark, J., 1976, On post-glacial sea level: Geophys. J., Roy. Astron. Soc., v. 46, p. 647–667.

Guilcher, A., 1963, Estuaries, deltas, shelf, slope: *in* Hill, M. N., ed., The sea, v. 3, New York, Interscience, p. 620–654.

Hillaire-Marcel, C., and Fairbridge, R. W., 1978, Isostasy and eustasy of Hudson Bay: Geology, v. 8, p. 117–122.

Holdahl, S. R., and Morrison, N. L., 1974, Regional investigations of vertical crustal movements in the U.S., using precise relevellings and mareograph data: Tectonophysics, v. 23, p. 373–390.

Lamb, H. H., 1972, Climate: Present, Past, and Future: London, Methuen, v. 1, 613 p. (also v. 2, 1976).

Morgan, J. P., 1970, Deltas - a resume: J. Geol. Educ., v. 18, p. 107–117.

Mörner, N. A., 1976a, Eustasy and geoid changes: J. Geol. v. 84, p. 123–151.

———, 1976b, Eustatic changes during the last 8000 years in view of radiocarbon calibration and new information from the Kattegat region and other northwestern European areas: Pal. Pal. Pal., v. 19, p. 63–85.

———, 1977, Climatic framework of the end of the Pleistocene and the Holocene: paleoclimatic variations during the last 35,000 years: Geogr. Phys. Quat., v. 31, n. 1, 2, p. 23–35.

Pavoni, N., and Green, R., 1975, Recent crustal movements: Amsterdam, Elsevier (also in: Tectonophysics, v. 29), 551 p.

Rice, A., and Fairbridge, R. W., 1975, Thermal runaway in the mantle and neotectonics: Tectonophysics, v. 29, p. 59–72.

Walcott, R. I., 1975a, Past sea levels, eustasy and deformation of the earth: Quat. Res., v. 2, p. 1–14.

———, 1975b, Recent and late Quaternary changes in water level: Eos, v. 56, n. 2, p. 62–72.

Walker, H. J., 1975, Coastal morphology: Soil Science, v. 119, n. 1, p. 3–19.

Wilson, A. T., 1969, The climatic effects of large-scale surges of ice sheets: Can. J. Earth Sci., v. 6, p. 911–918. (cf. Hughes, 1973, J. Geophys. Res., v. 78, p. 7884.)

II

FLUVIAL LANDFORMS

4

THE STREAM HEAD AS A SIGNIFICANT
GEOMORPHIC THRESHOLD

M. J. Kirkby

INTRODUCTION

In any landscape system, there are more processes acting than we can reasonably hope to know about, and certainly more than we can analyze. Theories or models of a system must therefore necessarily neglect many minor processes, and concentrate on the few most dominant. Most existing explanations in geomorphology relate landforms to one, or perhaps two dominant processes. Wherever the dominant process changes, the landforms respond, and generally become qualitatively different in at least some respects. Such changes in dominance are the thresholds of the landscape system.

Geomorphic thresholds may be studied in two crucial ways. The first is the establishment of domains for particular processes, within which that process is dominant. A model or explanation based on that process is then effectively valid within its domain and nowhere outside it. The second important aspect of thresholds is the nature of the transition between domains of dominance. The changeover is commonly progressive, but in several important cases involves an instability which sharpens the transition. This paper reviews the properties of dominance domains and the nature of transitions at thresholds in general terms, before examining the hillslope/channel threshold in more detail.

DOMINANCE DOMAINS

The measure of geomorphic effectiveness is generally sediment transport, which tends to act as a common unit in much the same role as the flow of money in economic systems. Sediment transport in which I include solution removal occupies this central position because it changes the landscape necessarily, and without sediment transport there can be no change. Figure 1a shows a set of three hypothetical processes, which vary as a single environmental control varies (others being held constant). To the left of point A, process 1 is dominant. Process 2 dominates from A to B, and to the right of C; while process 3 dominates between B and C. In practice this type of analysis must be extended to several dimensions corresponding to the dimensionality of the independent controlling factors. From it, dominance domains can be defined for the range of relevant processes, as shown conceptually in Figure 1b for two dimensions. In general for n independent factors, the surfaces defining process rates comparable to Figure 1a are n-dimensional; and their intersections which define the domains are surfaces of dimension $(n - 1)$.

Where the factors X and Y are directional coordinates, the dominance domains form one type of geomorphological map, but the factors are most commonly gross geomorphic controls such as slope gradient, overland flow discharge, soil depth, or stream width. Internal factors such as pH, pF, or C:N ratio are relevant in appropriate cases. A third

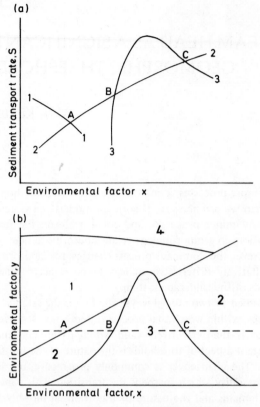

FIGURE 1 *(a) Concept of dominance defined in terms of sediment transport rate and a single environmental factor. (b) Dominance domains defined by two environmental factors.*

important set of factors are constraint factors largely uninfluenced by the system of interest, such as climatic, lithological, or tectonic controls. Figure 2 shows an example, for slope processes, based on simulation model results. In this form, it is closely related to the morphogenetic regions of Peltier (1950), Stoddart (1969), and others. Another familiar set of examples are phase diagrams for ionic dominance, which are commonly expressed in terms of pH and concentrations of parent solid phases. For phase diagrams, the threshold surfaces are linear (on logarithmic scales), so that the domains are polygonal (or polyhedral) in shape. Figure 3 shows the dominant ions for a mixture of SiO_2 and Al_2O_3, with slight curvature of the lines due to the fall in concentration of Al_2O_3 as SiO_2 approaches 100% of the solid mixture.

In the example of Figure 2 there is no feedback from process to climate, so that there is no tendency to migrate across the diagram with time. For the phase diagram of Figure 3, however, removal of ions in solution tends to modify the solid phases toward lower isolines on the diagram. Whether the extreme lowest point, P, is accessible depends on ex-

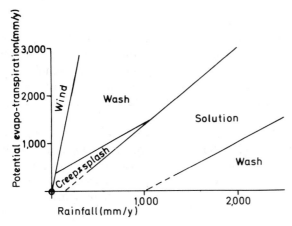

FIGURE 2 *Dominance domains for "slow" slope processes.*

ternal constraints on pH, for example through the availability of soil CO_2. For many phase diagrams, therefore, and in other cases where the controlling factors are themselves influenced by the denudation, there is a convergence toward a lowest point subject to external constraints. This convergence point can be analyzed using methods of linear or nonlinear programming, and will generally be at some point along a threshold between two or more process domains. It is then the balance between these processes which maintains the system in a dynamic equilibrium, so that a single dominant process is then not sufficient to provide an adequate explanation of the resulting landforms.

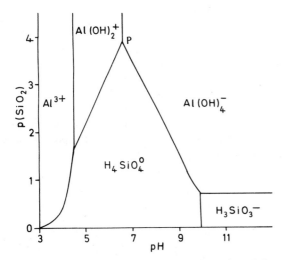

FIGURE 3 *Dominance domains for Al and Si ions for a simple mixture of solid Al_2O_3 and SiO_2.*

TRANSITION THRESHOLDS

In the thresholds and dominance domains described above, there is implicitly a steadily changing process rate as environmental factors change. At a threshold all that happens is that two such process surfaces intersect, and the resulting landforms show differences which are more or less sudden according to the angle at which they intersect. In the neighborhood of the threshold, each surface can usually be approximated by a linear function, so that the linear case has a fair degree of generality.

One of the simplest examples of a threshold is that for mass movements. The safety factor for landslide provides a clear threshold above which movement is rapid, and below it slight. Even such a conceptually simple transition is complicated by the existence of slower slope processes such as wash and soil creep (Fig. 4). The frequency distribution of, for example, soil water provides a second complication which requires that the threshold must be defined, for this example, in terms of hydrological as well as morphological factors.

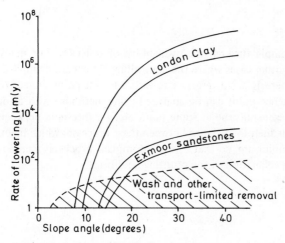

FIGURE 4 *Transitional threshold between rapid mass movement and "slow" slope processes.*

The balance between the slow slope processes changes downslope, but generally produces rather weak thresholds. For example, the ratio of soil creep is roughly proportioned to slope gradient, giving a rate of sediment transport per unit width:

$$S_c \propto -\frac{dz}{dx} \tag{1}$$

where z is elevation and x is distance from the divide. Wash process rates are most simply expressed in terms of distance from the divide, x, which acts as a surrogate for overland flow: and reasonable fits are found for the expression:

$$S_w \propto -x^2 \frac{dz}{dx} \tag{2}$$

Combining these empirical expressions, the total sediment transport

$$S_T \propto -\left(1 + \frac{x^2}{u^2}\right)\frac{dz}{dx} \tag{3}$$

where u is a constant distance which is small for areas where wash is significant, and large when it is not. A threshold occurs when $S_c = S_w$, at distance $x = u$ (Fig. 5a). For the

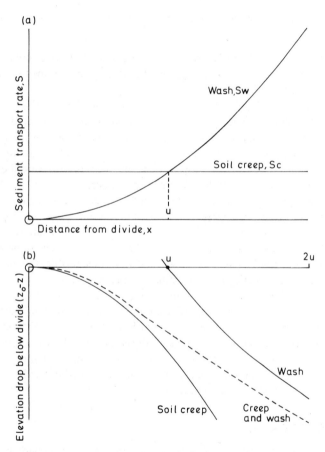

FIGURE 5 *(a) Sediment transport rates for a uniform slope gradient. The transition between creep and wash processes. (b) Forms resulting from transition threshold between creep and wash.*

simplest possible conditions, corresponding to equilibrium with a constant downcutting role, creep alone gives the convex slope profile:

$$(z_0 - z) \propto \tfrac{1}{2}x^2 \tag{4}$$

whereas wash alone gives the concave profile:

$$(z_0 - z) \propto u^2 \ln\frac{x}{u} \tag{5}$$

The combined processes lead to the convex-convexo profile:

$$(z_0 - z) \propto \tfrac{1}{2}u^2 \ln\left(1 + \frac{x^2}{u^2}\right) \tag{6}$$

which reduces asymptotically to the forms above as $x \to 0$ and $x \to \infty$, respectively. Comparison of the three forms in Figure 5b shows however a rather gradual transition in the neighborhood of the threshold at $x = u$.

The transition shown in Figure 5 between creep and wash downslope is less pronounced than that between landslides and wash in Figure 4, and produces a rather gradual transition of form between the two processes. The transitions between the slow processes of wash and solution with rainfall, modeled in Figure 2, are even more gradual, and produce landform and soil transitions that might not merit the name of thresholds were they not a member of this family of transitions between dominant processes.

In all these cases, and in the case of solutional thresholds exemplified by Figure 3, the landforms can readily be seen to differ appreciably across the threshold, even where it is a relatively weak transition. For rapid mass movements, their presence is indicated by a marked scar and toe morphology as well as a tendency to evolve toward a straight slope at the threshold angle. As creep and splash give way to wash down a slope profile, the morphology of the surface layer is diagnostic of the dominant process, at least for a short period after active transport, and the slope profile itself changes from convex to concave near the threshold point. In a similar way, the solution processes illustrated in Figure 3 show convergence on a weathering profile of very different Si:Al ratio, and so different mineralogy in practice, as the profile pH is constrained to different values by the vegetation and drainage conditions, leading to soils which may vary over the range from podsols to laterites. In other words, the landform response is generally more marked in qualitative terms than the threshold between the processes responsible for it.

INSTABILITY THRESHOLDS

For simple transition thresholds the change of process rate with environment can often be approximated as linear, and is more generally a steadily increasing (or decreasing) function. An important set of thresholds are, however, associated with a process rate which shows a pronounced maximum or minimum as a controlling factor changes. This is illustrated in Figure 6 for a one-dimensional case. Instability will tend to set in over one limb of the curve if there is any appreciable feedback from the process rate (S) to the controlling factor (x), and this feedback operates consistently in the same sense.

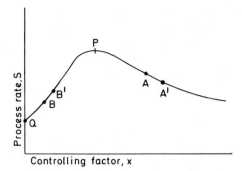

FIGURE 6 *Concept of an instability threshold.*

As an example consider the influence of soil depth (x) on the rate of bedrock weathering (S). The dependence of S on x is generally considered to take the general form shown in Figure 6. It is argued that, under very thin soils, water runs off too quickly to come to equilibrium with soil minerals, so that rock weathering is slow. Under very thick soils most of the rainfall never penetrates deeply, so that again rock weathering is slow. At some intermediate thickness where the residence time is of the same order as the time to equilibrium, weathering is at a maximum. At the same time, there is a feedback from weathering rate to soil thickness, in that a reduced rate of weathering allows the soil to become thicker over the whole range. Consider now point A in Figure 6, representing average conditions at a site. Suppose that a neighboring point (represented by A') has a locally deeper soil. Then its rate of weathering is less than at A, so that conditions at A and A' tend to become more alike. Local irregularities in the soil-bedrock boundary therefore tend to lessen over time, producing equilibrium soils at all points to the right of the maximum P. If a similar argument is applied to points B and B' to the left of P, it can be seen that the higher rate of weathering at B' causes a further increase in soil thickness at B' relative to B, so that the points tend to become less alike. Neighboring sites initially almost identical may therefore diverge to exposed bedrock tors and stable soil layers. The whole of the section PQ is unstable in this way, and soils will not remain in it unless externally constrained to do so.

This particular argument can be generalized to any process curve with a maximum or minimum, where there is a unidirectional feedback from process to controlling factor. According to the directions of the curves, one or other limb of the process curve is inherently unstable. If the feedback from S to x is positive in sense, the rising limb of the process curve produces an unstable positive feedback loop and so is unstable; and if the feedback is negative in sense, the falling limb is unstable. Before looking more closely at examples of this type of instability in geomorphic systems, it is as well to recall that not all peaked process curves give rise to instability. The best known example of such a process curve is perhaps Langbein and Schumn's (1956) relationship between sediment yield and annual rainfall. This produces no instability because there is no feedback from the sediment yield to the climate. It may also be noted that the instabilities, when they occur, do not depend on the peaked shape of the curve. The peak (or trough) guarantees that both stable *and* unstable states occur. Where only unstable states occur, morphological expressions of the process will essentially be absent, so that other processes limit and control the forms. It is the possibility of both stability and instability together which produces forms and thresholds relevant to geomorphology in practice.

To see how these instabilities are expressed at a threshold, the soil example can be pursued further. Suppose for simplicity that weathering takes place in the presence of a constant rate of erosional denudation. Then the rate of soil thickening is the rate of weathering less this constant rate.

Provided that the rate of erosion is not greater than the maximum rate of weathering, then two possible equilibria may develop, at points A and B in Figure 7a. Solving a suitable set of equations for the development of soil thickness over time, however, leads to a solution like that of Figure 7b. It shows that B is an unstable equilibrium, and that points near B will diverge either to zero soil or to the stable equilibrium at A. Points near A will, however, converge stably to A from both directions. At a variety of sites points in the

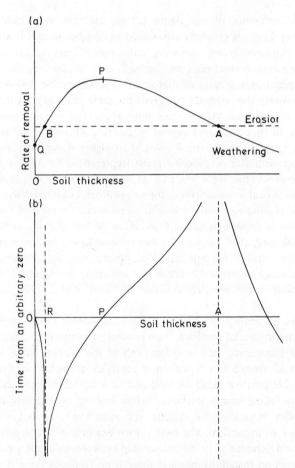

FIGURE 7 *(a) Instability threshold for soil thickness and its rate of weathering. (b) Evolution of soil thickness over time.*

range *PQ* will therefore be very sparsely represented in the landscape morphology, whatever the erosion rate.

The stable zone to the right of *P* is associated with an appreciable and rather uniform soil cover. Its evolution is therefore generally "transport-limited," in that soil mass movement and wash can proceed at its capacity, unconstrained by soil limitations. The unstable zone, *PQ*, is typically subjected to "weathering-limited" removal in which the rate of sediment transport is constrained to the available supply through weathering. The exact position of the changeover from weathering to transport-limited removal is generally within the unstable zone, so that weathering limited removal is mainly associated with bare rock. The instability therefore sharpens the transition into a rather clear dichotomy.

This familiar example of an instability threshold has been used as an introduction to two other important thresholds, which, it is argued here, can be considered in a very similar way—namely, the threshold between meandering and braiding, and that between smooth slopes and valleys.

The threshold between braiding and meandering has been discussed elsewhere by a number of authors (Leopold and Wolman, 1957; Henderson, 1961; Parker, 1976; Schumm and Khan, 1971) from different viewpoints. The present approach has been described more fully elsewhere (Kirkby, 1977), and is included here for comparison with other instability thresholds. The analysis arises from a suitable set of flow and bedload equations, for example the following:

$$Q = wrv = \left[\frac{2g}{f}\right]^{1/2} wr^{3/2} s^{1/2} \qquad (7)$$

for continuity and velocity;

$$f^{-1/2} = 1.77 \ln \frac{r}{d_{84}} + 2.0 \sim \left[\frac{r}{d_{84}}\right]^{1/6} \qquad (8)$$

for Darcy-Weisbach roughness, and

$$S = 8(g\Delta d^3)^{1/2} w \left[\frac{rs}{\Delta d} - 0.047\right]^{3/2} \qquad (9)$$

for bedload transporting capacity;

where Q = water discharge
$\quad S$ = sediment discharge
$\quad s$ = slope
$\quad w$ = channel width
$\quad r$ = depth or hydraulic radius
$\quad v$ = water velocity
$\quad f$ = Darcy-Weisbach friction factor
d_{84} = bed grain size
$\quad d$ = transported grain size
$\quad \Delta$ = ratio of submerged grain to fluid densities
$\quad g$ = gravitational acceleration.

For given values of slope and grain sizes, sediment discharge, S, may be expressed in terms of channel cross section A $(= wr)$ in the form

$$\frac{S}{Q} = \alpha \left(\frac{A}{Q}\right)^{\frac{1}{4}} \left[\beta - \left(\frac{A}{Q}\right)^{3/2}\right]^{3/2} \tag{10}$$

for appropriate constants α and β. This expression is shown qualitatively in Figure 8a, and

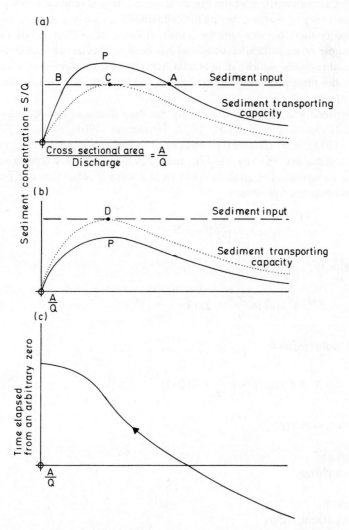

FIGURE 8 *(a) Stream bedload transport as a function of channel cross section on a fixed slope. The dotted curve is for a lower gradient (sinuous), toward which the river will tend. (b) Relationships where sediment input exceeds maximum capacity. (c) Evolution of channel cross section over time for case (b).*

is, of course, very similar to Figure 7a, except that $S = 0$ at $A = 0$. This left-hand part of the curve is, however, strongly dependent on the cross-sectional shape for low width/depth ratios. (As shown, it is correct for triangular cross sections.) The feedback from sediment discharge is positive, in that an increase in it scours the channel and so enlarges its cross section. The section *OP* of the curve is therefore unstable.

If a channel reach is modeled by its dominant discharge, then over a period its transporting capacity must reach an equilibrium with the sediment supplied from its catchment area. If this supply is assumed constant as for the broken line now added to Figure 8a, the evolution of the channel cross section over time will be qualitatively the same as in Figure 7b, with the soil thickness axis relabeled as channel cross-sectional area. Any equilibrium, *B*, to the left of *P* in Figure 8a is therefore unstable, and there will be convergence either to 0 or to the stable equilibrium at *A*.

The morphology associated with convergence to 0 implies the progressive choking of a channel, mainly through lateral deposition. This process must be eventually associated with a change of channel course, and it is argued that the region of instability is associated with a spatial separation into growing and closing channels—that is, with a braided channel pattern. This is a spatial partition into zero and stable forms which corresponds to the tor landscape associated with the soil weathering instability. It is plain, however, that the balance as shown in Figure 8a, with sediment input less than maximum transport capacity, will eventually tend to the stable equilibrium at *A*, so tht the braided pattern is short-lived, with a life which may perhaps be measured in decades (Schumm and Lichty, 1963). On a longer time scale, the equilibrium at *A* will gradually give way to a sinuous meandering course at *C*, corresponding to the peak of the dashed curve in Figure 8a for a lower channel gradient.

For conditions of aggradation, where a contemporary channel cannot cope with its sediment supply, the solid line in Figure 8b, the channel(s) will tend to choke whatever their size (Fig. 8c), giving much more widespread braiding than in the previous example. Braiding is then likely to survive until aggradation of the entire valley floor has built up the gradient sufficiently to allow equilibrium transportation (dashed curve and point *D* in Fig. 8b). The role of major storms in providing unusually large slugs of sediment may in this way initiate temporary braiding.

This analysis has ignored the frequency distribution of both flows and grain sizes. The suggestion above about the role of large storms may imply that highly variable flow regimes encourage braiding, and a similar argument may be applied to highly heterogeneous bed materials. It may be noted, finally, that this model for braiding is consistent with Leopold and Wolman's (1957) empirical discriminant. Analysis of equations 7 to 9 gives the maxima at

$$s \frac{d_{84}}{d} = \frac{0.115 \ln (r/d_{84}) + 0.208}{r/d_{84}} \tag{11}$$

Over reasonable values of r/d_{84}, the right-hand side of this expression behaves like $(r/d_{84})^{-1}$, and substituting the hydraulic geometry exponent for depth in terms of downstream variations in discharge $(r \propto Q^{0.4})$,

$$s \frac{d_{84}}{d} \propto \left(\frac{r}{d_{84}} \right)^{-1} \propto Q^{-0.4} d_{84}$$

or

$$s \propto dQ^{-0.4} \tag{12}$$

and the transported grain size, d, is relatively conservative, so this expression compares well with Leopold and Wolman's:

$$s \propto Q^{-0.44} \tag{13}$$

THE VALLEY-HEAD INSTABILITY THRESHOLD

For slope profiles, instability occurs when small hollows tend to grow in size with a positive feedback, so that they develop into valley heads. Near divides, the profiles are stable, in that small hollows tend to fill in. If a small hollow develops through local influences or external constraints, it concentrates both water and sediment within it. The hollow tends to grow unstably if the additional water can transport more than the additional sediment coming into it. This system has been analyzed by Smith and Bretherton (1972), and examined in the context of hillslope hydrology models (Kirkby, 1977). The instability threshold is analyzed explicitly here as a predictor for drainage density. In Figure 9, two neighboring flow lines are drawn down a hillslope orthogonal to the contours. At a distance x from the divide, the width of the flowstrip is denoted by Ω.

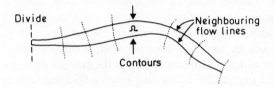

FIGURE 9 *Pair of neighboring flow lines, orthogonal to the contours, which define a flow strip, of varying width, Ω.*

Discharge and hence sediment transport may reasonably be expressed in terms of gradient, s, and area drained per unit contour length, a. The radius of curvature of the contours is denoted by ρ (positive in hollows), and it can be shown that these quantities are related by the geometrical relations:

$$a = \frac{\int_0^x \Omega \, dx'}{\Omega} \tag{14}$$

$$\Omega = \exp\left(\int_0^x -\frac{1}{\rho} dx\right) \tag{15}$$

$$\frac{da}{dx} = 1 + \frac{a}{\rho} \tag{16}$$

For a general (transport-limited) slope sediment process, the rate of transport, S, may be written in the general form

$$S = f(a, s) \tag{17}$$

The mass balance (storage) equation takes the form

$$\frac{\partial z}{\partial t} + \frac{1}{\Omega} \frac{\partial}{\partial x}(\Omega S) = 0 \tag{18}$$

when z is the elevation of the slope surface at time t. Expanding the second term in equation 18 gives, in turn,

$$
\begin{aligned}
-\frac{\partial z}{\partial t} &= \frac{1}{\Omega} \frac{d\Omega}{dx} \cdot S + \frac{da}{dx} \cdot \frac{\partial S}{\partial a} + \frac{\partial S}{\partial s} \cdot \frac{ds}{dx} \\
&= -\frac{1}{\rho} S + \left(1 + \frac{a}{\rho}\right) \frac{\partial S}{\partial a} + \frac{\partial S}{\partial s} \\
&= \frac{1}{\rho}\left(a\frac{\partial S}{\partial a} - S\right) + \left(\frac{\partial S}{\partial a} + \frac{\partial S}{\partial s} \cdot \frac{ds}{dx}\right)
\end{aligned} \tag{19}
$$

in which $\partial S/\partial a$ indicates differentiation at constant slope, etc. The first term on the right-hand side shows the response of the rate of erosion to a small hollow ($1/\rho > 0$). It may be seen that erosion rates will be higher in the hollow if and only if

$$a\frac{\partial S}{\partial a} > S \tag{20}$$

which is then the condition for instability.

If equations 17 and 18 are solved for conditions of equilibrium with constant basal downcutting at rate T, the slope form is given (relative to axes moving down with the slope) by

$$S = f(a, s) = Ta \tag{21}$$

Differentiating with respect to x,

$$\frac{\partial S}{\partial a} \cdot \frac{da}{dx} + \frac{\partial S}{\partial s} \cdot \frac{ds}{dx} = T\frac{da}{dx} \tag{22}$$

This form is concave in profile if and only if ds/dx is negative. Since $\partial S/\partial s$ and da/dx are always positive, a condition for concavity is therefore given by

$$T < \frac{\partial S}{\partial a} \tag{23}$$

Eliminating T from equation 21 gives

$$\frac{S}{a} < \frac{\partial S}{\partial a} \tag{24}$$

This recalls the result proved by Smith and Bretherton (1972, p. 1517) that the slope form is stable to a given process where and only where the constant downcutting form for that process is convex in profile. Where, however, as is common for mature drainage basins, the base of the slope is being reduced at a lower rate than the divides, the threshold of instability necessarily occurs within the concave part of a slope profile. The inequality 24 may be expressed in a third form, which allows a graphical representation. The average rate of downcutting above a given point is given by $\overline{T} = S/a$. Its rate of increase with increasing a (area drained per unit contour length) is

$$\frac{d\overline{T}}{da} = \frac{a(\partial S/\partial a) - S}{a^2} \tag{25}$$

Comparison with equations 20 and 24 shows immediately that the condition for instability is that $\overline{T}$ is increasing with a (keeping slope gradient constant). This leads to the graphical representation of Figure 10, which is drawn for a fixed gradient, and for a combination of creep and wash of the form

$$S \propto \left(1 + \frac{a^2}{u^2}\right) s$$

The graphical argument corresponding to Figure 10 runs as follows. At a point A to the right of the minimum, a local hollow corresponds to a locally higher value of a, which consequently leads to a higher average erosion rate in its catchment strip. This higher erosion rate must lower at least some part of its catchment relative to neighboring (non-hollow) areas. This means that the catchment area will grow, so that a increases. Thus a positive feedback is taking place, leading to instability for all points to the right of the minimum at D. Points to the left of this minimum, such as B, are similarly stable.

For a hillside which is approaching some sort of equilibrium, say with a constant rate of downcutting, the stable zone will give rise to smooth hillsides with little local variation, while the unstable zone will be characterized by some points moving into the stable zone, and others developing hollows and ultimately valleys, the size of which is eventually contained by channel processes and rapid mass movements of some kind. In other words, the hillside is dissected by a channel network.

For a unit length of channel bank, the area drained, a, must be less than the stable value. For the example shown in Figure 10, this corresponds to the value $a = u$ for all slopes, although this constancy is often too simple a view. Summing over the entire channel network of total length L,

$$\Sigma a = A$$

the drainage basin area. At every point $a \leqslant u$, so that

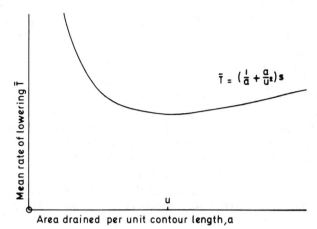

$$\bar{T} = \left(\frac{1}{a} + \frac{a}{u^2}\right)s$$

FIGURE 10 *Variation in mean denudation rate down-slope for a constant slope gradient, for sediment transport $S \propto (1 + a^2/u^2)$ is corresponding to the semiarid case $(M/r_o = 0)$.*

$$A \leqslant 2L \cdot u$$

or

$$DD = \frac{L}{A} \geqslant \frac{1}{2u} \tag{26}$$

Thus, the concept of stable areas can be directly linked to drainage density (DD), and it is thought that for an efficient network the relationship of equation 26 should approach equality. This approach thus provides a way of estimating drainage density, which will now be developed with particular reference to humid areas, for which existing work has not been very effective.

A DRAINAGE DENSITY MODEL

The model developed here is for creep, wash, and solution processes, which are thought to dominate in the determination of drainage density in areas where slopes are below landslide thresholds. Creep and rainsplash is modeled by a linear function of slope:

$$S_c = 10^{-3}s \quad m^2/yr \tag{27}$$

Wash processes are related to slope and overland flow by

$$S_w = 0.02Q_0^2s \quad m^2/yr \tag{28}$$

where Q_0 is the annual overland flow discharge per unit width in m^2/yr.

Solution is linked to subsurface flow, on a linear solubility model which is independent of gradient:

$$S_s = 10^{-6}KQ_s \quad \text{m}^2/\text{yr} \tag{29}$$

where K is an effective solubility in mg/liter and Q_s is the annual subsurface discharge in m^2/yr. Combining these, the total sediment and solute transport

$$S = 10^{-3}s(1 + 20Q_0^2) + 10^{-6}KQ_s \quad \text{m}^2/\text{yr} \tag{30}$$

In order to interpret equation 30 in terms of topographic variation, and so predict stability criteria, an underlying hydrological model is needed. This is built on the notion of an average soil moisture deficit level which is related to average subsurface flow, with overland flow predicted for days where the rainfall exceeds the average deficit. This model may need to be disaggregated into seasonal units, but this has not been done here. Subsurface flow at a point on a slope may be expressed approximately by the relationship

$$Q_s = q_0 \exp\left(-\frac{D}{M}\right)s \tag{31}$$

where q_0 and M are constants of the soil which are taken as invariant over a slope in this model, and D is the moisture deficit at a site. This exponential model is thought to describe Darcian flow in a soil of declining permeability with depth. $v_0 = q_0/M$ is the effective velocity of the surface layer at unit gradient, declining by the ratio e for each increment M of deficit. A more detailed version of this model has been explored fully for short-term hydrograph predictions (Beven and Kirkby, 1979). The average deficit is obtained for equilibrium with the net rate of runoff:

$$i = (R - E) \quad \text{m}/\text{yr} \tag{32}$$

where R is the annual rainfall and E the annual actual evapotranspiration (here assumed constant over the slope). At equilibrium

$$Q_s + Q_o = ia \tag{33}$$

so that

$$Q_s = ia - Q_o = q_0 e^{-D/M}s \tag{34}$$

and

$$D = M \ln\frac{q_0 s}{ia - Q_o} \tag{35}$$

Daily rainfalls approximate to the distribution

$$N(r) = N_e^{-r/r_0} \tag{36}$$

where $N(r)$ = number of days with rainfall $\geqslant r$
$\quad N = N(0)$ = number of rain-days
$\quad r_0 = R/N$ = mean rain per rain day

Summing overland flows produced on days for which $r > D$, the total overland flow produced at a site is

$$\int_{r=D}^{\infty} \frac{N}{r_o}(r - D) \exp\left(-\frac{r}{r_o}\right) dr = R \exp\left(-\frac{D}{r_o}\right)$$

$$= R\left(\frac{ia - Q_o}{q_o s}\right)^{M/r_o} \tag{37}$$

Finally, summing the overland flow contributions downslope:

$$Q_o = \frac{R}{\Omega} \int_0^x \Omega\left(\frac{ia - Q_o}{q_o s}\right)^{M/r_o} dx' \tag{38}$$

For most purposes of concern in the context of stability, the overland flow (Q_o) terms on the right-hand side of equations 35 and 38 may be neglected, simplifying them considerably. In the context of incipient stability for the sides of a ridge, the contour curvature may also be ignored, so that $\Omega = 1$ and $a = x$ throughout. With these simplifications, the final form for the sediment and solute transport is obtained by substituting the simplified version of equations 34 and 38 in 30:

$$S = 10^{-3} s \left\{ 1 + 20R^2 \ [\int_0^x \left(\frac{ix}{q_o s}\right)^{M/r_o} dx']^2 \right\} + 10^{-6} Kix \tag{39}$$

The simplest way to calculate the condition for stability is via the profile under a constant downcutting rate, T. The reason for choosing this approach is that a stability distance is gradient-dependent, and the constant downcutting profile gives an approximation to relevant gradients under other conditions. In this case $S = Tx$, and

$$(T - 10^{-6} Ki)x = 10^{-3} s \left\{ 1 + 20R^2 [\int_0^x \left(\frac{ix'}{q_o s}\right)^{M/r_o} dx']^2 \right\} \tag{40}$$

Divide through by s and set $x/s = p$ and $(T - 10^{-6} Ki)/10^{-3} = T'$. Near the divide, normal convexities have a near-constant value of x/s, designated by p_o. Then

$$p_o = \frac{1}{T'} \tag{41}$$

which immediately relates form to the rate of divide lowering. Setting

$$(20)^{1/2} R \left(\frac{i}{q_o} \ p_o\right)^{M/r_o} = \frac{1}{u} \tag{42}$$

where u is a parameter having the dimension of distance, equation 40 can be rewritten as

$$\frac{p}{p_o} = 1 + \left[\frac{\int_0^x (p/p_o)^{M/r_o} \ dx'}{u} \right]^2 \tag{43}$$

Rearranging,

$$\int_0^x \left(\frac{p}{p_0}\right)^{M/r_0} dx' = u\left(\frac{p}{p_0} - 1\right)^{\frac{1}{2}} \tag{44}$$

Differentiate with respect to x and regroup to

$$\frac{dx}{u} = \frac{1}{p_0} \int_1^p \frac{dp}{(p/p_0)^{M/r_0} (P/p_0 - 1)^{\frac{1}{2}}} \tag{45}$$

Setting $p/p_0 = 1 + w^2$:

$$\frac{x}{u} = \int_0^w \frac{dw}{(1 + w^2)^{M/r_0}} \tag{46}$$

The family of solutions to this integral are shown in Figure 11, and the corresponding patterns of slope as it varies downslope in Figure 12.

Interpreting these results in terms of drainage density, it can be seen that the distance to maximum slope for the constant-downcutting profiles, which is the same as the length of stable slope, can be written in the form

$$x_c = u\phi\left(\frac{M}{r_0}\right) \tag{47}$$

for a function ϕ which is shown in Figure 13. For the case of $M/r_0 = 0$, which is the limiting arid case, this model simplifies to that shown in Figure 10. In other cases, it models a range of minimum drainage densities as $1/2x_c$.

Two ways of understanding what this approach to drainage density may mean are illustrated in Figure 14 for lowland United Kingdom (U.K.) conditions. In the figure the surface velocity of soil water on unit gradient, r_0, has been taken fixed at 200 m/day (0.23 cm/s), as this is considered relatively conservative. Maximum slope has been calculated from the summit convexity and slope length (given by drainage density). If a soil and vegetation covered surface is incised by streams, initially on a gently sloping surface, then as the streams cut down, drainage density will fall as slope gradients increase, along one of the lines for fixed M. This process of channel infilling will continue until slopes rise to an upper limit corresponding to the onset of landslide instability. The value of drainage density at this point is thought to be most characteristic of the landscape as a whole. Subsequent erosion will gradually lower gradients again, causing drainage density to rise very slightly, and perhaps to eliminate any residual disparity between its actual and theoretical minimum. In this first view of the relationships in Figure 14, there appears to be a negative relationship between drainage density and both soil permeability (M) and stable slope angle. These sensitivities should be very much greater than for more arid areas where M/r_0 is low.

Figure 14 also shows predicted values of the average deficit below saturation, D, on the summit convexity. It may be seen that this variable on its own is a very fair predictor of drainage density, as is shown by the low gradients of the isolines. In a field study, measurement of this deficit may be the most convenient practical index. Under humid

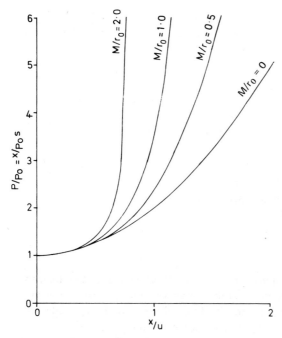

FIGURE 11 *Variation in $p = x/s$ downslope in humid areas as the exponent M/r_0 varies.*

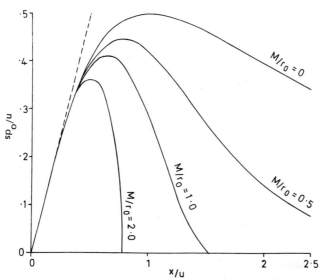

FIGURE 12 *Variation in gradient downslope for humid areas as M/r_0 varies.*

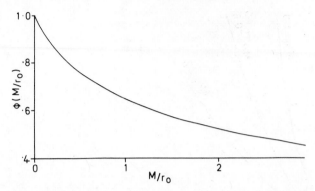

FIGURE 13 *Effect of the exponent M/r_0 in reducing the critical distance $x_c = u\phi(M/r_0)$.*

conditions, the drainage density indicated by instability criteria is greater than that estimated from perennial channel networks (at the margin of which $D = 0$). This is in accordance with field observations (Gregory and Walling, 1968); and the instability network may be shown to correspond to flow on only a few days of the year.

Although this study is only at the theoretical stage to date, it seems that an approach to drainage density based on the instability of hillslopes as they evolve promises to explain significant sources of variation within humid areas, as well as contrasting their relationships effectively with those of semiarid areas.

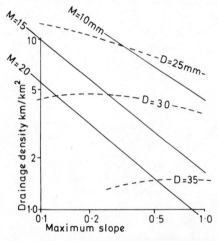

FIGURE 14 *Factors influencing drainage density in a humid area of given climate ($R = 700$ mm, $r_0 = 4$ mm rain/day, $V = 200$ m/day, $i = 1$ mm/day).*

CONCLUSIONS

As more complex models are constructed of interacting landscape processes, the need to define domains of validity and the threshold at their boundaries will become increasingly important. It is already plain that no usable and realistic model has more than a narrow range, and planetary landscapes, with their very different processes, illustrate this point in its most extreme form.

It has been argued above that a number of the most important thresholds in geomorphology can be interpreted in terms of a stable and unstable regime, and that the morphological expression of potentially unstable equilibrium is by means of an areal split between two distinct forms. For the soil weathering case, these were soil-covered sites and tors; for the river braids, they were the channels and island bars; and for the hillslopes, valleys and interfluves. At a simple level, the graphical type of analysis presented here allows analogies to be drawn among diverse landscape systems and with other systems unconnected with landforms. One way in which these morphological arguments might naturally lead is into catastrophe theory, although the connections with it have not been made explicit above.

REFERENCES CITED

Beven, K. J., and Kirkby, M. J., 1979, A physically based, variable contributing area model of basin hydrology: Hydrol. Sci. Bull., v. 24, p. 43–69.

Gregory, K. J., and Walling, D. E., 1968, The variation of drainage density within a catchment: Bull. Int. Assoc. Sci. Hyd., v. 13, p. 61–68.

Henderson, F. M., 1961, Stability of alluvial channels: Proc. A.S.C.E., J. Hyd. Div. HY6, p. 109–138.

Kirkby, M. J., 1977, Maximum sediment efficiency as a criterion for alluvial channels: *in* Gregory, K. J., ed., River channel changes: New York, Wiley, Chap. 27, p. 429–442.

Langbein, W. B., and Schumm, S. A., 1956, Yield of sediment in relation to mean annual precipitation: Trans. Amer. Geophys. Union, v. 39, p. 1076–1084.

Leopold, L. B., and Wolman, M. G., 1957, River channel patterns—braided, meandering, and straight: U.S. Geol. Survey Prof. Paper 282B, p. 39–85.

Parker, G., 1976, On the causes and characteristic scales of meandering and braiding in rivers: J. Fluid Mech., v. 76, n. 3, p. 457–482.

Peltier, L. C., 1950, The geographic cycle in periglacial regions as it is related to climatic geomorphology: Ann. Ass. Amer. Geog., v. 40, p. 214–236.

Schumm, S. A., and Khan, H. R., 1971, Experimental study of channel patterns: Nature, v. 233, p. 407–409.

——, and Lichty, R. W., 1963, Channel widening and flood plain construction along Cimarron River in s.w. Kansas: U.S. Geol. Survey Prof. Paper 352D, p. 71–88.

Smith, T. R., and Bretherton, F. P., 1972, Stability and the conservation of man in drainage basin evolution: Water Resources Res., v. 8, n. 6, p. 1506–1529.

Stoddart, D. R., 1969, World erosion and sedimentation: *in* Chorley, R. J., ed., Water, earth and man: London, Methuen, p. 43–64.

CONCLUSIONS

REFERENCES CITED

5

EROSIONAL DEVELOPMENT OF VALLEY-BOTTOM GULLIES IN THE UPPER MIDWESTERN UNITED STATES

Joe M. Bradford and Robert F. Piest

INTRODUCTION

A gully is an incised drainage channel, is usually steep-sided, and transmits ephemeral flow, often with a steeply sloping and actively eroding head scarp. There are no widely agreed upon dimensions for distinguishing gullies from rills. Generally, a gully is an obstacle to farm machinery and is too deep to be obliterated by ordinary tillage; whereas a rill is of lesser depth and can be smoothed over by ordinary tillage. Gullies are classified on the basis of topographic location as valley-bottom, valley-side, or valley-head gullies (Brice, 1966). A valley-bottom gully becomes a valley-head gully as its head scarp migrates into the valley head. Valley-side gullies usually result from less concentrated flow coming from diverse directions.

Gullying is important to the geomorphologist as a mechanism for slope retreat. Depending upon the type of material under consideration, slope retreat progresses at a rate dependent upon the materials competence and the energy conditions of the local environment. The study of gullies is essential not only to the understanding of natural landscapes, but also as a practical means for controlling erosion and sedimentation. To the agriculturalist and the environmentalist, gullying is of concern because of its effects on land destruction and water pollution. In this chapter we discuss the dynamics and processes of gullying that occur on a short-term basis, with some implications for the long term.

Why certain landforms respond to external influences and result in rapid gullying is still somewhat puzzling. Is the erosion and sedimentation of a gully cyclic? If so, how does one identify or predict which physical parameters will trigger the onset of erosion? Vegetative cover, climate, and other variables change with time; the terrain merely responds to these alterations.

A landscape is an energy regime with a delicate balance between the form of the system and inflow and outflow of energy. The system is continuously importing energy, and an erosion threshold may be reached if the mode of energy utilization changes. The way in which this added potential energy is stored or utilized changes when a change of state threshold is reached. An example is a bank of dry loess standing vertically along a gully. It slowly absorbs water from surface infiltration or base wetting, increasing its potential energy, and, finally, reaching the threshold where the potential energy is transferred or converted into kinetic energy. The mass of soil slumps. Thus, the addition of potential energy triggers the entire system at the threshold point. The gully wall advances.

The bank that failed is now at the base of the gully wall, and, apparently, will remain there until it is disaggregated and entrained by runoff—or enveloped by runoff—with enough energy to transport the soil particles. This transport will continue until the supply diminishes or the energy level of the stream goes below the threshold required to do this work.

Gullying results from a complex array of many processes; any one of which may operate for limited duration or landscape position. Changes in the rates of gully development can be attributed not only to changes in external forces but also, in part, to thresholds within the geomorphic system. Schumm (1973) and Patton and Schumm (1975) recognized geomorphic thresholds in the process of gully erosion. They observed critical threshold valley slopes above which alluvium was entrenched. However, in small drainage basins (less than 10 km^2) in Colorado, variations in vegetation prevented the recognition of a critical threshold slope. In this chapter we trace gully development and examine the concept of geomorphic thresholds as applied to the processes of valley-bottom gullying in small, agricultural drainage basins in the loessial soils regions of western Iowa and Missouri.

THE STUDY AREA

The gully study area consists of four small watersheds, from 0.302 to 0.608 km^2 in size, in Pottawattamie County in western Iowa (Fig. 1). They are located about 26 km east of the Missouri River, near the small town of Treynor, within Soil Conservation Service Land Resource Area M-107 (Agric. Handb. 296, 1965), Iowa and Missouri deep loess hills. The topography of each watershed is shown in Figure 2. Adjoining watersheds 1 and 2 are about 5 km south of adjoining watersheds 3 and 4.

Pleistocene Geology

The study area has been subjected to two major glaciations, the Nebraskan and Kansan. Each resulted in emplacement of till and was followed by periods of soil formation and erosion. Glacial deposits and driftless areas were subsequently covered by two major loess sheets—Loveland loess of Illinoian age, and Wisconsinan loess. Wisconsinan loess deposition began about 25,000 years ago and continued to about 14,000 years ago. After loess deposition, erosion removed materials from hillsides; the resulting downstream deposition is herein described as loess-derived alluvium. This alluvial fill, defined by Daniels and Jordan (1966) as the DeForest Formation, occupies most drainageways (Fig. 3).

The Pleistocene history of the region was described by Daniels and Jordan (1966) and Ruhe et al. (1967). The Pleistocene and Recent deposits in watershed 3 were described in detail by Allen (1971) and are shown in Figure 4.

The Kansan till is not exposed at the land surface and is found at 4 to 9 m below the surface in the drainageways and is about 18 m thick. A Yarmouth paleosol about 3 to 5 m thick is present in the upper part of the till and is overlain by about 9 to 12 m of Loveland loess on the hilltops. A 3- to 5-m-thick Sangamon paleosol is blanketed on the ridges by about 8 to 12 m of Wisconsin-age loess.

The sequence of post-Wisconsinan alluvial fill in drainageways was established by Daniels and Jordan (1966) and Allen (1971). The oldest unit of the DeForest alluvial

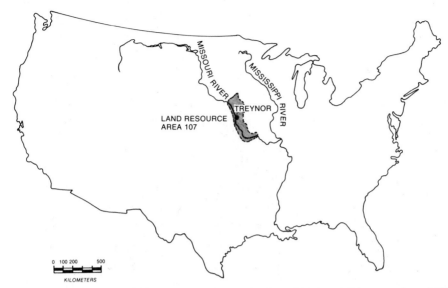

FIGURE 1 *Location of experimental watersheds within Land Resource Area 107 near Treynor, Iowa.*

formation, the Soetmelk member, lies over both Kansan till and the Yarmouth paleosol. Daniels and Jordan (1966) placed the end of this period of deposition about 11,000 years before present (YBP). Next is the Watkins member, which ranges from 11,120 ± 400 YBP at the base to 2020 ± 200 YBP at the top.

Above this is the Hatcher member dated at 2020 ± 200 YBP to 1800 ± 200 YBP. The next youngest fill is the Mullenix, 250 to 1800 years old with most of the accumulation

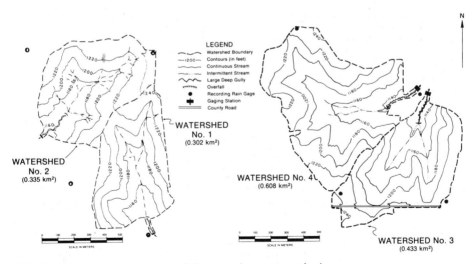

FIGURE 2 *Topographic map of Treynor, Iowa, watersheds.*

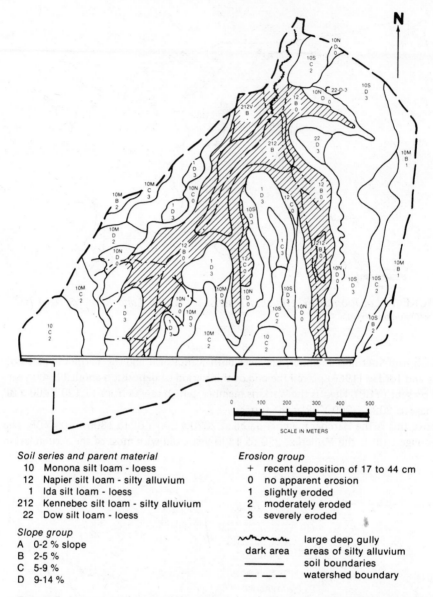

Soil series and parent material
  10   Monona silt loam - loess
  12   Napier silt loam - silty alluvium
   1   Ida silt loam - loess
  212 Kennebec silt loam - silty alluvium
  22   Dow silt loam - loess

Slope group
  A   0-2 % slope
  B   2-5 %
  C   5-9 %
  D   9-14 %

Erosion group
  +   recent deposition of 17 to 44 cm
  0   no apparent erosion
  1   slightly eroded
  2   moderately eroded
  3   severely eroded

  ᴧᴧᴧᴧ   large deep gully
  dark area areas of silty alluvium
  ──────   soil boundaries
  ── ── ──   watershed boundary

FIGURE 3 *Areas of natural loess and local alluvium within watershed 3, Treynor, Iowa.*

within 1100 years. The youngest fill is the Turton member dated at 95 to 250 years old. In addition, there is post-settlement alluvium that has accumulated since about 1850.

Vegetative Cover

The research watersheds have been managed so as to compare the effects of contoured corn, pasture grass, and level-terraced corn on hydrology and erosion. Table 1 lists

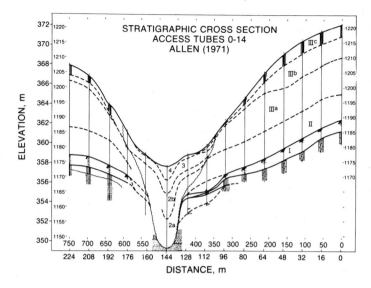

ELEVATION, m

DISTANCE, m

-1200	Mean sea level elevation (feet)
150	Distance from access tube 0
⟋	Ground surface
⟋	Recognized boundary
- -	Proposed boundary
		Observation site
4	Access tube number
⌇⌇	Unconformity

DeForest formation alluvium

2a	Watkins member
2b	Hatcher member
3	Mullenix member
4	Turton member

Wisconsinan loess

⊤⊤⊤	Zone IIIc (Solum)
III^b	noncalcareous increment of Zone III
III^a	calcareous increment of Zone III
II	calcareous Zone II
I	"basal" Wisconsin loess

⊤⊤⊤⊤	basal Wisconsin paleosol
XXX	Sangamon paleosol in Loveland loess
⊔⊔⊔	late Sangamon paleosol
≡≡≡≡	Yarmouth silty clay
░░░	Kansan till

FIGURE 4 *Stratigraphic cross section of watershed 3, Treynor, Iowa (Allen, 1971, Fig. 9, p. 78).*

the size, crop, and conservation practice of each of the four watersheds. Each watershed is in a single land use, and treatments are those commonly used in the area. The "nonconservation" watersheds are farmed on the contour with conventional tillage and planting. In 1972, watershed 3 was changed from bromegrass to continuous corn by planting directly into the grass sod. Thereafter, a mulch tillage system with minimum cultivation has been used. The effect of the sod cover has lasted for several years. Before fall 1972, watershed 4 was level-terraced with 92% of its area above terraces. The terraces had a storage capacity of about 5 cm of surface runoff. In 1972, the number of terraces was reduced by about one-half; the terrace system was made more parallel; and a complete pipe outlet system was installed to avoid excessive water pondage in the terrace channels.

TABLE 1 *Watershed Descriptions*

Watershed number	Area (km²)	Year	Crop	Conservation practice
1	0.302	1964–77	Continuous corn	Field-contoured
2	0.335	1964–77	Continuous corn	Field-contoured
3	0.433	1964–72	Bromegrass	Rotation-grazed
		1972–77	Continuous corn	Mulch tillage
4	0.608	1964–72	Continuous corn	Level-terraced, conventional tillage
		1972–77	Continuous corn	Revised terraced system, mulch tillage

Climate and Hydrology

Climatic information before 1964 is available from records at Omaha, Nebraska, 38 km northwest of the watersheds. The mean annual temperature is 10.8°C with mean maximum and mean minimum of 16.0 and 5.2°C, respectively. The mean annual snow is 72 cm and the mean annual precipitation is 69.9 cm.

The 13-year mean precipitation (Table 2) over the four watersheds was 80.6 cm. Annual precipitation varied considerably from year to year but was similar for the four watersheds for each year. Certain years were characterized by low-intensity, long-duration rainfall, while high-intensity rainfall was characteristic of 1965 and 1967. For the study area, 1 storm every 10 years should have rainfall in excess of 10 cm/25 hr. The 1-year frequency for a storm duration of 24 hr is 7.1 cm.

TABLE 2 *Water and Sediment Yield Summary of Treynor, Iowa, Watersheds, 1964–76*

Year	Watershed number	Annual precip. (cm)	Runoff (cm) Base	Runoff (cm) Surface	Runoff (cm) Total	Sheet rill (kg/m²)	Gully (kg × 10³)	Total (kg/m²)
1964	1	90.45	4.88	11.58	16.46	5.60[a]	608[a]	7.62
	2	89.31	5.46	10.21	15.67	5.60[a]	300[a]	6.50
	3	85.06	5.99	1.07	7.06	0.07	58	0.20[b]
	4	88.39	14.38	2.01	16.38	0.16	9	0.18[b]
1965	1	115.19	9.04	26.97	36.02	9.86	1050	13.36
	2	112.65	7.54	27.13	34.67	8.16	599	9.95
	3	112.47	11.73	11.68	23.42	0.09[a]	78[a]	0.27
	4	113.97	26.82	6.38	33.20	0.20[a]	15[a]	0.22
1966	1	51.61	6.45	1.65	8.10	1.50	84	1.77
	2	52.15	6.10	2.24	8.33	1.93	161	2.40
	3	55.91	6.45	0.97	7.42	0.02[a]	9[a]	0.04
	4	55.58	15.01	0.48	15.49	0.13	13	0.16

TABLE 2 (Continued)

Year	Watershed number	Annual precip. (cm)	Runoff (cm)			Sediment yield		
			Base	Surface	Total	Sheet rill (kg/m²)	Gully (kg × 10³)	Total (kg/m²)
1967	1	97.16	5.77	29.39	35.15	22.22	1320	26.56
	2	95.53	6.35	26.54	32.89	16.86	1250	20.53
	3	86.94	8.38	6.73	15.11	0.13	109	0.38
	4	87.76	18.49	1.85	20.35	0.65	−21[c]	0.61
1968	1	82.04	4.24	2.92	7.16	0.83	93	1.14
	2	82.55	4.62	2.87	7.49	0.92	40	1.03
	3	78.99	4.04	2.59	6.63	0.04	12	0.07
	4	81.74	10.74	0.30	11.05	0.07	2	0.07
1969	1	79.81	8.08	6.43	14.50	0.40	107	0.76
	2	80.11	7.54	5.97	13.51	0.22	50	0.38
	3	77.83	8.36	4.39	12.75	0.02	17	0.07
	4	77.98	15.52	0.69	16.21	0.02	−5	0.02
1970	1	80.04	5.61	5.44	11.05	2.65	161	3.14
	2	78.28	5.97	4.55	10.52	1.66	155	2.13
	3	73.30	5.56	0.94	6.50	<.02[d]	5	0.02
	4	73.13	10.13	0.33	10.46	0.02	<0.9[d]	0.02
1971	1	73.81	5.23	12.55	17.78	4.48	362	5.69
	2	74.09	6.65	9.75	16.41	2.98	219	3.63
	3	75.74	7.21	3.86	11.07	0.09[a]	27[a]	0.13
	4	76.40	14.02	1.73	15.75	0.34[a]	5[a]	0.36
1972	1	86.21	6.78	3.84	10.62	1.68	328	2.78
	2	86.46	7.65	3.89	11.53	1.77	109	2.08
	3	95.22	15.80	2.11	17.91	0.27	−29	0.20
	4	95.25	14.66	10.72	25.37	1.46	44	1.52
1973	1	105.94	20.78	6.63	27.41	0.22	91	0.54
	2	104.60	25.55	7.49	33.05	0.11	61	0.29
	3	103.20	37.03	2.72	39.75	0.02	−0.9	0.02
	4	102.39	30.51	8.48	38.99	0.22	45	0.31
1974	1	63.04	16.43	1.37	17.81	0.11	28	0.20
	2	62.15	21.77	1.42	23.19	0.07	34	0.18
	3	56.01	20.70	0.20	20.90	<0.02[a,d]	<0.9[a,d]	<0.02[d]
	4	53.80	18.90	0.61	19.51	<0.02[d]	5	0.02
1975	1	78.26	11.99	2.62	14.61	0.36	33	0.25
	2	78.66	19.79	2.11	21.89	0.18	34	0.07
	3	74.35	16.74	0.33	17.07	<0.02[a,d]	<0.09[a,d]	<0.02[d]
	4	73.76	16.00	3.05	19.05	0.04	5	0.04
1976	1	53.98	10.06	0.46	10.52	<0.02[a]	4	0.02
	2	53.92	12.32	0.38	12.73	<0.02[a]	2	<0.02
	3	60.71	12.62	0.97	13.59	0.25[a]	0.9	0.25
	4	64.14	10.44	1.83	12.27	0.16	12	0.18
Averages 13 years 1964–76	1	81.36	8.89	8.59	17.48	3.86	328	4.91
	2	80.80	10.59	8.05	18.62	3.09	231	3.79
	3	79.68	12.37	2.95	15.32	0.07	23	0.13
	4	80.34	16.59	2.95	19.53	0.27	10	0.29

[a]Division between sheet-rill and gully erosion estimated.
[b]Total and component erosion values estimated.
[c]Negative value indicates channel fill.
[d]Soil losses are less than 0.01 kg/m² from sheet-rill source or less than 0.4 kg from gully source.

Partitioning the total annual precipitation into its hydrologic components gives insights both into modification of vegetative cover and conservation treatments upon the erosion processes within each watershed and into the energy conditions of the local environment (Saxton and Spomer, 1968; Saxton et al., 1971; Saxton et al., 1974).

The annual runoff for each watershed and the base and surface flow components are given in Table 2. Paired, contoured-corn watersheds 1 and 2 had nearly the same precipitation and runoff; their total stream flow differed by less than 7%. The relative yearly amounts of base flow and surface runoff for watersheds 1 and 2 depended on antecedent moisture and rainfall intensity and duration. Base flow for watersheds 3 and 4 exceeded surface runoff by 300%, on the 13-year average.

Grass watershed 3 had the lowest total runoff of all four watersheds before 1972, for example, 38% less than that of watershed 1. However, 63% of its total was base flow. This lower water yield reflects the longer growing season and greater annual evapotranspiration of grass as compared with corn (Saxton et al., 1974). Surface runoff was less for grass than from corn, because infiltration rates were greater for grass. The dense grass prevented the soil surface from sealing, and the increased evapotranspiration from grass, especially from the early- and late-growing season, depleted soil moisture more than did with corn.

Total water yields from level-terraced corn watershed 4, before 1973, and contoured-corn watersheds 1 and 2 have been similar. There was a significant difference, however, in water source. Base flow accounted for most of the total stream flow from the level-terraced watershed. The large base-flow component from the terrace area was due to increased infiltration and percolation caused by ponding behind the terraces.

Gully Erosion

Gully erosion rates at the Treynor watersheds were measured, during periods of surface runoff, by sampling the sediment content of stream flow at two locations—immediately upstream and about 100 m downstream from each gully headcut (Fig. 5). With allowance for time of travel and minor added runoff between sampling locations, it was possible to monitor downstream sediment accretion due to gully scour throughout the storm.

The sediment yield from the outlet gully of each of the four Agricultural Research Service watersheds are summarized on an annual basis in Table 2. Large amounts of soil debris were transported. Gullies of watersheds 1 and 2, respectively, lost 3.3 and 2.3 × 10^5 kg of soil per year during the 13 years (1964–1976). Significantly smaller amounts were eroded from conservation watersheds 3 and 4.

Valley-bottom gullies within the four watersheds are characterized by near-vertical headwalls 2.7 to 4.3 m high and a total depth of 4.0 to 6.7 m. The sidewalls of the plunge pool area also have vertical walls but only after complete debris cleanout; normally, much debris is left at the base of the walls. Downstream from the headcut, the upper 0.5 to 1.0 m of the sidewall is vertical, and the lower sidewall has angles typically between 40 to 70°. Figure 6 gives the cross section of the gully approximately 14 m downstream from the headcut at watershed 1. The lower slopes, composed mainly of materials slumped from the sidewalls, are stable as long as runoff is insufficient to transport the debris from the base.

Gully voiding was sporadic. Although loosely correlated with storm runoff, the voiding rate varied according to the availability of soil debris which accumulated in the

FIGURE 5 *Aerial view of gully headcut at watershed 1 outlet drainageway, showing sampling footbridges and measuring weir.*

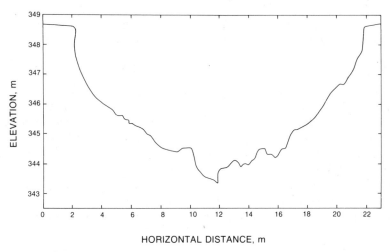

FIGURE 6 *Cross-profile of gully 14 m downstream from headcut at watershed 1, October 1973.*

channel from various weathering processes (Piest et al., 1975a). The 1967 gully erosion at watersheds 1 and 2 was 31 and 41%, respectively, of total 13-year gully erosion. At watershed 2, the single event of June 20, 1967, caused 21% of the 13-year gully erosion, principally by removing 1200 kg/m of soil, from each bank in the 213-m long reach of gully between upstream headcut and downstream measuring weir.

The gullies within watersheds 3 and 4 were actively eroding before the establishment of conservation practices. By 1964, these gullies had advanced about 210 to 250 m upslope. Before 1964, the four watersheds differed little in gully growth rate. The causes for present-day equilibrium in the conservation watersheds will be discussed in light of threshold processes.

INSTABILITY PROCESSES OF MASS WASTING

The dominant mode of gully growth within the thick loessial soils study area is by mass wasting of the walls, with most of the soil loss coming from the vicinity of the retreating headscarp. Little soil is eroded from the standing banks by tractive forces of flowing water.

No single mechanism can be used to describe the mass wasting process. Stability or instability is essentially a matter of whether or not driving forces exceed resistances and depend upon such factors as the amount of debris at the slope base, the slope angle, and moisture and soil conditions. The material begins to fail only when the forces involved become greater than the resistances. In mathematical terms, this can be expressed as

$$F_s = \frac{\tau_f}{\tau} \tag{1}$$

where F_s = factor of safety
τ_f = shear strength along some shear surface
τ = equilibrium shear stress along the same shear surface

A value of $F_s > 1$ implies stability, and a $F_s \leqslant 1$ implies instability. $F_s = 1$ is the threshold value at which the material is in a state of incipient movement.

Driving Forces

Changes in the driving forces require energy, and this energy is either associated with climate or gravity. Climate adds energy to the system in many ways through moisture and temperature changes. However, since gully wall slumping is mainly due to energy associated with water, other energy sources will not be discussed. Slope failure due to the addition of water can be attributed to (1) the weight of water added to the soil mass by surface infiltration and/or by a rise in the water table, and/or (2) increased seepage forces as water in the saturated zone exits through the bank as water table discharge or as drawdown of bank storage after passage of a flood wave through the gully, and/or (3) reduction in apparent cohesion of unsaturated soil, through reduction in capillary tension (or negative pore pressure).

Higher water table levels at the headcut are responsible, in part, for the geometry of the massive slumping near the headcut. The water table level along the gully wall, thus the seepage force at the base of the gully wall, decreases downstream from the headcut and causes little slumping of the walls. However, incipient bank failure is not directly related to the magnitude of the exit gradients near the headcut because headcut failure depends also on slope geometry, soil conditions, and runoff volume and duration.

Figure 7 shows the elevation of the phreatic surface in May 1977 at a cross section of (1) the nonincised upstream drainageway 3 m upstream from the gully plunge pool or headcut of watershed 1 and (2) about 1 m downstream from the vertical headcut. Seepage zones occur on the faces of the gully wall about 1.2 to 2.4 m below ground surface, depending on watershed and antecedent precipitation. These data, combined with water surface profiles in the gullies, indicate that rapid drawdown of water in the stream bed does not contribute to gully instability in the four watersheds; the height of flowing water at the gully head never reaches the level of the free water seepage surface in the gully banks. Stream stage in the gully is seldom greater than 0.6 m. For watershed 1, with a 5.5 to 6.0-m deep gully, the peak stage would normally be about 1.2 m below the seepage surface along the gully wall.

Resisting Forces

Forces that resist slope failure are due to the shear strength of the soil mass. Soil shear strength is usually expressed by Coulomb's equation:

$$\tau_f = c' + (\sigma_n - \mu) \ \tan \phi' \tag{2}$$

where c' = cohesion expressed in terms of effective stresses
ϕ' = friction angle expressed in terms of effective stresses
σ_n = total normal stress on the failure plane at the time of failure
μ = pore water pressure

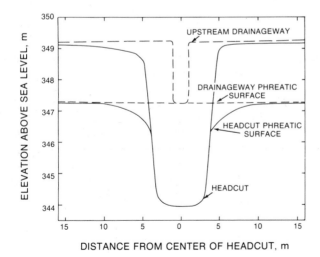

FIGURE 7 *Single-plane projection of cross sections of upstream drainageway and headcut and their phreatic surfaces.*

The valley-bottom gullies within each of the four watersheds are incised within silty alluvium. In general, the properties of this loess-derived silty alluvium are similar to the natural loess; however, specific differences and their influence on the mechanical behavior are not documented in the literature. Much of the literature deals specifically with the properties of loess.

The mechanical behavior of silty alluvium and loess depends largely on its internal structure and interparticle bonding. The structure of loess is typically a loose cluster arrangement of clay particles with a silt grain at the center (Rahman, 1973). Densities of undisturbed loess may be as low as 1.2 g/cm³. Loess cohesion has been attributed not only to clay as a binder, but also to cementation by calcium carbonate. Strength of loessial deposits seems to be more sensitive to moisture changes than deposits of other origins; however, to our knowledge, no comparisons have been published. Turnbull (1968) presented a graph showing a radical change in shear strength with changing moisture content on undisturbed Mississippi loess soil.

The macrostructural features of loess and silty alluvium also contribute to its failure patterns. Loess cleaves on vertical planes; the cause of these planes of weakness is mere conjecture. In their review of this problem, Lohnes and Handy (1968) stated that vertical cleavage has been attributed to vertical root holes (Von Engeln, 1949), secondary carbonate filling vertical root holes (Schultz and Cleaves, 1955), or shrinkage cracks resulting from compaction (Strahler, 1963). Vertical faces in loess are quite stable under dry conditions; but when saturated, their stability is greatly decreased.

Undisturbed soil cores were taken within each DeForest Formation unit (Fig. 4) with an 8.9-cm-diameter Shelby thin-walled sampler. The samples were taken in the drainage-way at the same location (Access Tube 9) that Allen (1971) used in watershed 3, which is at the toeslope position with a 4.5% slope. Consolidated, drained direct shear tests were run on 2.00-cm-long, 6.37-cm-diameter hand-trimmed samples from each unit. The samples were saturated for 3 days before consolidation and shearing. The rate of shearing was 0.0005 cm/min. Sampling depths and shear strength parameters (eq. 2) are given in Table 3.

Table 3   *Direct Shear Strength Test Results of DeForest Formation Alluvium at Access Tube, AC 9 (Allen, 1971), Watershed 3, near Treynor, Iowa*

Stratigraphic unit	Horizon depth (cm)	Sample depth (cm)	Cohesion (g/cm²)	Friction angle (deg)	Bulk density (g/cm³)
Turton	0–152	80–100	138	15	1.24
Mullenix	152–274	190–210	70	20	1.34
Hatcher	274–503	400–420	124	16	1.43
Watkins	503–	740–760	77	23	1.39

Mode of Bank Failure

A full understanding of the processes of mass wasting of gully walls cannot be achieved without identifying the manner in which failure occurs. Three modes of failure have been identified in gully growth in silty alluvium in the loessial study area. The first is a deep-seated, circular arc failure. The second is slab failure. The third is a combination of base collapse or "popout," followed by slumping of overhanging material.

Deep-Seated Slides

Downstream from the near-vertical headcut, lower-angle slopes (possibly about 51° or lower, according to Lohnes and Handy, 1968) may fail along a circular arc or a loga-rithmic spiral (Fig. 8) under prolonged wet conditions or stream downcutting. Stability analyses of such failures can be conducted from conventional slope stability methods, such as the Swedish circle method (Fellenius, 1936), simplified Bishop method of slices (Bishop, 1955), or the Morgenstern and Price method (Morgenstern and Price, 1965). Under the simplified Bishop method of slices (Bailey and Christian, 1969) failure criteria, the slope heights necessary to cause instability (FS = 1) for 90°, 77°, and 51° slopes, are given in Table 4. For each calculation, the slope was assumed to be a uniform alluvial de-posit with the shear strength parameters, c', and ϕ', as given in Table 3. Soil unit weights were calculated from the bulk densities in Table 3, assuming a soil particle density of 2.70 g/cm^3 and a degree of saturation of 100%. Pore water pressures along the slip surface were determined by assuming a flow-net system similar to that determined by Bradford and Piest (1977) for the Treynor study area. The phreatic surface in each case was taken to be 180 cm below the soil surface 15 m from the slope face. Slope angles of 77° and 51° were used due to Lohnes and Handy's (1968) discussion of slope-angle frequencies for loess in western Iowa. Their field observations had indicated that 77° slopes, however, usually fail on a plane, whereas 51° slopes can also fail along a circular arc.

FIGURE 8 *Deep-seated circular slide along gully channel.*

TABLE 4 *Critical Slope Heights for DeForest Formation Alluvium*

Soil deposit	Slope angle (deg)	Calculated slope[a] height for FS = 1 (cm)	Maximum slope height[b]	
			$Z = 0$ (cm)	$Z = Z_0$ (cm)
Turton	90	350	404	202
	77	386		
	51	579		
Mullenix	90	180	217	108
	77	235		
	51	350		
Hatcher	90	300	349	175
	77	342		
	51	495		
Watkins	90	210	248	124
	77	261		
	51	352		

[a]Calculated by the simplified Bishop method of slices (Bailey and Christian, 1969).
[b]Calculated from equations 3 and 4.

Slab Failure

Bradford and Piest (1977) have shown that a curved failure surface is inappropriate to describe the failure near the headcut in alluvium and that conventional slope analysis methods are not applicable. However, the failure surface can be approximated by two planes forming a trapezoidal mass of soil (Lohnes and Handy, 1968). Slab failure similar to that in a gully headcut in silty alluvium is seen at a road cut exposing Wisconsinan loess (Fig. 9). Intact slabs are seldom seen in the gully study area; upon falling, they break up because of the high moisture conditions. Vertical tension cracks form one of the planes, and the other plane is assumed to pass through the slope base at an inclination from the horizontal of $45° + \phi/2$ (Lohnes and Handy, 1968). Thus, the failure mode is similar to an unconfined compression test in that the slab fails when the weight of soil exceeds the compressive strength of the underlying soil. The vertical cracks caused by tensile stresses in the upper layers of the slope and the macrostructural features typical of loess seem to be responsible for this failure mode.

From Lohnes and Handy (1968), the maximum height H_c of the cut before slab failure is

$$H_c = \frac{4c}{\gamma[\cos \phi - 2 \cos^2 (45 + \phi/2) \tan \phi]} - z \tag{3}$$

where z is the depth of the tension crack and γ the unit weight of the soil mass. Maximum vertical slope heights were calculated (Table 4) for the case when $z = 0$ and

$$z = z_0 = \frac{2c}{\gamma} \tan \left(45° + \frac{\phi}{2}\right) \tag{4}$$

The slopes were assumed to be uniform, with zero tensile strength and other soil properties defined in Table 3.

75 cm

FIGURE 9 *Slab failure at roadcut near Missouri River.*

FIGURE 10 *Popout failure that develops at the base of the gully wall.*

Base Failure

Bradford et al. (1978) identified the typical failure sequence of gully headwalls in alluvium in the thick loessial area of western Iowa as (1) a popout or alcove failure near the toe of a near-vertical wall (Fig. 10), (2) columnar sloughing of the overhanging material, and, finally, (3) the transport of the eroded material downstream.

As described by Lutton (1969), both alcoves and popouts consist of pyramidal blocks freed from the wall by one fracture, inclined into the slope, and a second fracture below, inclined out of the slope. Popouts as compared with alcove failures are localized low on cut faces but above the base.

The initiating failure at the base of the wall is not thoroughly understood; however, three explanations can be offered.

1. The soil is, to a certain degree, collapsible (Handy, 1973). Handy defined collapsibility in loess as a state of underconsolidation related to apparent cohesive strengths of perennially unsaturated soils. Dudley (1970) stated that two prime requirements for collapse are a loose soil strength and a moisture content less than saturation. When water is added, the soil volume is reduced rapidly at some degree of saturation.

Reginatto and Ferrero (1973), however, distinguish between "truly collapsible" soils and "conditionally collapsible" soils. Conditionally collapsible soils are those able to support a certain level of stress upon saturation.

Collapsible loess occurs extensively in Iowa and in the study region (Handy, 1973) and collapsibility was proposed by Handy as a possible contributing factor to the development of valley-side and valley-head gullies within natural loess where no seepage water exists at the base of gully walls. Since the study area gullies are in alluvium and have a high zone of seepage above the base, collapsibility, as defined by Dudley (1970) and Handy (1973), could not be responsible for the base failure. To determine if the silty alluvium could meet the "conditionally collapsible" criterion, consolidation tests are run on undisturbed soil cores taken from each soil layer given in Table 3, and K_0 tests (Kane, 1973) were run on the most compressible alluvial member, the Mullenix unit.

K_0, the lateral stress ratio at rest, is the ratio of the total lateral stress σ_r to the total axial stress σ_n, where there has been no lateral strain. The K_0 tests were run on saturated samples in a standard triaxial cell on specimens with diameters of 7.0 cm and heights of 15 cm. The K_0 conditions were maintained by increasing the radial and axial stresses so that the volume change, as indicated by a burette, corresponded to the vertical deformation of the sample. The behavior of the soils was analyzed by the stress path method of representing the states of stress (Lambe and Whitman, 1969; Kane, 1973). In this method the state of stress is represented by a stress point on a $p–q$ diagram, where $p = (\sigma_v + \sigma_H)/2$ and $q = (\sigma_v - \sigma_H)/2$. σ_v and σ_H represent the total vertical (normal) and horizontal (radial) stresses, respectively. An abrupt change occurred in the K_0 value upon collapse of the soil structure. For the Mullenix unit, K_0 was 0.420 before collapse and increased to 0.806 after collapse (Fig. 11).

Oedometer test results also indicated that a certain vertical pressure was required to collapse saturated samples. Little to no volume change occurred upon saturation at a vertical stress of 0.15 kg/cm^2.

The results of the K_0 and oedometer tests give further evidence that the collapse phenomenon due to wetting does not seem to be responsible for the base failure. However, a slow migration of soil particles from zones adjacent to the vertical wall and below

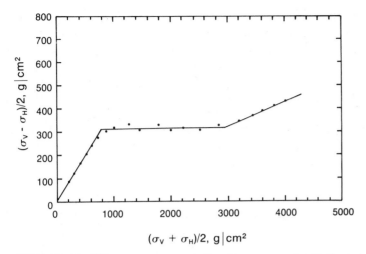

FIGURE 11 *Effective stress path for K_0 tests on the Mullenix member of the DeForest Formation Alluvium.*

the seepage level might trigger collapse. The particles would move along lines of concentration of seepage. The discussion by Brink and Kantey (1961) would support this conclusion. They stated that a marked characteristic of the collapse phenomenon is that it seems to be confined to slopes where the soils are well drained, and no recorded case has been found on flat plateaus or in depressions where the internal drainage is impeded. Visual observations of the exposed soil mass below the seepage zone and adjacent to the headcut walls indicated an orientation of drainage paths toward the walls.

2. Stress relief at the base of the slope causes bulging. The removal of debris from the slope base results in a change in the state of stress, with corresponding strains and displacements. For soils with a brittle structure in drained shear, slight deformations can result in a large decrease in the shearing resistance of the soil. This strength reduction may be due to the opening of cracks in the alluvium proceeding from the face into the wall. Since a zone of stress concentration occurs adjacent to the base of the slope, localized failure occurs there.

3. Reduced soil cohesion results from increased water content. Since undisturbed loess is a loose, open-structured soil composed of silt particles separated by clay coatings, and the silty alluvium of the DeForest Formation may be similar, as water contents increase, strength will decrease. Bradford et al. (1978) found that base failure can be initiated by a slight increase in the phreatic surface along a vertical wall of alluvium. Limit equilibrium slope stability analyses indicated that seepage forces exerted by the flowing groundwater had little effect on slope stability at the test site.

No satisfactory theoretical procedures have been developed for estimating conditions at time of base failure, because the mechanism of failure has not adequately been defined. Thus, even though mass wasting is responsible for gully growth, we are not able to predict response rates to external environmental changes because of the lack of quantitative description of processes. However, we do know that process thresholds occur in the development of gullies.

Thresholds of Instability

From the calculated maximum slope heights in Table 4, thresholds of instability can be defined within the processes of gully mass wasting. As a valley-floor gully advances up-valley, the scarp cuts into different materials, according to the geomorphology and stratigraphy of the drainage basin, and may gradually increase in height due to surface-erosion deposition. The additional gravity force caused by increased height and the differences in strength of the incised soil layers result in alteration of the stability of the gully scarp. For example, if we visualize the gully advance as shown in Figure 12, as the

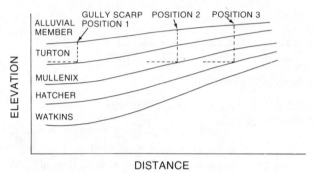

FIGURE 12 *Schematic representation of valley-floor gully advance.*

gully incises into the different alluvial soil units, the stability of the entire gully system is abruptly changed. Using the calculations in Table 4 and assuming the mode of failure to be slab-type, a vertical slope of Turton alluvium could reach a height of 404 cm before becoming unstable (position 1, Fig. 12). However, as the gully advances to position 2 and cuts into the Mullenix member, instability results because the Mullenix member can withstand only a 217-cm vertical wall. As the gully head moves from position 2 to position 3, stability is increased. Further gully growth up the drainageway into the Watkins unit again decreases stability.

This concept of abrupt changes in gully wall instability due to strength differences in the stratigraphic layers can be readily seen in the channel failures in the easily identifiable sandy alluvium in Mississippi (Grissinger, personal communication, 1977) and in gully failures in thick zones of sandy glacial outwash below loess in central Missouri. The gullies or channels seem to erode very little for many years; and then, in a matter of a few days, gullying can create huge voids in the land surface. This accelerated erosion is most often associated with eroding through to a weak layer and the alteration of the resistance forces of the system.

DEBRIS TRANSPORT

As previously stated, the production of soil debris by gully bank and head-scarp failure is a necessary, but not sufficient, condition for gully growth in this loessial region. In order to maintain gully erosion, runoff energies are required to entrain and transport soil debris. That is,

$$G = f(S, Q) \tag{5}$$

where G = gully growth rate
    S = debris supply function
    Q = runoff variable

Figure 13 (Piest et al., 1975a) shows gully sediment transport rates for May 25, 1972, at watershed 1, based on 30 streamflow samples. The supply of soil debris in the gully was exhausted (during a period near the runoff peak), and the transport of gully materials was essentially zero. Temporary gully cleanout for this storm occurred on the hydrograph

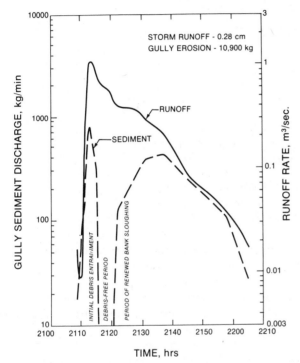

FIGURE 13 *Gully erosion rate, storm of May 25, 1965, at watershed 1.*

recession prior to a runoff rate of about 0.84 m³/sec. Piest et al. (1975a, 1975b) reported the same processes in several other storms; that is, gully sediment discharge was maximum soon after runoff began but declined sharply during subsequent periods of high runoff. The large reduction in transport rate of eroded soil from the gully verifies that the tractive forces of runoff along the channel boundary do not play a major role in gully erosion but that soil debris from gully wall failure is the primary material removed from gullies of loessial regions.

If tractive force on the channel boundary had been the predominant eroding agent, either by direct shear on the channel boundary or by bank undercutting, a discharge of gully sediments would have been proportional to the stream velocity, and transport of gully sediments at significant runoff rates would not have been minimal. Piest et al. (1975a) measured runoff velocities exceeding 3m/sec in drainageways that showed little deterioration. If we assume that the soil debris accumulated in gullies is essentially co-

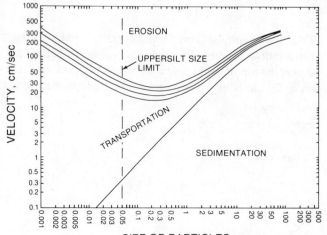

SIZE OF PARTICLES, mm
THE CURVES FOR EROSION AND DEPOSITION OF
A UNIFORM MATERIAL-E HJULSTROM

FIGURE 14 *Hjulstrom's (1935) critical drag curves, separating the flow regions where entrainment, deposition, and steady carriage occur.*

hesionless, Figure 14 (Hjulstrom, 1935) shows that nearly all discrete silt particles can be entrained and transported at velocities exceeding 0.61 cm/sec; indeed, most silt aggregates are transportable by the velocities commonly encountered.

Once the intermittent nature of debris supply and the transport effectiveness of runoff are known, a predictive model can be developed. Although the intricate input relations to equation 5 are poorly defined at present, the relative import can be postulated. For example, moderate storm runoff velocities are commonly found whenever flow depths exceed some minimum value (Fig. 15). The Manning formula was used to compute hydraulic radius and flow velocity for a typical gully cross section (Fig. 15):

$$V = \frac{K}{n} S^{1/2} R^{2/3} \tag{6}$$

where K = constant ($K = 1$, in SI units)
    V = mean channel velocity, m/sec
    n = roughness factor
    S = hydraulic gradient
    R = hydraulic radius, m (stream cross-sectional area/wetted perimeter)

Figure 15 summarizes hydraulic geometry and flow characteristics for several depths of flow and two approach slopes. Typically, the hydraulic radius, and therefore the

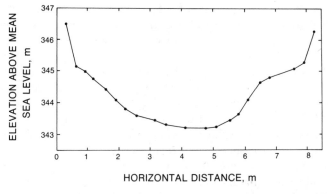

Flow depth (m)	Stream cross-sectional area (m²)	Wetted perimeter (m)	Hydraulic radius (m)	n	Vs_1[1] (m/sec)	Vs_2[2] (m/sec)
0.30	0.16	2.68	0.02	0.1	0.48	0.59
0.61	1.52	3.96	0.38	0.1	0.75	0.91
0.91	2.71	4.69	0.58	0.1	0.98	1.20
1.22	4.04	5.55	0.73	0.1	1.14	1.40
1.52	5.53	6.40	0.86	0.1	1.28	1.57
1.83	7.21	7.38	0.98	0.1	1.39	1.70

[1]s_1 = a hydraulic gradient of 0.02
[2]s_2 = a hydraulic gradient of 0.03

FIGURE 15 *Hydraulic geometry—discharge relationships for typical cross section (top figure) for watershed 3 near Treynor, Iowa.*

velocity, increase rapidly with depth. Although it would be difficult to select any threshold runoff or velocity that would initiate erosion, Figure 16 attempts to define a threshold runoff volume that will sustain gully erosion. Cumulative storm runoff volume for

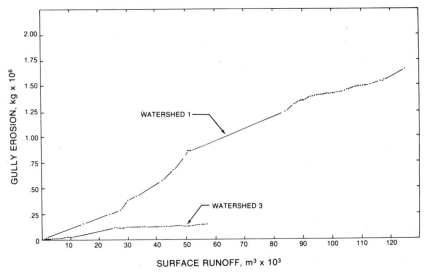

FIGURE 16 *Cumulative gully erosion and surface runoff at watersheds 1 and 3, Treynor, Iowa, for 1967–70.*

1967–70 is plotted versus cumulative gully erosion for 59 storm events at watershed 1 and 44 events at watershed 3. Since watershed 3 was in a conservation management (grass), runoff for comparable storms was less than at rowcropped watershed 1, and 15 of the 44 rainfall events caused insignificant runoff.

If we assume that watershed 1/watershed 3 are representative nonconservation/conservation watersheds that drain through raw and active/healed and vegetated gully systems, then several conclusions can be made from Figure 16. Nearly any volume of runoff is adequate for removing gully debris from watershed 1, whereas the only appreciable gully erosion at watershed 3 required a storm runoff volume of 15,000 m^3 with a peak discharge of 6.1 m^3/sec. We can conclude that this large runoff exceeded the threshold necessary to initiate erosion in a vegetated gully.

Figure 17 describes several geomorphic stages or thresholds that pertain to a raw and active gully. At stage I, runoff is insufficient ($<Q_1$) to erode a gully, regardless of the supply of loose soil debris that is available for transport. Gully erosion under stage II circumstances is dependent upon debris supply and independent of runoff. Although the supply has been shown to vary within and between storm runoff events, the long-term supply can be considered essentially constant. Therefore, the erosion rate for a particular gully would define a horizontal line somewhere along plane II. Runoff values greater than Q_2 represent the class of catastrophic landform changes that cause an unlimited supply of

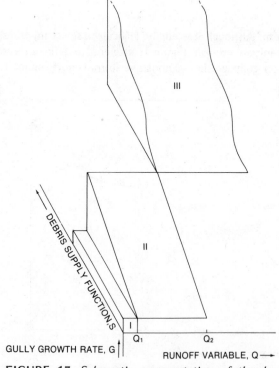

FIGURE 17 *Schematic representation of thresholds related to surface runoff and gully growth.*

debris to become available. This may have been caused by a "change of state," like the "washouts" near Fargo, North Dakota, in 1975 (Spomer et al., 1978) or by debris flows (Brown and Ritter, 1971).

PERSPECTIVES ON MODERN GULLYING AND GULLY CONTROL

Let us examine causes for the apparent rejuvenation of drainage systems, including gullying, in the context of Schumm's (1973) theories of geomorphic thresholds and the complex response of drainage systems. He cautioned against interpreting dynamic geomorphic phenomena on the basis of external causative variables when, in fact, some intrinsic erosion thresholds were exceeded. However, as Schumm implied, recognizing these erosion thresholds for areas of extensive agricultural activity can be extremely difficult.

The complexity of the relationship among climate, drainage area, and geomorphic characteristics may also be responsible for lack of establishing thresholds inherent to the landscape. We cannot be certain that the present cycle of gully erosion is due to land-use changes. Ruhe et al. (1967) recognized cycles of cut-and-fill (due to shifts in climate) in western Iowa during geologic time. We are uncertain about the influence of minor climatic shifts on gully formation (in the postsettlement period). A sensitive balance exists between the moisture regime (or hydrology) of the watershed and land surface instability. Tillage and cropping systems complicate the response of the valley floor to the processes causing gully erosion.

In this chapter we have attempted to relate gully erosion in agricultural watersheds to thresholds of mass wasting and debris transport. In the strictest interpretation of a geomorphic threshold, neither threshold is geomorphic. Neither threshold, using the words of Schumm, is inherent in the manner of landform change. That is not to say that true geomorphic thresholds are nonexistent; limiting the study to four watersheds would have, most likely, prevented such a finding.

The thresholds encountered in the processes of mass wasting of the gully walls are related to the stratigraphy of the land, and only indirectly related to the shape of the land surface. The shear strength samples represent only a limited location; however, the test results quantify what has been previously observed—that some soil layers are weaker than others. Not only is the strength per se of each successive stratigraphic layer critical to erosional processes, but also the relative permeabilities of each layer. Potentially unstable areas of the drainage basin can result from the concentration of flow lines, resulting in greater decreases in soil strength because of higher pore pressure gradients. Thresholds of the gully erosion processes are then reached.

The thresholds encountered in the debris transport process are directly related to climate and vegetation. Prolonged wet periods increase the likelihood of gully bank failure, but only when runoff exceeds some critical value does the rate of gully development rapidly accelerate. This can be seen in the data of Table 2 and Figure 16. The years with above-normal rainfall and periods of high intensity result in increased sediment loss from gullies. Land use and vegetation overshadow climate's effect on present-day gullying. Referring again to Table 2, we see that mulch tillage or level terraces greatly reduce gully growth. This is due, in part, to the prevention of debris cleanout because of not exceeding some critical value of runoff. It is also due to the action of the water in the plunge pool

FIGURE 18 *Scarp 2 m high in channel reach of Keg Creek near Underwood, Iowa.*

area of the gully headscarp. Increased flow volumes and velocities over the headcut supply energy to the plunge pool region of the gully. This causes local scour in this part of the gully; slope height and steepness in the vicinity of the wall are increased. A reduction in energy at the headcut of watersheds 3 and 4 is the probable cause of the decrease in gully erosion since 1964, when the conservation practices were implemented.

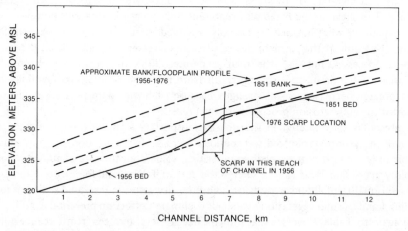

FIGURE 19 *Profile changes from 1851 to 1976 in an unstraightened channel reach of Keg Creek near Underwood, Iowa.*

Gullies that drain small upland fields are affected somewhat by the general degradation of downstream drainageways (valley trenches) to which they are tributary. Historic records throughout the loess hills region show a trend of base lowering that began after European settlement in mid-nineteenth century and has accelerated in the past half century. Piest et al. (1976) traced the development of the Tarkio River system, where some of the channel has degraded more than 6 m since dredging and straightening about 1920, with commensurate widening. Causes include changed base levels, coupled with channel alteration and drastically increased surface runoff brought about by agricultural land use.

Keg Creek, to which Treynor watersheds 1 and 2 are tributary, is typical of degrading streams that are causing upland dissection. Figure 18 shows a 2-m-high scarp migrating upstream in an unstraightened channel reach of Keg Creek near Underwood, Iowa. Figure 19 approximates the 1852 and the existing channel profile through the reach, as derived from the 1852 original land survey, the 1956 U.S. Geological Survey 1:24,000, 7 ½-min topographic map, and current measurements.

There has been an interaction of **extrinsic/intrinsic** stresses to accelerate gullying beyond any natural geomorphic cycle. Exceeded extrinsic thresholds result from anthropic land-use changes (that increased surface runoff) and overt channel alteration downstream. Intrinsic thresholds result from exceeding a particular soil or geomorphic parameter; this changes slopes and base levels of drainageways.

It is not possible to quantify the complex response of these drainage systems according to individual stimulus. We know, for example, that the higher flood runoff regime, generated by land-use changes, would alone have been sufficient to have altered (straightened) meanders and enlarged channels. Whatever the causes of deteriorating drainageways, however, the geomorphologist, engineer, and environmentalist should have one common goal—to understand the complex variables that are operating so that we can best manage this resource for the common good.

REFERENCES CITED

Allen, W. H., 1971, Landscape evolution and soil formation: Treynor, Iowa: Ph.D. thesis, Iowa State University, Ames, Iowa, 676 p.

Bailey, W. A., and Christian, J. T., 1969, ICES-LEASE-1, a problem oriented language for slope stability analysis: R69-22, Soil Mechanics Publ. 235, Dept. Civil Eng., Massachusetts Inst. Tech., Cambridge, Mass., 56 p.

Bishop, A. W., 1955, The use of the slip circle in the stability analysis of earth slopes: Geotechnique, v. 5, p. 7–17.

Bradford, J. M., and Piest, R. F., 1977, Gully wall stability in loess-derived alluvium: Soil Sci. Soc. Amer. J., v. 41, n. 1, p. 115–122.

——, Piest, R. F., and Spomer, R. G., 1978, Failure sequence of gully headwalls in western Iowa: Soil Sci. Soc. Amer. J., v. 42, n. 2, p. 323–328.

Brice, J. C., 1966, Erosion and deposition in the loess-mantled Great Plains, Medicine Creek drainage basin, Nebraska: U.S. Geol. Survey Prof. Paper 352-H, p. 255–335.

Brink, A. B. A., and Kantey, B. A., 1961, Collapsible grain structure in residual granite soils in southern Africa: Proc. Fifth Int. Conf. Soil Mechanics and Foundation Engineering, Paris, v. 1, p. 611–614.

Brown, W. H., and Ritter, J. R., 1971, Sediment transport and turbidity in the Eel River basin, California: U.S. Geol. Survey Water-Supply Paper 1986, 70 p.

Daniels, R. B., and Jordan, R. H., 1966, Physiographic history and the soils, entrenched stream systems, and gullies, Harrison County, Iowa: USDA Tech. Bull. 1348, 133 p.

Dudley, J. H., 1970, Review of collapsing soils: Journ. of the Soil Mechanics and Foundations Division, ASCE, v. 93, p. 925–947.

Fellenius, W., 1936, Calculation of the stability of earth dams: Trans. 2nd Congress on Large Dams (Washington), v. 4, p. 445.

Handy, R. J., 1973, Collapsible loess in Iowa: Soil Sci. Soc. Amer. Proc., v. 37, n. 2, p. 281–284.

Hjulstrom, F., 1935, Studies of the morphological activity of rivers as illustrated by the River Fyris: Bull. Geol. Inst. Univ. Uppsala, v. 25, p. 221–527.

Kane, H., 1973, Confined compression of loess: Proc. Eighth Int. Conf. Soil Mechanics and Foundation Engineering, Moscow, v. 2, p. 2, p. 115–122.

Lambe, T. W., and Whitman, R. V., 1969, Soil mechanics: New York, Wiley, 553 p.

Lohnes, R. A., and Handy, R. L., 1968, Slope angles in friable loess: J. Geol., v. 76, p. 247–258.

Lutton, R. J., 1969, Fractures and failure mechanics in loess and applications to rock mechanics: Res. Report S-69-1, U.S. Army Eng. Waterways Exp. Stn., Corps Eng., Vicksburg, Miss. 53 p.

Morgenstern, N. R., and Price, V. E., 1965, The analysis of the stability of general slip surfaces: Geotechnique, v. 15, p. 79–93.

Patton, P. C., and Schumm, S. A., 1975, Gully erosion, northwestern Colorado: a threshold phenomenon: Geology, v. 3, n. 2, p. 88–90.

Piest, R. F., Bradford, J. M., and Spomer, R. G., 1975a, Mechanisms of erosion and sediment movement from gullies: in Present and prospective technology for predicting sediment yields and sources, ARS-S-40, ARS-USDA, p. 162–176.

——, Bradford, J. M., and Wyatt, G. M., 1975b, Soil erosion and sediment transport from gullies: J. Hydraulics Div., ASCE, v. 101, p. 65–80.

——, Beer, C. E., and Spomer, R. G., 1976, Entrenchment of drainage systems in western Iowa and northwestern Missouri: Proc. Third Federal Interagency Sedimentation Conf., Denver, Colo., p. 548–560.

Rahman, A. U., 1973, The behavior of loess in confined compression: Ph.D. thesis, University of Iowa, Iowa City, Iowa, 234 p.

Reginatto, A. R., and Ferrero, J. C., 1973, Collapse potential of soils and soil-water chemistry: Proc. Eighth Int. Conf. Soil Mechanics and Foundation Engineering, Moscow, v. 2, pt. 2, p. 177–183.

Ruhe, R. V., Daniels, R. B., and Cady, J. G., 1967, Landscape evolution and soil formation in southwestern Iowa: USDA Tech. Bull, 1349, 242 p.

Saxton, K. E., and Spomer, R. G., 1968, Effects of conservation on the hydrology of loessial watersheds: Trans. Amer. Soc. Agricultural Engineers, v. 11, n. 6, p. 848–853.

——, Spomer, R. G., and Kramer, L. A., 1971, Hydrology and erosion of loessial watersheds: J. Hydraulics Div., ASCE, v. 97, n. HY11, p. 1835–1831.

——, Johnson, H. P., and Shaw, R. H., 1974, Modeling evapotranspiration and soil moisture: Trans. Amer. Soc. Agricultural Engineers, v. 17, n. 4, p. 673–677.

Schultz, J. R., and Cleaves, A. B., 1955, Geology in engineering: New York, Wiley, 592 p.

Schumm, S. A., 1973, Geomorphic thresholds and complex response of drainage systems: *in* Morisawa, M., ed., Fluvial geomorphology: New York State Univ. at Binghamton Pubs. Geomorphology, p. 299–310.

Soil Survey Staff, 1965, Land resource regions and major land resource areas: Agric. Handbook 296, SCS-USDA, Washington, D.C., U.S. Government Printing Office, 82 p.

Spomer, R. G., Piest, R. F., and Poggensee, R. L., 1978, Erosion on the Sheyenne River delta, North Dakota: submitted for presentation at the Tenth Int. Cong. Sedimentology in Jerusalem, Israel, July 9–14, 1978.

Strahler, A. N., 1963, The earth sciences: New York, Harper & Row, 681 p.

Turnbull, W. J., 1968, Construction problems experienced with loess soils: Highway Res. Record, v. 212, p. 10–27.

Von Engeln, O. D., 1949, Geomorphology: New York, Macmillan, 655 p.

6

THRESHOLDS AND VALLEY WIDTHS IN THE SOUTH RIVER BASIN, IOWA

Neil E. Salisbury

INTRODUCTION

Big rivers carve big valleys. Small streams create small valleys. This makes sense, and so it has seemed since Playfair first stated the most basic law of landform geometry.

> Every river appears to consist of a main trunk, fed from a variety of branches, each running in a valley proportioned to its size, and all of them together form-ing a system of vallies, communicating with one another, and having such a nice adjustment of their declivities, that none of them join the principal valley, either on too high or too low a level; a circumstance which would be infinitely improb-ably, if each of these vallies were not the work of the stream that flows in it (Playfair, 1802, p. 102).

While rather little testing of Playfair's statement has been done, considerable attention has been directed toward the underfitness of stream valleys, or valleys too wide for their present streams, particularly by Dury (1958, 1960, 1964a, 1964b, 1965). Aside from the underfit condition, however, scant attention has been paid to the geometric relationships between the sizes of streams and their valleys.

In an integrated stream network, the major valleys receive increments of discharge and sediment from their tributaries. What is the response of the main stream valleys to these increments? Is the whole (the major valley) equal to the sum of the parts (the tri-butary increments)? Does every tributary act as a kind of threshold, forcing the major stream to slowly and continuously attack its confining bluffs? Or is there a threshold value of tributary size, below which no impact can be seen? Do certain subbasins contri-bute most of the stream flow and therefore the impact upon valley geometry? If a stream basin is sufficiently compact areally and uniform in its geologic base, extrinsic thresholds can be eliminated from consideration and the answers to these questions would seem to involve geomorphic thresholds (Schumm, 1977, p. 7–8).

Climatic change will affect all portions of a relatively compact stream basin. While individual intensive storms may initiate abnormally high stream discharges in some tri-butary basins of a system and not in others, these should average out over a period of time. Moreover, intensive storms are part of the normal suite of climatic conditions in most parts of the world and are not indicative of major climatic change unless their patterns of frequency of occurrence changes. Similarly, the impact of diastrophic changes can usually be identified in compact basins.

Rzhanitsyn (1960, p. 3) notes that a stream confluence represents a jump in flow, and the result is a new river, not a continuation of the old stream. This is particularly true when the streams effecting a confluence are nearly equal in discharge. He also states that the basic law of structure of the river net is spasmodic change in stream and channel characteristics (1960, p. 97). However, such changes as width of channel smooth out and become continuous downstream.

Many stream basins in the Middle West have experienced repeated episodes of storage and flushing of sediments from the landscape. The evidence has been reported in the work of Knox and associates in Wisconsin and Ruhe and associates in Iowa (Knox, 1972; Knox and Johnson, 1974; Knox et al., 1975; Ruhe, 1967; Ruhe, 1969; Ruhe and Cady, 1967; Daniels and Jordan, 1966). While such episodes add complexity to explanations of the underlying causes of variations in valley geometry, they also result in situations that conform closely to the classic descriptions of geomorphic thresholds as expressed by Schumm (1977, p. 8).

In examining the role of thresholds in valley geometry I will restrict my attention to variations in the width of the valley bottom, a choice of dependent variable that has been rationalized earlier (Salisbury et al., 1968, p. 10–20). The focus will be on the South River basin of Iowa, which possesses characteristics that favor generalization of the conclusions to broader environments. Like most stream valleys, the South River and its major tributaries exhibit width variations which tend to confirm Rzhanitsyn's ideas concerning spasmodic change and allow us to examine the question of whether jumps in flow as a consequence of tributary entrances result in new valleys or merely continuations of the old.

The general question to be examined is: Do tributary confluences impose a threshold effect upon main stream valley width?

THE SOUTH RIVER BASIN OF IOWA

The basin of the South River (Fig. 1) was chosen for study because an earlier investigation revealed that the valley of the main stem manifests a high distance-decay relationship (Salisbury et al., 1968, p. 62–64). A perfect "normal" stream valley with narrow headwaters and a broad mouth would exhibit a relationship between valley width and distance from the mouth of $r = -1.00$. Mapping of the South River valley with measurements taken every 5 mi (8 km) revealed a relationship of $r = -0.96$. Thus, the South River is so close to perfect that it is an excellent choice for testing whether tributary contributions to main stem discharge have an impact upon valley geometry, specifically, the width of the valley bottom.

Critical to such a test is the matter of localized constraints upon valley-widening processes. The valley walls in the South River basin are gentle and often irregularly sloping with what Ruhe describes as a stepped profile, due to pedimentation of the Pleistocene deposits (Ruhe, 1967, p. 101). The lower slopes are concave upward, and choked with slope wash and colluvium. No rock crops out along the interface between valley wall and stream bottom. Bedrock must be searched for in the rare cutbank of the lowermost elevations, and in gully heads which transect the bluffs. The influence of rock, then, is subtle at best, and probably long forgotten by the major streams. Perhaps during erosional periods long past (Ruhe, 1969, p. 129–168), the contest between constricting rock layers and meandering streams was active. Today the streams encounter mostly soft materials along their bank and bluffs.

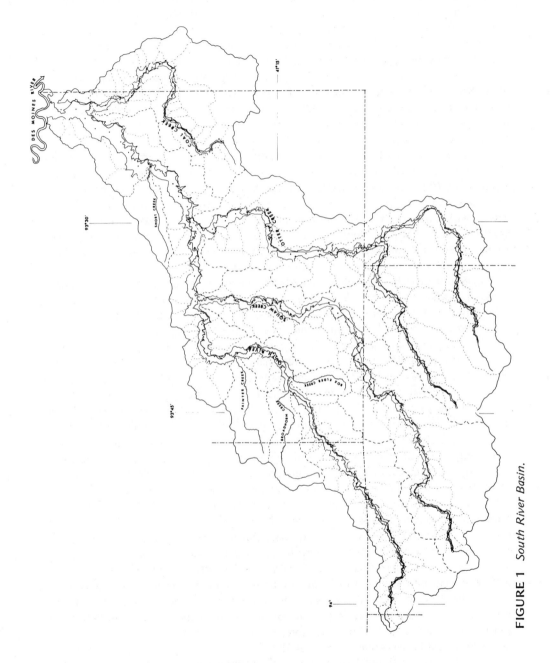

FIGURE 1 *South River Basin.*

105

While bedrock control of valley widths is minimal under present conditions, some constrictions may exist in association with buried bedrock high (Salisbury et al., 1968, p. 63). Where valley widening has probably been episodic and associated with more general erosional periods, it is possible that relict constrictions lie buried beneath more recent sediment storage sites. Mapping of the bedrock topography is at too coarse a scale to reveal the details (Fig. 2), but the highest elevations of the bedrock surface are in the headwaters of the basin. Here the present streams frequently ignore the configuration of the bedrock surface, including the buried Decatur Channel. In downstream reaches of the basin the streams are confined to bedrock gorges, and it is here that buried relict constraints may be most important in defining valley widths. Details of the association between the bedrock surface and valley widths will be considered later.

The South River basin has an area of 590 mi². The South River empties into the Des Moines River at an elevation of slightly less than 750 ft above sea level. The headwaters are at about 1225 ft above sea level, creating a basin relief of 475 ft. Three major tributaries and many smaller ones join the main stem (Fig. 1). Squaw Creek is farthest upstream, with a basin area of 136 mi² and a relief of 415 ft. Otter Creek, with two main branches, joins a few miles downstream, adding its 168 mi² area to the total basin. It has a relief of 360 ft. Coal Creek, with an area of 75 mi² and relief of 225 ft, joins the main stem a few miles upstream of its confluence with the Des Moines River.

The South River basin's well-integrated stream network is carved into a loess-mantled Kansan glaciation drift plain. Underlying the drift are hundreds of feet of Pennsylvanian sediments—Des Moines series sandstones, shales, and coal beds near the mouth of the basin, and Missouri series limestones and shales to the southwest. Drift thickness varies from zero at stream cutbanks and gully heads near the mouth of the basin to over 200 ft in the headwaters and overlying the Decatur Channel (Fig. 2). The loess mantle is Wisconsinan in age and has been stripped off the lower slopes of hillsides by episodes of erosion similar to those that created the stepped profiles of the interfluves.

According to Ruhe (1969, p. 156) the major valleys of the region were formed during the later Yarmouth (post-Kansan) interglacial episode. The stepped profiles of the interfluves developed in several stages. The upland surface of the Kansan drift plain was sufficiently stable to permit intensive soil development. These flattish upland tracts also retain the Loveland loess deposited during the Illinoian glacial episode. Subsequent soil development during the Sangamon interglacial period cannot be distinguished easily from Yarmouth soils in this region, and the resulting weathering profile is known as the Yarmouth-Sangamon paleosol. This forms the highest-level erosion surface, which is mantled and protected by Wisconsinan loess. During Late Sangamon time an erosion surface was cut by pedimentation below the Yarmouth-Sangamon surface, truncating the latter. This is the intermediate level. Finally, during Early Wisconsin time an erosion surface was cut into Kansan drift below the Late Sangamon surface, forming the lowest level (Ruhe, 1967, p. 102). All three surfaces—Yarmouth-Sangamon, Late Sangamon, and Early Wisconsin—have been subject to erosion and sedimentation during Late Wisconsin and Holocene time. Valley slopes, therefore, cannot be older than Late Wisconsin and are probably Recent in age (Ruhe, 1967, p. 149).

Gully cutting and filling developed on these surfaces commencing about 6800 radiocarbon years ago. This episode has been associated with climatic change which resulted in a vegetation cover shift from forest to grassland (Ruhe, 1967, p. 153–155). Valley fills

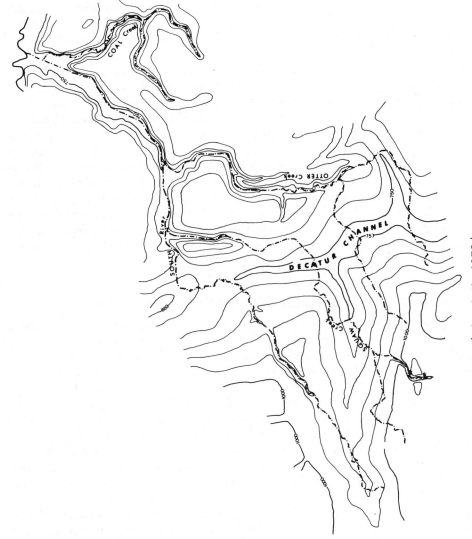

FIGURE 2 Bedrock topography. (After Cagle, 1973.)

certainly received sediment during all these erosional periods, but are obscured by even more recent periods of alluviation. Daniels and Jordan, reporting on their work in Harrison County (to the west of the South River basin), suggest that five periods of erosion and alluviation have occurred in the past 15,000 years, three of these in the past 2000 years (1966, p. 36–37). Erosion rates varied from 0.23 ft/100 yr to 0.36 ft/100 yr, resulting in valley alluviation rates of 0.85 ft/100 yr to as much as 5 ft/100 yr. Their evidence led them to conclude that "The alluvial history of these watersheds indicates that landscapes do not develop through continuous slow erosion and alluviation but through periods of erosion and alluviation, probably followed by a period of stability" 1966, p. 38).

Thus, the valleys of the region in which the South River basin is located were first formed tens or even hundreds of thousands of years ago. The valley walls have been subjected to at least four major periods of erosion in the last few tens of thousands of years. And the valley bottoms, whose surfaces are no more than few thousand years old, have experienced several periods of erosion and alluviation in Holocene time. This geomorphic history has been episodic in nature, with periods of relative stability or sediment storage followed by periods of sediment flushing. Postsettlement alluviation in the Middle West has been followed by a more recent episode of renewed gully cutting and sediment flushing (Daniels and Jordan, 1966; Knox, 1972; Knox et al., 1975). Certainly, the complexity of this episodic erosion has afforded the streams of the South River basin ample opportunity to broaden their valleys in response to the varying inputs of discharge and sediment. Explanations of valley width variations, however, may be obscured by the episodic nature of the processes involved.

MEASUREMENT OF VALLEY WIDTH AND BASIN AREAS

The boundaries of valleys in the South River basin were interpreted from USGS topographic maps of varying scales, dates of publication, and levels of accuracy. The South River main stem, Otter Creek, and South River's two major branches, Squaw Creek and Coal Creek, were the streams whose valleys were mapped. The downstream end of the basin, including the entire Coal Creek basin, is covered by 1:24,000 scale maps with a contour interval of 10 ft which comply with National Map Accuracy Standards. The lowermost 6 mi of Squaw Creek valley and the South River main stem valley from Mile 15 to Mile 30, as well as short segments of Otter Creek valley are on the Indianola 15-minute (1:62,500 scale, 20-ft contour interval) quadrangle, which was published in 1931 and does not conform to National Map Accuracy Standards. A small segment of Otter Creek valley is on the Chariton 15-minute quadrangle (20-ft contour interval), published in 1918. This segment could not be interpreted accurately from the topographic map and more recent soil maps (1:31,680 scale) were used as the data source. Most of the Otter Creek, Squaw Creek, and upstream portions of the South River main stem basins are on the New Virginia and Osceola 15-minute quadrangles (20-ft contour interval) which do conform to National Map Accuracy Standards.

After mapping the limits of the valley bottoms, sample sites for width measurements were selected on a nonrandom, ronregular spacing basis (Fig. 3). An attempt was made to include the range of values within the population of valley widths that make up a given reach. The broadest and narrowest parts of the valley where tributary crenulations of

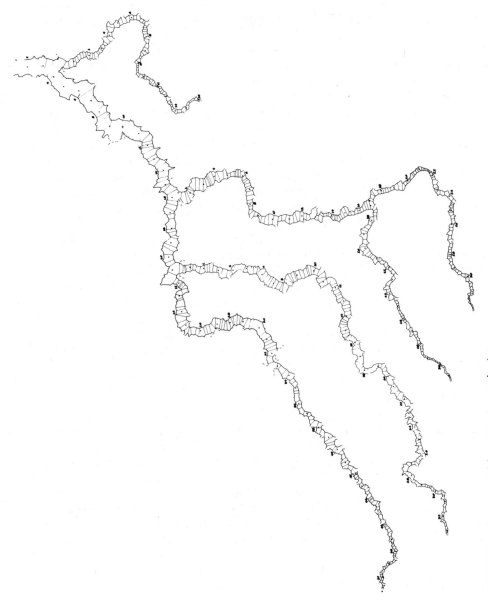

FIGURE 3 *Valley width measure sample sites.*

valley walls did not interfere were chosen as sample sites. Thus, valley width measuring sites cross the valley approximately normal to its trend at successive wides and narrows. No attempt was made to weight the number of sites chosen according to the length of a wide or narrow reach. Rather, my purpose was to preserve the spatial integrity of widening or narrowing of the valley downstream of tributary entrances. The distance of the width measuring site from the mouth of the valley was determined by a map measure or opisometer.

Valley form and width variations are revealed in Figures 1 and 3. As Dury has pointed out, underfitness is not always evident manifestly (1964a). The South River valley is a good example of a meandering stream (before channelization several decades ago) in a relatively nonmeandering valley. Evidence of former high flows can be sought by sub-surface exploration, and for precise, final conclusions concerning their magnitude this is what is needed. But valley geometry (especially width) can also reveal underfitness. Former high levels of flow ought to have had an impact on the intensity and persistence of valley wall erosion and should be revealed in systematic downstream modifications. The major valleys of the South River basin have bends and are occasionally sinuous, but rarely display a meandering pattern for more than one or two bends in succession. Stream channels are obviously meandering at smaller wave lengths, but because of the lack of meandering valleys must be classified as Dury's Type 5 underfitness pattern. Regardless of whether valleys are meandering or straight, thresholds of tributary impact should be revealed. Valleys are conservative; they represent discharge conditions at the period when underfitness was established. Even present-day floods which fill valley bottoms are in-effective in widening valleys. As Dury points out, "field observation shows that quite exceptionally high floods do not erode the valley walls except where the present channel impinges upon them" (1964a, p. A41). The actual work of widening the valley is thus accomplished by stream channels which do not reflect relict conditions, but are adjusted to equilibrium conditions based on more recent discharges. The impact of tributary con-fluences upon channel form deserves a fuller treatment than can be included here, and is reported elsewhere.

Modern stream discharges may not be a good indicator of the amount of water that was contributed by tributary basins under past environmental conditions. Moreover, discharge figures are not available for all the tributaries of a stream basin, except in rare, short-term research situations. For purposes of discharge prediction for small basins in bridge building and dam or road construction, engineers employ equations which include a variety of factors. The relevant equations for this part of Iowa consider only basin area and channel slope to be significant (Schwob, 1966, p. 7). Because tributaries of the major streams of the South River basin have similar slopes, basin area alone is used as a sur-rogate for discharge. This assumes that small subbasins within a compact, low-relief basin should share equally in precipitation inputs over an extended period of time; con-sequently, discharge contributions will be proportional to subbasin area. The areas of upland portions of tributary basins were measured with an Imlac PDS-4 Graphic Display System with a Graf-pen GP3 Sonic Digitizer.

HYPOTHESIS FOR EXPLANATION OF VALLEY WIDTH VARIATIONS

The "Valleys of Iowa" model of stream valley form for "normal" humid environment or effluent streams proposes a fairly constant decline or decay in width upstream from

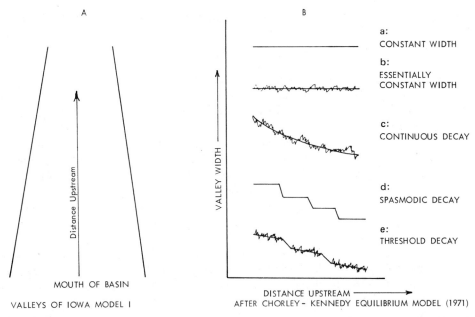

FIGURE 4 *Valley width models.*

the mouth of the basin as shown in Figure 4A (Salisbury et al., 1968, p. 48). As noted earlier, the distance decay relationship for the South River valley is $r = -0.96$. Even as close a fit to the model as this may obscure the pattern of how downstream width increases are achieved. Are these constant or continuous increases, or are these spasmodic jumps downstream of tributary confluences, as Rzhanitsyn implies (1960, p. 97)? To restate the general question asked by this research: Do tributary confluences exert a threshold effect upon valley widths, and if so, with what pattern?

Variations in valley width exist in every valley which is not straight-jacketed by intransigent rock walls or concrete. For an idea of how form variations may occur with distance, we can tip the Chorley-Kennedy equilibrium model (1971) (as shown in Fig. 1-2 in Schumm, 1977, p. 5) on its side and consider the curves as spatial changes rather than time relationships (Fig. 4B). Curves a and d can be eliminated from further consideration at this time because these are distance-decay relationships that would obtain only in special circumstances—a concrete-lined channelized stream or one unusually constrained by rock walls. Curve b, with local width variations obscured by an overall constancy in distance decay, exists in some natural streams but is not of concern here because we have already established that there is, indeed, a decay of valley width with distance upstream from the mouth of the South River basin ($r = -0.96$). This leaves as our most important question whether the decay in valley width as seen in Figures 1 and 3 is essentially spatially continuous (whether the function is linear or curvilinear is not our primary concern) as in curve c—the *continuous decay* case—or if certain tributary confluences conbribute sufficient discharge that a jump in valley width occurs—curve e, or *threshold decay*. In addition to the overall pattern of decay we need to consider possible explanations for variations in valley width. Local variations in width become general if they are repeated, or if they are sufficiently large so that they provide the spasmodic jumps of curve e. The explanations considered here are posed as hypotheses, not all of which are tested in this study.

Hypothesis 1: Valley width variations have spatially unique solutions. Most obvious is an association with the underlying bedrock surface. Valley wides are related to buried bedrock channels. Valley narrows are constrained by rock gorges.

Hypothesis 2: Valley width variations are spatially random—their patterns in space cannot be related to causative factors.

Hypothesis 3: Valley width variations are spatially regular or cyclical. This would suggest that while width variations are indeed spasmodic, they occur at regular pulses in the decay relationship. It is conceivable that such regularities have hydraulic explanations, not unlike those underlying meander bends or pool and riffle sequences.

Hypothesis 4: Valley width variations show a continuous downstream response to increments in discharge from tributary confluences without regard to tributary size. Essentially, this is the pattern of curve c, continuous decay.

Hypothesis 5: Valley width variations show a discontinuous downstream response to increments in discharge from tributary confluences that is a function of tributary size. This is the pattern of threshold decay in curve e, with tributary confluences associated with the steps or jumps in valley size.

Hypothesis 6: Valley width variations show a discontinuous downstream pattern which is related to slope variations in the valley and channel which are associated with narrow and wide reaches of the valley. This hypothesis assumes that sediment storage and flushing within the valley is more important in explaining width variations than tributary confluences per se.

Hypothesis 7: Valley width variations downstream show a response to relict widening processes that is associated with variable precipitation inputs to different tributary basins. This can be called the "Big Thompson Canyon" effect. Do catastrophic storms affecting only certain tributaries leave their impact for millenia afterward? In effect, we have already disposed of this hypothesis with the general assumption of tributary discharge increments being a function of tributary basin area, but it must remain as a possible explanation for deviations from the general case.

TESTS OF THE HYPOTHESES

Distance-Decay Relationships

The first step in examining the role of tributary discharge increments in valley width variations is to establish the general distance-decay relationship for each of the valleys in the South River basin. Valley widths were measured at the sample sites shown in Figure 3 and simple regressions were calculated with distance of the site from the mouth of each stream as the independent variable (Table 1). Because Otter Creek has two nearly equal-sized branches joining halfway up its length, these are shown separately ("above junction") and as though each is the main stem of Otter Creek, the other serving as a tributary ("total").

Spatial autocorrelation is ignored. Each successive measure is treated as an independent value, which may or may not be a rational assumption. It is irrational in the sense that a small creek is not likely to possess a narrow valley created by discharges of less

TABLE 1 *Regression Results, Valley Width[a] (Y) on Distance (X) (Widths and Distances in Miles)*

Stream	Basin area[b] (mi^2)	Sample points	Regression equation	r	r^2	S_{yx}
South River-main stem	590	192	$Y = 0.99 - 0.017X$	−0.85	0.73	0.15
Coal Creek	75	76	$Y = 0.37 - 0.017X$	−0.72	0.51	0.08
Squaw Creek	136	162	$Y = 0.58 - 0.013X$	−0.82	0.66	0.09
Otter Creek,-below junction	168 (66)	78	$Y = 0.59 - 0.018X$	−0.47	0.22	0.16
North Otter Creek,-above junction	55	69	$Y = 0.70 - 0.021X$	−0.82	0.67	0.06
North Otter Creek,-total	121 (+47)	147	$Y - 0.58 - 0.016X$	−0.76	0.57	0.12
South Otter Creek,-above junction	47	83	$Y - 0.43 - 0.010X$	−0.57	0.32	0.06
South Otter Creek,-total	113 (+55)	161	$Y - 0.55 - 0.015X$	−0.74	0.55	0.12

[a]Valley width measurements from topographic maps.
[b]Basin areas from Larimer (1957).

than 1000 cfs at one point in space succeeded by a valley width as broad as the Missouri River's at the next measurement point. It is rational in that considerable width variations in a reach have been noted in the field, and except for those situations where bedrock constrains the attack on valley walls, streams apparently are free to adjust their meander traces just as channels vary considerably in width. In short, while valley widths do not vary by several orders of magnitude within short distances, they often vary by 100 to 200%. The assumption is made that this range of variation suffices to indicate relative independence in the width values. Spatial autocorrelation is not forgotten; the question is simply deferred to future research.

Note that the South River main stem valley has an r of −0.85 compared to the value of $r = -0.96$ recorded in the earlier study (Salisbury et al., 1968, p. 62 and Fig. 15). The present study examined widths at many more sample sites, each valley wide and narrow, rather than simply at 5-mi intervals. The regression equation for the South River main stem in the earlier study was "$Y = 0.89 - 112.41X$," with an S_{yx} of 0.09 mi. Intercept values are comparable in these equations, but b values differ because of measurement differences.

The slopes of the regression equations in Table 1 cluster in the vicinity of −0.01 to −0.02. All b values are statistically significant at the 0.01 level. These values describe the rate of distance decay, which is fairly uniform among the valleys overall. Anomalies from the regression equations may be explained by the hypotheses listed earlier. The correlation coefficients and standard errors give some indication of the degree of variation in each valley. Visually, Otter Creek-below junction and South Otter Creek-above junction appear to have the most variable valley widths (Fig. 1).

The plots of valley width with distance from the mouth reveal considerable variations in the distance-decay relationships when examined spatially (Fig. 5). Regression lines and

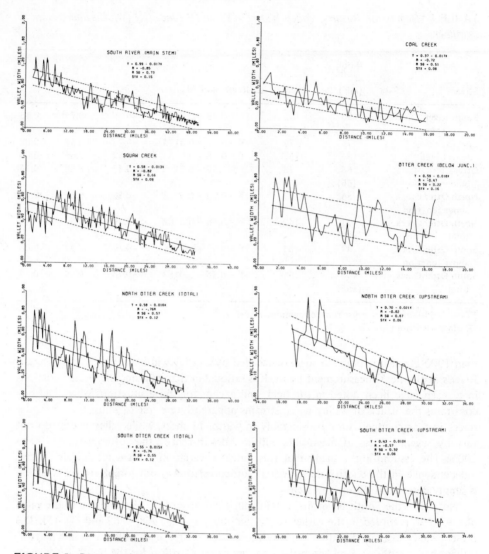

FIGURE 5 *Distance decay plots.*

standard error bands have been added to the plots to highlight the anomalies, or residuals from regression. All of the valleys have actual widths which are more than one standard error from the predicted widths. These frequently occur in series, as near the mouth of the South River main stem or the lower 10 mi of Squaw Creek, which displays both an overpredicted and underpredicted series in this reach. The most common characteristic of all these graphs, however, is that even when wides or narrows occur in series, extreme fluctuations in valley widths are evident. This rapid reversal of trends on valley widths complicates the task of interpretation of the possible role of tributary increments in valley widening. Statistical analysis is necessary to sort out actual from apparent impacts.

One of the more obvious relationships to examine is the rate of change of valley widths downstream from the headwaters. This is given by the *b* values or regression coefficients in Table 1. The uniformity of *b* values, however, is not promising as a means of revealing whether tributary discharge increments have an effect on valley width. The South River main stem was divided into segments in order to determine whether regression analysis would be more revealing for individual reaches of a stream than as an aggregate for the entire valley. The results are shown in Table 2. Only segments C and H had significant *b* values. Pairs of segments were subjected to chi-square tests to determine whether their mean valley widths were significantly different. In upstream order, segments B-C, C-D, E-F, and F-H were significantly different at the 0.01 level, as were C-F, A-C, and A-H.

TABLE 2 *South River Main Stem: Relationship Between Valley Width in Miles (Y) and Distance from Mouth in Miles (X), by Segments*

Segment	*N*	Regression equation	r^2	*r*	S_{yx}	$\bar{Y}$	S_y
A. Below Coal Creek (mile 2.1 to 3.5)	5	$Y = 0.80 + 0.104078X$	0.16	0.40	0.13	1.10	0.14
B. Coal Creek to Short Creek (mile 4.9 to 9.0)	17	$Y = 1.16 - 0.023727X$	0.04	−0.19	0.20	1.01	0.20
C. Short Creek to Otter Creek (mile 9.9 to 14.7)	18	$Y = 1.12 - 0.031293X$	0.21	−0.46	0.09	0.74	0.10
D. Otter Creek to Squaw Creek (mile 16.2 to 20.6)	14	$Y = 0.73 - 0.008415X$	0.01	−0.08	0.15	0.57	0.16
E. Squaw Creek to Painter Creek (mile 21.2 to 26.6)	25	$Y = -0.06 + 0.022782X$	0.10	0.31	0.11	0.49	0.12
F. Painter Creek to Box Elder Creek (mile 26.8 to 31.8)	26	$Y = 0.18 + 0.13708X$	0.08	0.28	0.07	0.58	0.08
G. Box Elder Creek to Broadhorn Creek (mile 32.3 to 32.7)	2					0.39	
H. Broadhorn Creek to head (mile 33.0 to 50.8)	82	$Y = 0.965 - 0.016799X$	0.49	−0.70	0.09	0.24	0.12

This approach was abandoned because valley widths do not increase uniformly in a downstream direction. Segment F is significantly wider than segment E and is approximately the same width as segment D. Except for this aberration, however, there is a systematic broadening of the valley downstream, although the width differences between segments are not always statistically significant. Three of the segments (A, E, and F) actually have positive *b* values, disclosing a tendency toward an increase upstream in valley width rather than the expected decline.

Because regression analysis for segments downstream of major tributaries had mixed results, it was decided to concentrate upon the individual width measures and their relationship to drainage basin increments of area. Both valley widths at the point of tributary entrances and width differences from above and below tributary entrances were considered.

The relationship between valley width at-a-point and drainage area increment at-a-point was not revealing. For Coal Creek $r^2 = 0.0044$ and for the South River main stem r^2

= 0.049. There is no reason to believe that there should be a strong relationship between these variables, actually, because small increments in drainage area will occur not only in narrow headwater reaches but also in wide downstream reaches.

The relationship between valley width at-a-point and the total drainage area upstream of that point is much stronger (Table 3). This confirms the belief that total contributions of discharge must result in wider valleys in the aggregate, but the relationship fails to reveal whether individual tributary confluences have an impact. For this we must turn to "width differences downstream," or the difference in valley width upstream and downstream of a tributary confluence.

TABLE 3 *Relationship Between Valley Width At a Point (Y) and Drainage Area Upstream (X)[a]*

Stream	Sample points	Regression equation	r^2	r
Main stem	65	$W = 0.001328A + 0.273099$	0.66	0.81
Main stem upstream of major tributaries	33	$W = 0.004806A + 0.111562$	0.46	0.68
Coal Creek	22	$W = 0.003794A + 0.123354$	0.65	0.80
Squaw Creek	42	$W = 0.002296A + 0.16976$	0.73	0.85
North Otter Creek, to junction	19	$W = 0.004451A + 0.07720$	0.64	0.80
South Otter Creek, to junction	14	$W = 0.003472A + 0.151633$	0.37	0.61
Otter Creek, downstream of junction, no upstream	18	$W = 0.00415A + 0.2670$	0.38	0.62
Otter Creek, downstream of junction, upstream as headwaters	19	$W = 0.002773A + 0.04892$	0.17	0.41
Otter Creek, North Fork through main stem downstream junction	36	$W = 0.002203A + 0.136823$	0.54	0.74

[a]Uncorrected for opposite valley wall low-order increments.

As Table 4 indicates, there is essentially no difference between valley widths downstream of a confluence ("drainage area increment") and the width of the valley upstream of the tributary. The results of this particular regression analysis are most damning to the general question posed by this research. Only one stream, Otter Creek downstream of the junction of its branches, shows tributary increments in discharge to be significant in affecting valley widths (technically speaking, at least, this is the only stream for which the relationship is significantly greater than 0, at the 0.05 level).

The question may be raised as to whether a single measuring point or width difference can be expected to respond to an increment of discharge. Figure 5 disclosed that width values ordinarily fluctuate greatly from one site to the next. More meaningful would be the change in width for the reach downstream of a tributary entrance. If discharge increments fail to make an impact in an entire reach of a valley, this is much more conclusive than when it occurs at a single point in the valley. As Table 5 shows, this is precisely the case. Considered as entities, tributary confluences make little impression upon valley width.

TABLE 4 *Relationship Between Width Difference Downstream (Y) and Drainage Area Increment at a Point (X)*[a]

Stream	Sample points	Regression equation	r^2	r
Main stem	64	$W_{diff} = 0.000076A + 0.034010$	0.00036	0.0189
Main stem upstream of major tributaries	32	$W_{diff} = 0.008487A + 0.050282$	0.0282	0.1680
Coal Creek	20	$W_{diff} = -0.002529A + 0.017369$	0.0136	0.1166
Squaw Creek	41	$W_{diff} = 0.004804A + 0.004421$	0.0397	0.1993
North Otter Creek, to junction	17	$W_{diff} = -0.000739A + 0.004539$	0.0033	0.0572
South Otter Creek, to junction	13	$W_{diff} = -0.005488A + 0.060650$	0.0482	0.2197
Otter Creek, main stem downstream junction	17	$W_{diff} = 0.037078A - 0.154970$	0.3184	0.5642
Otter Creek, North Fork through main stem downstream junction	34	$W_{diff} = 0.007928A - 0.040062$	0.0534	0.2312

[a]Uncorrected for opposite valley wall low-order increments.

TABLE 5 *Relationship Between Width Difference Downstream—Mean for Section and Drainage Basin Increment at Head of Section*

Stream	Sample points	Regression equation	r^2	r
South River, main stem	63	$\overline{W}_{diff} = 0.000011A + 0.010999$	0.000024	0.0049
South River, main stem upstream of tributaries	31	$\overline{W}_{diff} = -0.014508A + 0.040253$	0.13	-0.37
Coal Creek	20	$\overline{W}_{diff} = -0.001532A + 0.008371$	0.05	-0.22
Squaw Creek	41	$\overline{W}_{diff} = 0.001208A + 0.002933$	0.01	0.12
North Otter Creek, to junction	17	$\overline{W}_{diff} = 0.001518A + 0.006905$	0.08	0.28
South Otter Creek, to junction	13	$\overline{W}_{diff} = 0.001356A - 0.00230$	0.14	0.38
Otter Creek, downstream of junction	17	$\overline{W}_{diff} = 0.014651A - 0.053373$	0.08	0.29
Otter Creek, North Fork through main stem downstream of junction	35	$\overline{W}_{diff} = 0.000202A + 0.001412$	0.0004	0.02

Although aggregate statistical relationships fail to support the hypothesis that tributary increments in discharge have an impact on valley widening, the possibility remains that larger tributaries could have such an effect. Is there some drainage area increment level which serves as a threshold for main stem valley widening downstream? Three possibilities exist. If basin area increments do not in the aggregate serve as threshold for valley widening: (1) it may be that larger tributary increments are submerged in the statistical relationship by the great number of small tributary increments, (2) it may be that in the South River system larger tributaries are not yet at this threshold level, or (3)

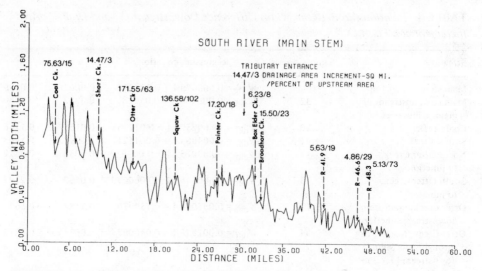

FIGURE 6   *Distance decay plot, South River main stem.*

the buried bedrock surface may be related to surface valleys in such a way as to overwhelm a normal threshold response. Individual valleys were examined to determine their response to large tributaries.

The valley of the South River main stem (Fig. 6) belies the notion that there is a threshold of tributary importance associated with valley widening. This is true, at least, if we examine tributary increments in discharge relative to upstream drainage area (note that Figs. 6–11 indicate both drainage area increment at a point in square miles, and tributary incremental area as a percentage of upstream basin area). Only Coal Creek, which

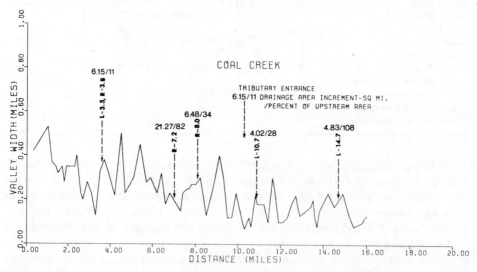

FIGURE 7   *Distance decay plot, Coal Creek.*

adds a basin increment 15% of the main stem drainage basin, serves to significantly increase valley size. Other tributaries have little or no effect, except for the combined contributions of Box Elder and Broadhorn Creeks. Often even the largest increments, such as
Squaw Creek (which more than doubles the size of the basin) are accompanied by narrowings of the valley immediately downstream. It should be noted that Figures 6–11
show only the large tributary entrances, either greater than 5 mi² or an increment of 10%
of the upstream drainage area. Smaller tributary entrances reveal similar patterns of valley
width increase and decrease downstream.

Coal Creek valley produces similar results (Fig. 7). Two of five tributary entrances
increase valley width, three are accompanied by decreases. From upstream, an approximate doubling of drainage area near the headwaters is accompanied by a decrease in
valley width. At mile 10.7 a tributary with 28% of the area of the upstream main basin
produces a slight widening, but this is immediately reversed and width does not "recover"
for nearly 2 mi, and then with only minor tributary confluences and low-order stream
increments (to mile 9.1, W = 0.40 mi). A significant tributary at mile 8 is followed downstream by continued narrowing for nearly 1.5 mi. A very large tributary entrance at mile
7.2 is followed downstream by an erratic but persistent widening of nearly 2 mi. A
smaller tributary entrance downstream is followed by pronounced valley narrowing.
Thus, in the Coal Creek valley, only one tributary entrance, the largest increment downstream of the headwaters, acts as a threshold in increasing valley width. Four out of five
large tributary confluences have essentially no positive impact on valley width.

In Squaw Creek valley the evidence is mixed (Fig. 8). Four relatively modest downstream tributaries yield varying results. Tributaries R-2.5 and L-3.6 both incur small
valley widenings, but these are reversed. Tributaries L-7.6 and L,R-10.5 are followed by
valley narrowings downstream, but then by erratic widenings. Upstream, four significant
tributaries are all accompanied by immediate valley widenings, which are fairly persistent
in the case of those tributaries that make important percentage increments to basin area.

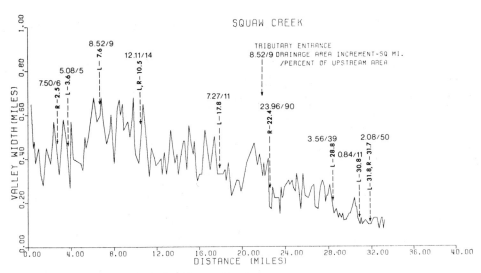

FIGURE 8 *Distance decay plot, Squaw Creek.*

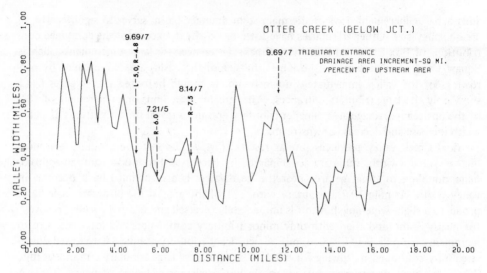

FIGURE 9 *Distance decay plot, Otter Creek below junction.*

But the example farthest upstream illustrates that significant increments can also be accompanied by essentially no change in width.

Otter Creek valley below the confluence of its major branches has no large tributaries (Fig. 9). Three modest tributary entrances from mile 7.5 to mile 4.8 are associated with valley widenings, but narrowings also occur downstream of these increments. The valley narrows within little more than a mile downstream of the junction of North and South Otter Creeks to less than one-fourth its size at the confluence.

The North Otter Creek valley is characterized by very few large tributary increments, and by many quite small tributaries which are not shown on Figure 10. At mile 24.5 a

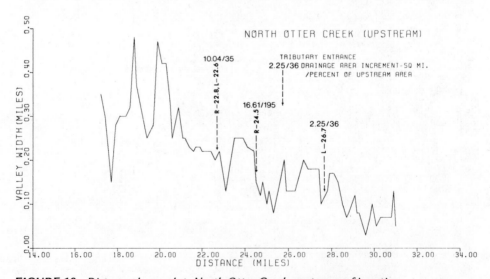

FIGURE 10 *Distance decay plot, North Otter Creek upstream of junction.*

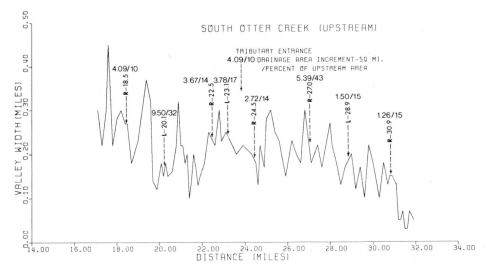

FIGURE 11 *Distance decay plot, South Otter Creek upstream of junction.*

shorter stream with a basin area twice that of the North Fork itself effectively widens the valley for a distance of more than a mile. Two tributaries entering at about mile 22.6 have no impact upon valley width. Upstream the largest of the smaller tributaries is followed by an immediate narrowing of the valley, and then by a significant broadening.

Unlike North Otter Creek, South Otter Creek has numerous moderately sized tributaries contributing discharge to its main valley (Fig. 11). A number of these are associated with immediate, short-lived valley widenings (miles 18.5, 22.5, 23.1, 24.5, 27.0, and 30.9). Others are followed downstream by valley narrowings (miles 20.1 and 28.9). In either case, the immediate response of the valley rarely remains consistent for as much as 0.5 mi. For this reason valley width fluctuations appear to be independent of drainage basin increments.

Thus, for all valleys the impacts of larger tributary increments do not produce consistent valley widenings. Narrowing of the valley floor is just as likely to occur. Considered simply in plan view, then, either tributary confluences do not have an effect upon valley widths or tributaries in the South River basin fail to reach threshold size. The latter conclusion is unlikely because several tributary confluences effectively double basin area and presumably discharge.

To this point we have considered solely the plan view of the valley. Perhaps the slope of the valley bottom or channel operates in conjunction with valley width in producing a threshold. Whether or not this is true, the combination of slope and width leads to some assumptions concerning sediment storage and flushing (Fig. 12).

In situation 1, valley slope exceeds channel slope. This is an obvious relict situation. Valley slopes were produced by past flow conditions and probably exist as low terraces. High terraces are conspicuously absent in the South River basin, at least on a large scale. The channel is adjusted to present flow conditions, which require a lower slope than at some past time, the period in which valley slopes were established as the floodplain. This is a zone of sediment storage with no thresholds imminent.

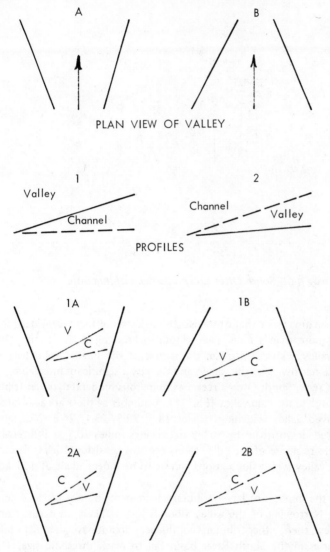

FIGURE 12 *Models of valley and channel slopes and plan view.*

In situation 2 channel slopes exceed valley slopes. Valley slopes were established as floodplains during past flow conditions. Present flow conditions are such that, perhaps due to greater sediment loads, channels have adjusted by establishing steeper slopes. If a considerable disparity exists between channel and valley slopes, a threshold of erosion of the valley bottom and tributary basins may be imminent. This is a zone of sediment flushing.

In situation A the valley is widening. This may be the result of greater discharge through tributary increments; or it may be because of a lowering of resistance of the valley walls, perhaps because the stream has encountered a buried bedrock valley.

In situation B the valley is narrowing. This is possibly the result of a lowered discharge (an unlikely event in a humid stream), the stream encountering highly permeable valley fills which permit or encourage greater throughflow or subsurface flow, or the greater resistance of the valley walls.

In combination, zone 1A has a widening valley with channel slopes less than valley slopes. This is a zone of sediment storage with no imminent threshold of erosion. Zone 1B has a narrowing valley with channel slopes less than valley slopes. Although the stream is not downcutting relative to the valley, under natural flow (meandering) conditions it is more likely than in situation 1A to encroach upon valley walls in cutbanks. The downcutting of the channel bed is not in a threshold situation, but is a function of slope. The more steeply sloping channels are capable of erosion. This is a zone of general sediment storage with localized flushing.

In zone 2A channel slopes exceed valley slopes in a widening valley. Despite a lowered probability of encountering valley walls in cutbanks, the channel is downcutting to achieve a new equilibrium and may set off a train of erosion upstream in the main channel and its tributaries, and eventually on valley wall or upland slopes. The valley widening is probably relict and thus unrelated to current channel erosional conditions. It represents some past threshold. This is a zone of general valley wall storage of sediment, but of channel flushing.

In zone 2B channel slopes exceed valley slopes in a narrowing valley. The channel is in the position of setting off a chain of erosion in alluvial valleys upstream, in both the main stem and tributaries, and on contributing upland slopes. Moreover, there is a higher probability of cutbank erosion in the narrowed valley. This is a zone of sediment flushing and the optimal future threshold situation.

To isolate reaches of the South River basin streams that would conform to the situations posed in Figure 12 and the assumptions or hypotheses of threshold and sediment storage of flushing conditions, channel and valley profiles were prepared from topographic maps (Fig. 13). Detailed graphs were also prepared illustrating each reach, but are not reproduced here. While the graphs are subject to considerable error, because of the poor quality of some of the topographic maps used as the source of data, they are nevertheless superior to those used in determining frequency and magnitude of flooding in Iowa stream basins (Schwob, 1966).

In the South River main stem between the entrance of Short Creek and Coal Creek, valley widths fluctuate greatly but are mainly average to much above predicted widths. Channel slopes exceed valley floor slopes from miles 6–10. Abrupt widening reaches here are zone 2A; a short, critical zone of 2B exists. It should be pointed out that this entire reach of the South River has been channelized for several decades, and the channel presently appears to be relatively stable. From Otter Creek to Short Creek valley widths do not fluctuate so greatly as downstream, and are generally below average in width with a overall widening trend. Here the valley slope exceeds channel slope and we can consider most of this reach to be a zone of sediment storage. Shortly downstream of Squaw Creek's entrance the valley broadens, continuing a trend that obtains upstream to about mile 23. This is a 1A zone with sediment storage. But at about mile 19 the valley narrows abruptly to two standard errors below the predicted width. This is a 2B zone with considerable potential for sediment flushing. Above the entrance to Squaw Creek, from mile 24 to mile 22–23, is a second zone of pronounced narrowing; but here valley slopes exceed channel slopes, producing a 1B zone of only localized sediment flushing.

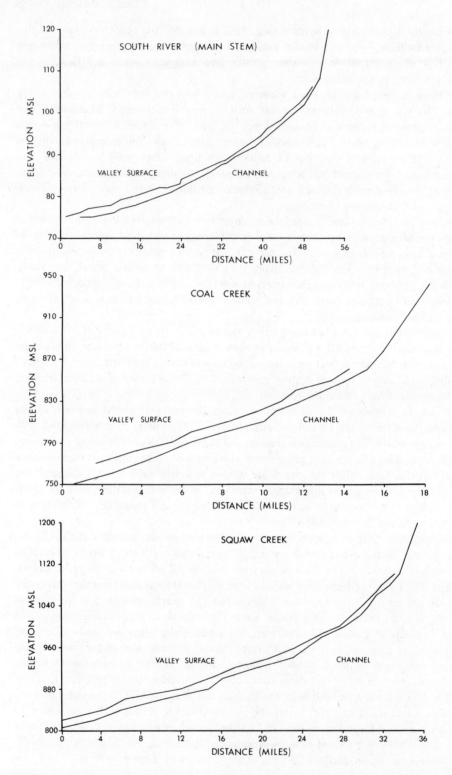

124

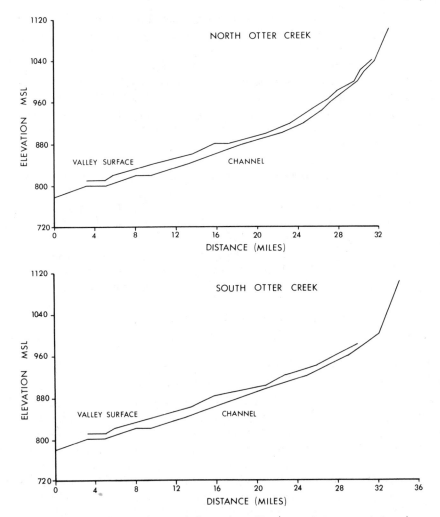

FIGURE 13 *Valley surface and channel profiles* (opposite page and above).

From mile 24 upstream to mile 35 channel slopes exceed valley slopes, with a general narrowing between miles 30 and 25. Only limited flushing could be expected here because valley narrowings are not pronounced and the disparity between channel and valley slopes is not large. Although there is an abrupt narrowing above the entrance of Broadhorn Creek, this tributary and Box Elder Creek are associated with a pronounced widening of the valley and a zone 2B situation is restricted to the mile or so above Broadhorn. Above mile 35 valley slopes are highly variable but generally exceed channel slopes until the headwater reaches. The only exception is between miles 41 and 43, where the general widening trend of the upper main stem is in evidence, lessening the chances of widespread sediment flushing.

Coal Creek is characterized by a number of abrupt and short-lived shifts in valley width. Consequently, stable and threshold situations do not persist for as long distances as on the South River main stem. Moreover, along much of the middle and lower course of the creek there is not a great disparity in valley and channel slopes. Zone 2B situations exist for half a mile or so in the vicinity of mile 3.5 and miles 5–5.5. The reach between these points fluctuates between zones 2A and 2B as a result of abrupt changes in valley widths. Upstream, a 2A zone exists above mile 10, where the valley is narrow and valley slopes are appreciably greater than channel slopes from miles 10.9 to 11.6, and again from miles 13.4 and 14.2. Perhaps the fact that Coal Creek is incised in a bedrock valley (Fig. 2) protects it from potential erosion surges, but the sites mentioned above are the most likely reaches for such potential activity.

Although Squaw Creek has some abrupt changes in valley width, reaches that are appreciably above or below average or predicted dimensions tend to persist for several miles. It is not a valley that is characterized by persistence in threshold or flushing situations, however. There is a 2-mi reach of a 1B zone from above mile 4 to above mile 6. No other persistent valley narrowing zone exists; sediment flushing situations alternate with conditions that we might expect are more stable.

The Otter Creek basin has several reaches that can be characterized as zone 2B or in the threshold or sediment flushing state. These are downstream of the junction from miles 5.2–5.6, 9.5–10, and 10.5–11.6; along the North Fork from miles 17.8–18.6 and 28.6–29.2; and in the South Fork from miles 24.5–25. All other reaches are either stable or fluctuate rapidly between stable and threshold conditions.

THE HYPOTHESES REVISITED

A useful way of summarizing the diverse results of this investigation is to reexamine the hypotheses that were put forth earlier. Hypothesis 1 suggested that valley width variations can be explained by spatially unique solutions, the most germane in this instance being as association with the buried bedrock surface. While some narrowings of the valley might be explained by the mechanism of constraints imposed by rock walls, most cannot. Except for Coal Creek, only the lower courses of the valleys conform to the bedrock surface. Even in these lower reaches both wides and narrows occur, and bedrock outcrops are rare. The buried Decatur Channel intercepts the major valleys as follows: South River main stem, miles 27–31; Squaw Creek, miles 11.5–14 and 18–20; North Otter Creek, miles 21–24; and South Otter Creek, miles 19–21 and 26–29. The absence of bedrock constraints in the valley walls in these reaches should result in widened valleys as it does elsehwere in Iowa. With the possible exception of the South River main stem there is nothing unusual about the widths of the valleys in the vicinity of the buried channel. Most of them have narrows in this zone. Thus, while hypothesis 1 can be called upon to account for certain widenings or narrowings, it fails to explain most of them.

Hypothesis 2 and 3 dealt with the randomness or regularity of valley width variations. While both can be supported by casual visual inspection of Figure 6–11 for some reaches, the hypotheses were not subjected to a rigorous statistical test, such as near-neighbor or time-series analysis.

Hypothesis 4 suggested that width variations would display a continuous downstream response to increments in discharge without regard to tributary size. While there is an overall increase in valley widths downstream, it is discontinuous in many instances. The statistical tests on an aggregate basis, summarized in Tables 3–5, failed to provide support for the notion that tributaries make an impact upon valley widths.

Hypothesis 5 stated that tributaries above a threshold level of discharge contribution would create steps or jumps in valley size. Analysis of Figures 6–11 revealed that half or more of the larger tributaries either had no impact on width or else were actually followed downstream by valley narrowings.

Hypothesis 6 suggested that channel and valley slopes are related to valley wides and narrows. While there are significant relationships between slope and mean valley widths for sections of the stream (Tables 6 and 7), comparison of detailed valley and channel longitudinal profiles with Figures 6–11 revealed that the associations were mixed. For the moment this hypothesis cannot be rejected conclusively. Detailed field work may yet provide support for it.

TABLE 6   *Relationship Between Mean Valley Width for Section and Channel Slope*

Stream	Sample points	Regression equation	r^2	r
South River, main stem	19	$\overline{W} = 0.644258 - 119.98S_{ch}$	0.35	-0.59^a
Coal Creek	11	$\overline{W} = 0.355366 - 76.87S_{ch}$	0.35	-0.60
Squaw Creek	15	$\overline{W} = 0.474648 - 79.95S_{ch}$	0.60	-0.77^a
Otter Creek, downstream of junction	4	$\overline{W} = 0.750876 - 315.62S_{ch}$	0.35	-0.59
North Otter Creek, to junction	9	$\overline{W} = 0.322337 - 56.49S_{ch}$	0.70	-0.83^a
South Otter Creek, to junction	7	$\overline{W} = 0.365376 - 92.46S_{ch}$	0.52	-0.72
Otter Creek, North Fork through main stem downstream of junction	13	$\overline{W} = 0.449620 - 91.43S_{ch}$	0.64	-0.80^a

[a]Significant at the 0.01 level.

TABLE 7   *Relationship Between Mean Valley Width for Section and Valley Slope*

Stream	Sample points	Regression equation	r^2	r
South River, main stem	18	$\overline{W} = 0.938856 - 312.04S_v$	0.53	-0.73^a
Coal Creek	11	$\overline{W} = 0.334177 - 58.49S_v$	0.20	-0.45
Squaw Creek	14	$\overline{W} = 0.507722 - 99.74S_v$	0.71	-0.84^a
Otter Creek, downstream of junction	5	$\overline{W} = 0.615714 - 158.67S_v$	0.18	-0.43
North Otter Creek, to junction	8	$\overline{W} = 0.251596 - 29.16S_v$	0.47	-0.68
South Otter Creek, to junction	5	$\overline{W} = 0.310443 - 59.93S_v$	0.88	-0.94^b
Otter Creek, North Fork through main stem downstream of junction	13	$\overline{W} = 0.409549 - 61.79S_v$	0.43	-0.66^b

[a]Significant at the 0.01 level.
[b]Significant at the 0.05 level.

Hypothesis 7 implied that anomalous valley widenings might be explained by high discharge catastrophic storms impacting only certain tributaries within each valley. It seems reasonable that only the largest subbasins would experience such an event exclusive of its neighbors. While there are several candidates for such a relationship (Coal Creek for the South River main stem, tributary R-7.2 for Coal Creek, tributary R-22.4 for Squaw Creek, tributaries L-5.0 and R-4.8 for Otter Creek downstream of the junction, and tributary R-24.5 for North Otter Creek) all these confluences are followed downstream by valley narrowings. If there ever was such a "Big Thompson Canyon effect" in the South River basin, its impact was soon lost in the normal fluctuations of valley widths downstream.

CONCLUSIONS

What makes a valley widen? Obviously, more water is too simple an answer, because all too often more water has not resulted in a positive impact upon valley widths. Valley widening may be a random or spatially stochastic process. Over the long term valleys widen downstream. But at any given moment we may see the process incompletely revealed, and the complexities of geometry imposed by little change or even negative impact may be an essentially temporary feature in the longer term history of the valley.

At any rate, the threshold effect of adding more water to a valley does not necessarily provoke an immediate and rational response to the valley's geometry. Similarly, valley narrowing may not be in any way related to physical obstacles in the path of the water flowing downstream, such as resistant rock walls.

That increments in discharge in a stream cause changes in valley width is undeniable. Downstream portions of all the valleys are broader than their upstream or headwater segments. But thresholds of impact resulting from substantial increments of discharge are harder to observe. Local unique variations such as differences in the strengths of valley walls obscure the pattern.

Generally speaking, it appears that stream valleys, conservative as they are in their response, manage to accommodate substantial increments in flow without immediate responses to the thresholds of flow implied by these increments.

REFERENCES CITED

Cagle, J. W., 1973, Bedrock topography of south-central Iowa: U.S. Geol. Survey Misc. Geol. Inv. Map I-763.

Chorley, R. J., and Kennedy, B. A., 1971, Physical geography: A systems approach: London, Prentice-Hall, 370 p.

Daniels, R. B., and Jordan, R. H., 1966, Physiographic history and the soils, entrenched stream systems, and gullies, Harrison County, Iowa: U.S. Soil Conserv. Service Tech. Bull. 1348, 116 p.

Dury, G. H., 1958, Tests of a general theory of misfit streams: Inst. Brit. Geogr. Trans. and Papers, p. 105–118.

——, 1960, Misfit streams: Problems in interpretation, discharge, and distribution: Geogr. Rev., v. 50, p. 219–242.

——, 1964a, Principles of underfit streams: U.S. Geol. Survey Prof. Paper 452-A, 67 p.

———, 1964b, Subsurface exploration and chronology of underfit streams: U.S. Geol. Survey Prof. Paper 452-B, 56 p.

———, 1965, Theoretical implications of underfit streams: U.S. Geol. Survey Prof. Paper 452-C, 43 p.

Knox, J. C., 1972, Valley alluviation in southwestern Wisconsin: Assoc. Amer. Geogr. Ann., v. 62, p. 401–410.

———, and Johnson, W. C., 1974, Late Quaternary valley alluviation in the Driftless Area of southwestern Wisconsin: *in* Knox, J. C., and Mickelson, D. M., eds., Late Quaternary environments of Wisconsin: Amer. Quat. Assoc. Guidebook 3rd Bienn. Meeting, Madison: p. 134–162.

———, Bartlein, P. J., Hirschboeck, K. K., and Muckenhirn, R. J., 1975, The response of floods and sediment yields to climate variation and land use in the Upper Mississippi Valley: Madison, Univ. of Wisconsin, Inst. Envir. Studies Rept. 52, 75 p.

Latimer, D. J., 1957, Drainage areas of Iowa streams, Iowa Highway Research Board, Ames, Bull. no. 7, 437 p.

Playfair, J., 1802, Illustrations of the Huttonian theory of the earth: New York, Dover, facsimile reprint, 1956, 328 p.

Ruhe, R. V., 1967, Geomorphology of parts of the Greenfield quadrangle, Adair County, Iowa: *in* Landscape evolution and soil formation in southwestern Iowa: U.S. Soil Conserv. Service Tech. Bull. 1349, p. 93–161.

———, 1969, Quaternary landscapes in Iowa: Ames, Iowa State Univ. Press, 255 p.

———, and Cady, J. G., 1967, The relation of Pleistocene geology and soils between Bentley and Adair in southwestern Iowa: *in* Landscape evolution and soil formation in southwestern Iowa: U.S. Soil Conserv. Service Tech. Bull. 1349, p. 1–92.

Rzhanitsyn, N. A., 1960, Morphological and hydrological irregularities of the structure of the river net: Leningrad, Gidrometeoizdat, 380 p.

Salisbury, N. E., Knox, J. C., and Stephenson, R. A., 1968, The valleys of Iowa-1: Valley width and stream discharge relationships in the major streams: Iowa City, Univ. Iowa, Iowa Studies in Geography, no. 5, 107 p.

Schumm, S. A., 1977, The fluvial system: New York, Wiley, 338 p.

Schwob, H. H., 1966, Magnitude and frequency of Iowa floods: Iowa Highway Research Board Bull. 28, 47 p.

7

SEDIMENT DEFORMATION AND TRANSPORT ON LOW-ANGLE SLOPES: MISSISSIPPI RIVER DELTA

H. H. Roberts, J. N. Suhayda, and J. M. Coleman

INTRODUCTION

River systems that carry large suspended loads and discharge into marine basins distinguished by low marine energy conditions generally build broad, clay-rich subaqueous platforms over which coarser grained subaerial and shallow-marine deposits prograde. Low-angle subaqueous slopes that typify these deltas are commonly characterized by processes and resulting geomorphic response features that indicate the disruption of "normal" sedimentary patterns and the displacement of sediments to deeper areas of the shelf and slope. Of the world's major river systems that fit into this category, the Mississippi River and its delta have been the most thoroughly studied. Much of the impetus for investigation of this area has been provided by industries interested in petroleum and mineral exploration. Sediment instability and deformational processes, however, have received attention only in recent years. Damage to and occasional complete loss of bottom-mounted structures by catastrophic soil movements have made it economically imperative to understand more about these processes, especially as exploration and production move steadily into deeper water, where operational costs increase tremendously.

The best documented example of the effects of sediment instability and slope failure on an offshore structure is given by Sterling and Strohbeck (1973) and Bea et al. (1975) with regard to a platform loss in Shell Oil Company's South Pass, Block 70. They demonstrate from data collected immediately before and after severe hurricanes that intense storm-induced wave activity is accompanied by substantial topographic changes in the delta-front environments, some of which result in the destruction of man-made installations. Although perturbation by storm waves commonly produces forces that exceed soil stability thresholds and cause failures that are accompanied by translation of bottom sediments, not all such sea-floor changes can be attributed to storm-related forces. Repeated bathymetric surveys run by the U.S. Coast and Geodetic Survey, as well as similar but generally more localized surveys run by industrial groups, have shown that shelf topography of the Mississippi River delta is in a constant state of change. It is obvious, from the magnitude of these changes and the vast areas of sea floor affected, that processes other than classical deltaic sedimentation are involved. Even though there is always a question of accuracy in comparing bathymetric data between two or more surveys because of positioning problems, bottom changes are commonly so great as to clearly fall outside the margin of interpretive error. The temporal history of many of these bottom changes indicates that distinct alterations of bottom topography can occur during non-storm periods. These observations suggest that sediment deformation is initiated by multiple forcing mechanisms and that geomorphic response forms are possibly controlled by various levels or thresholds of sediment instability. Coleman and Garrison (1977) suggest that from a geological point of view the following major factors influence the stability of marine sediments:

1. *Rates and location of sediment accumulation.* Sedimentation in the Gulf of Mexico varies from millimeters of sedimentation per century to rates in excess of 1 m/yr near the mouths of the major rivers. This extreme variability, coupled with variations in the gradient of the areas of deposition, is exceedingly important in determining the strength of the material and its vulnerability to failure. In those areas where rates of sedimentation are extremely high, the sediments contain a large amount of water, display loose packing arrangements, and are normally underconsolidated. Such sediments are highly prone to failure even when minimum stress is applied. In areas where rates of deposition are lower, pore waters are allowed to escape during sedimentation and normal consolidation can proceed; these sediments are less prone to failure.

2. *Size and composition of sediment.* Size distribution in continental shelf sediments can be extremely variable, ranging from coarse, sandy clastics to fine-grained clays. Deposition of large volumes of dense, coarse-grained, sandy clastics can result in differential loading of underlying weaker silts and clays. This initial stress-inducing mechanism can result in a variety of types of failures. The composition of the sediment, especially in the finer-grained clays, strongly influences early diagenesis and cementation of the sediments.

3. *Dynamics of marine and riverine processes.* In regions adjacent to active river deltas, hydrodynamic processes that operate at the river mouths control the outflow patterns and fundamentally determine the pattern of sediment dissemination and accumulation. Through geologic time different areas become the sites of maximum deposition, and offshore marine deposits are thus constantly subjected to differing sets of stresses. Currents within the water mass, driven by wind, tide, density, and semipermanent oceanic currents, induce spatially and temporally varying stresses on bottom sediments across the continental shelf and upper continental slope. Surface waves under the influence of tropical storms and hurricanes produce gradients in bottom pressure on the ocean floor to which pore waters and gases within the sediment must respond. This cyclic-type loading plays an important role in influencing the stability of marine sediments.

4. *Geochemical environment existing at time of deposition.* On a regional scale, geochemical environments across the continental shelves display high variability, ranging from those regions that are entirely in a reducing state to those that are highly oxidizing. On a microscale, variations in pore-water chemistry control the rate and type of early diagenesis and cementation and thus can impart to clays much greater strength than normally would be present. Biochemical processes degrade organics, and various sedimentary gases, primarily methane and carbon dioxide, are produced. All these features strongly influence the initial strength of the deposited sediment, as well as changing sediment strengths with time and depth of burial.

5. *Tectonic or structural history of the region.* The subsurface structure of the northern Gulf of Mexico is extremely complex, and throughout Tertiary and Quaternary times sites of deposition have constantly changed and a wide variety of structural readjustments have taken place. Deep-seated faults cut the modern sediment surface, pro-

ducing scarps and slopes on the sea floor; these are sites of potential sediment instability that were produced by phenomena originating in an earlier time framework. Diapiric intrusions, both salt and shale, strongly influence the topography of the continental shelf and slope and produce basins and highs on the sea floor in which marine processes can redistribute sediment. The high rates of accumulation that characterized the Tertiary and Pleistocene depocenters produced highly variable subsidence rates.

Interactions of these major factors on various temporal and spatial scales are important in producing deformational response forms in marine sediments that vary in size, shape, rate of change, and causative mechanisms. The vast importance of these processes as mechanisms for sediment transport in the context of the delta-building scheme has only recently been realized. New advances in offshore survey technology, including navigation systems with repositioning capability and accuracies of a few meters, side-scan sonar, and high-resolution geophysical profiling equipment, have enabled marine geoscientists to map with a reasonable degree of accuracy both the plan-view configuration and, to a lesser extent, the three-dimensional geometry of deformational features. Furthermore, repeated surveys allow collection of important data on rates of change of these features. On the basis of numerous sources of published and unpublished data on deformational processes and response forms, this chapter attempts to summarize the types of mass-movement features that are associated with the Mississippi River delta from the inter-distributary bays to the shelf margin. In addition, the thresholds of sediment instability for these depositional environments are discussed using conventional stability analysis in conjunction with recently acquired data.

ENVIRONMENTAL SETTING: MISSISSIPPI RIVER DELTA

Frazier (1967), and in earlier papers Kolb and Van Lopik (1958 and 1966), have shown that the modern Mississippi River delta complex (deposits of the past 7000 yr) is composed of a series of overlapping delta lobes representing distinct episodes of sedimentation. As each major lobe develops the distributary network becomes overextended, gradient advantages are lost, and the trunk stream begins diverting floodwaters to a course that offers a more favorable or steeper gradient. Eventually, this process promotes abandonment of one locus of sedimentation for another.

Within the past 600–800 yr the modern "bird-foot" delta (Fig. 1) has developed in a position between former active delta lobes, the St. Bernard subdelta to the east and the Lafourche subdelta to the west (Frazier, 1967). As is common of subdelta development in the Mississippi River complex, progradation of the coarse-grained facies was preceded by blanket deposition of prodelta clays. In general, the greater the depth of the water into which the distributaries are depositing their sediments, the slower the rate of progradation and the thicker the fine-grained prodelta facies. The coarsest sediment, which is fine sand in the case of the modern Mississippi River, is deposited near the distributary mouths as river-mouth bar deposits. Finer suspended-load sediments (estimated at 65% clay and 35% silt/fine sand for the Mississippi River) are carried beyond the immediate vicinity of the distributary mouths and are spread over the shelf (Fig. 1) as prodelta and shelf deposits. As distributaries grow seaward, thick, relatively coarse grained bar sediments are deposited over a thick sequence of highly incompetent prodelta clays. This natural process of deltaic evolution triggers deformation of underlying clays.

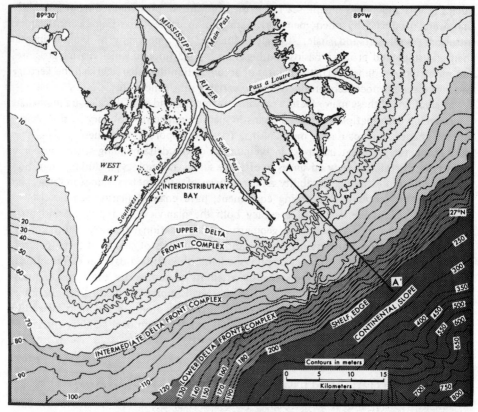

FIGURE 1 *Modern Mississippi River delta offshore topography. Deformational features along transact A–A' (interdistributary bay to shelf edge) are discussed in this chapter.*

The modern Mississippi River delta lobe is somewhat anomalous with regard to previous Holocene subdeltas because it has remained relatively confined in subaerial extent and has prograded onto the continental shelf to a position generally less than 20 km from the shelf edge. The subaerial extent of the bird-foot delta is approximately 1900 km², whereas the average area of older lobes was approximately 6200 km² (Coleman and Garrison, 1977). Localization of sedimentation on the continental shelf, coupled with progradation into relatively deep water, has promoted the deposition of a thick sequence of prodelta clay, over which coarser grained deposits of the modern delta have prograded. Figure 1 illustrates the rather distinctive shelf topography that has developed as a function of the delta-building process under these conditions. Between the major distributaries are areas of extremely low sloping bottom (generally between 0.1 and 0.3°) that receive very little sediment except during floods. Despite the flat bottom conditions and low sedimentation rates in these interdistributary bays, localized surface sediment deformation is taking place. Immediately seaward of and adjacent to the distributary mouths are the smooth-bottomed regions of rapid, coarse-grained sedimentation known as the river-mouth bars. Where the distributaries are in close proximity, the bar complexes merge, forming at any given time elongate sand bodies oriented at nearly right angles to the distributary trends. Seaward of these shallow areas of smooth bottom lie the delta-front deposits that are characterized by a sculptured sea-floor topography.

For convenience of discussion the delta-front deposits that extend between water depths of approximately 10 and 200 m (shelf edge) have been subdivided into three zones on the basis of gross shelf topography (Fig. 1). The shallow or upper delta-front complex extends from about 10 to 60 m and is characterized by highly crenulated bathymetric contours that reflect a concentrated network of small-scale troughs which radiate seaward from the distributary mouths. Bottom topography is extremely complicated within this zone. The intermediate and lower (deep) delta-front zones display somewhat more regular bathymetric contours than the upper delta-front zone. Although these lower two zones are crossed by linear troughs and accompanying depositional lobes, as is also typical of the shallow zone, these features tend to be of larger scale and to occur less frequently. This trend is particularly true of the intermediate zone, where slopes flatten slightly. The frequency of occurrence of linear sea-floor depressions increases again in the deep delta-front complex.

Within these depositional environments extending from the interdistributary bays to the shelf edge (Fig. 1), deformation of bottom sediments is taking place on different spatial scales and at various rates. Resulting geomorphic response features evolve in a variety of shapes and from several forcing mechanisms. By using high-resolution geophysical profiles and side-scan sonar data, coupled with accurate positioning, deformational features in these deltaic environments may be described. Side-scan sonar has been particularly important in deciphering the morphological characteristics of these deformational features. As described by Belderson et al. (1972), Newton et al. (1973), and Flemming (1976), the side-scan sonar record is simply a sonic representation of the bottom; it is much like an oblique aerial photo, which records reflected sound rather than reflected light. Shape distortion is inherent in the sonograph images for a variety of reasons. Sonographs used as illustrations in this report exhibit distortions resulting primarily from ship speed. Both horizontal and vertical scales are provided on each illustration so that shape corrections can be made.

DISTRIBUTION OF SEDIMENT DEFORMATIONAL FEATURES— INTERDISTRIBUTARY BAYS TO SHELF EDGE

Numerous types of sediment deformational features occur in the region from the extremely low sloping floors of the interdistributary bays to the relatively inclined sea floor of the continental slope. Figure 2 schematically illustrates the distribution of these deformational features that occur across the Mississippi River delta-front deposits and out to the shelf edge. Each major type of deformational feature is discussed with regard to probable initiating mechanisms and the deltaic environment where it most frequently occurs. It is understood that in reality most forms do not constitute discrete categories, but are transitional from one category to the next.

Interdistributary Bays

Interdistributary bays are situated between the major distribuatires of the modern bird-foot delta and are characterized by low sedimentation rates (<0.1 m/yr), shallow water (generally less than 10 m), and an extremely low sloping bottom (approximately

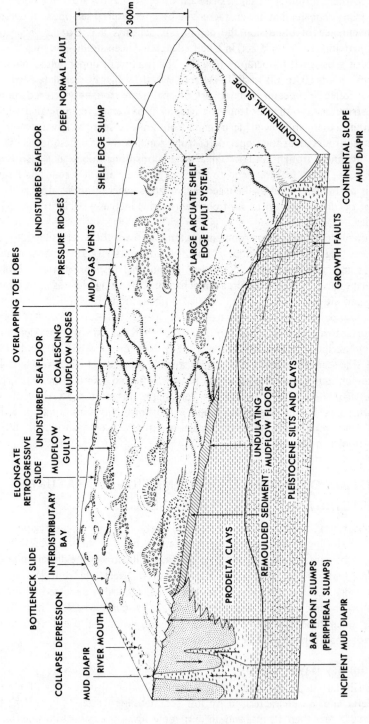

FIGURE 2 *Schematic distribution of deformational features from shallow interdistributary bays to the shelf edge.*

136

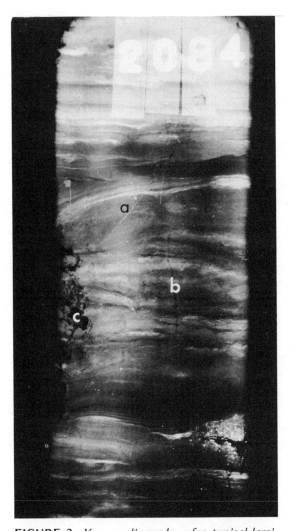

FIGURE 3 *X-ray radiography of a typical laminated interdistributary bay sediment illustrating alternating silts and clays probably related to flood cycles (depth 5.9 m). Both filled straight burrows (a) and open branching burrows (b) are apparent. Gas-related structures (c) are present, but not in such abundance as to obscure bedding. Dark areas of the X-ray radiograph are regions transparent to X-radiation, while light areas (silt-rich laminations) are more dense and do not allow radiation to pass freely to the film plane. The core diameter is 5.6 cm. (Roberts et al., 1976.)*

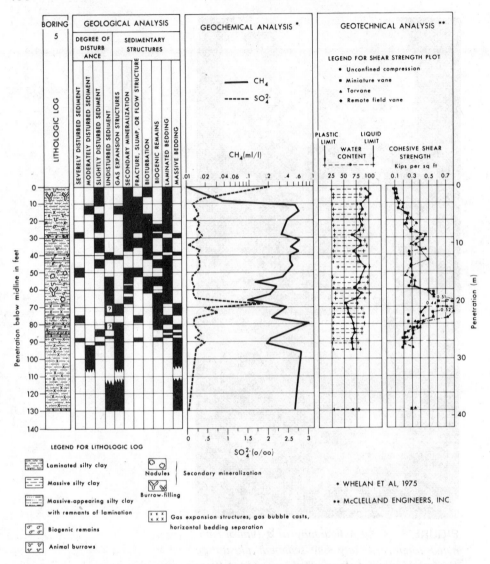

FIGURE 4 *Data summary log for a boring taken in East Bay approximately 8 km west of South Pass. Interdistributary-bay sediments interpreted to a depth of approximately 20 m from sedimentary structures analyzed from X-ray radiographs.*

0.0–0.2°). Bathymetric contours in Figure 1 illustrate a break in slope for these environments at about the 10-m contour. At this depth the regional slope steepens to maximum values of near 0.5° as the interdistributary-bay environment merges with the shallow delta-front complex.

Because of the sheltered position of these environments with respect to the main point sources of sediment, the river mouths, bottom sediments in the interdistributary bays are composed of clays and alternating silt-rich laminations (Fig. 3). Introduction of the coarser particles is generally related to the river's annual flood cycle. During nonflood times primarily clay-sized particles are deposited in these sedimentary environments. Tidal pumping of river water rich in clay-sized suspended load and wind forcing of the river-mouth plume into the bays are mechanisms of fine-grained sediment introduction. During floods both clay-sized and silt-sized suspended load is forced overbank and through the numerous crevasse systems in the lower reaches of the distributaries. Localized deposits of fine sand are associated with these breaks in the natural levees, forming crevasse-splay deposits which build into the interdistributary bays. The open bay bottoms, however, are composed of soft clays and silts containing a great deal of water and biogenic gas. Roberts et al. (1976), Suhayda et al. (1976), and Whelan et al. (1976) have described the properties of an interdistributary-bay sedimentary sequence from a bore hole in East Bay. These studies illustrated that plastic and liquid limits, ranged from 25 to 30% and 75 to 100%, respectively, and water content remained above 70% in the upper 18 m of the boring. In addition, biogenically derived sedimentary gas (methane) was found to occur in high concentrations. A variety of shear strengths were found throughout the core. However, in the upper 18 m of the section, which was interpreted as interdistributary sediment (Roberts et al., 1976), shear strengths were generally very low. Figure 4 summarizes these data and indicates sedimentary structures interpreted from analysis of X-ray radiographs.

Bottom Features. Sea-floor deformations are present at all water depths within the interdistributary bays. As a general rule, however, these features tend to be smaller and more equidimensional in the shallower bay areas (<15 m), becoming larger and more lineated as water depths and bottom slopes increase. A continuum of forms that range from small circular depressions to extremely elongated, trough-like features is represented within the confines of the interdistributary bays. It is also recognized that compound forms exist.

The biogenic gas (methane) content within interdistributary-bay sediments is large and has been reported to be as much as 15% (Whelan et al., 1976). These gas-charged conditions make acoustic penetration of the subsurface sediments very difficult. Many of the deformational features, however, constitute areas of acoustic returns, or "acoustic windows," that are surrounded by otherwise acoustically turbid sediments. These observations suggest sediment degassing during the deformation process and possibly a relative increase in soil density.

An instability type common to the low sloping bottoms of the interdistributary bays is the rather circular depression characterized by a blocky or coarsely "textured" interior that is bounded by low-relief scarps (<3 m). Prior and Coleman (1978a) refer to these features as "collapse depressions" (Figs. 5 and 6). Collapse depressions occur primarily in the very low sloping (0.1–0.2°) portions of the bays and in water depths that are generally less than 15 m. Coleman and Garrison (1977) point out that such deformational forms are quite variable in size and that diameters range between 30 and 90 m. In comparison to other mass-movement features associated with the delta-front environments, these collapse depressions are extremely small.

The upslope margins of collapse depressions commonly exhibit the most distinct escarpments bounding the feature. Curved crack systems, or "crown cracks," extend away from the main head scarp into the adjacent undeformed sediments. As Figure 5

COLLAPSE DEPRESSION

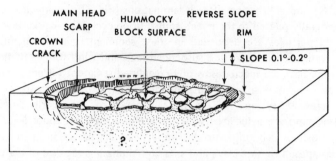

FIGURE 5 *Schematic representation of a collapse depression and its morphological characteristics. The depth of instability is unknown. (Prior and Coleman, 1978a.)*

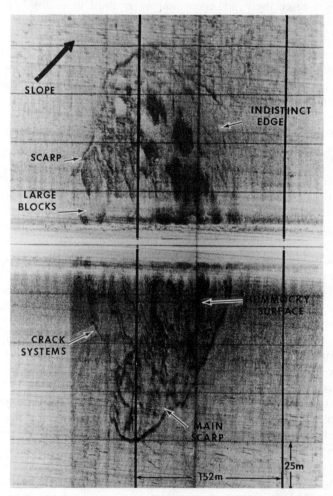

FIGURE 6 *Side-scan sonograph of a collapse depression. Note the irregular, "coarsely textured" internal topography and the low-relief escarpments that bound the feature.*

BOTTLENECK SLIDE

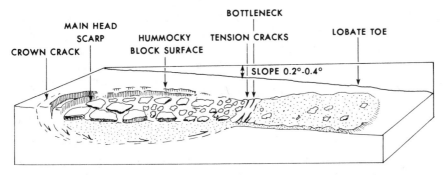

FIGURE 7 *Schematic diagram of a bottleneck slide and its salient features. (Prior and Coleman, 1978a.)*

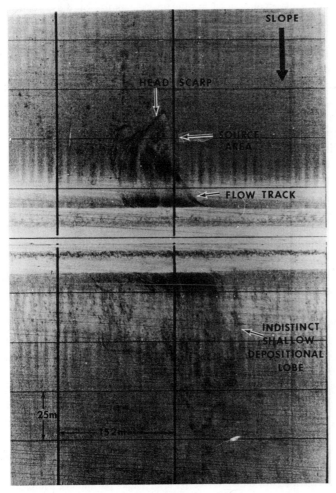

FIGURE 8 *Side-scan sonograph of a bottleneck slide showing distinct failure zone and thin, indistinct depositional lobe.*

indicates, the downslope limits of collapse depressions commonly display a reverse slope, indicating a tendency toward downslope movement. Collapse depressions are interpreted as features resulting from localized subsidence, possibly caused by the loss of sediment gas and pore water at the time instability is initiated. The forcing mechanisms for this instability are not clearly understood. Excess biogenic gas and water pressures in the sediment certainly create conditions that favor soil failures. Wave perturbation of these shallow sea-floor areas, as discussed by Suhayda et al. (1976), is also a probable initiating force.

Deformational forms that are intermediate between collapse depressions and well-defined elongate subaqueous slide features have been described as bottleneck landslides (Prior and Coleman, 1978a). These features display discrete escarpment and curvilinear cracks in the upslope source area, a blocky or textured sea floor composed of well-defined blocks, and generally thin, indistinct depositional lobes (Figs. 7 and 8). Although escarpments and cracks are most pronounced at the upslope limits of these features, systems of cracks commonly occur along lateral margins. The material forming the depositional lobe is most commonly discharged through a relatively narrow opening at the downslope end of a bowl-shaped feature similar to the collapse depression described above. Some slides of this general description may display wide discharge openings. Tension cracks (Fig. 7) can commonly be observed at the narrow opening where deposition begins. Bottleneck slides vary in length from approximately 150 to 600 m and occur on slopes of 0.2–0.4°. Occasionally, two or more bowl-shaped source areas will discharge into a common depositional lobe, forming a compound feature.

Low-angle slope landslides of the bottleneck variety are generally characterized by subsidence in the source area as material moves downslope and by retrogression or migration of the source area in an upslope direction. These failure types commonly involve remolding or liquefaction of the source material at the time of instability. High sedimentary gas and water pressures appear to set the stage for such slides. Degassing and dewatering during the process of failure is strongly suggested by the improved acoustic properties of sediments within both the depression and the depositional lobe as compared to surrounding materials.

In the continuum of deformational forms that occur in the interdistributary-bay environment, retrogressive elongate slides (Fig. 9) form one end member. These features occur on slopes of 0.2–0.5° and range in length from 500 to 1500 m; length/width ratios are up to 15, whereas the near-circular collapse depressions have ratios of from 1.0 to 1.5 (Prior and Coleman, 1978b). Axial slopes of these elongate features, as well as the more circular ones, are somewhat less than the regional slope, generally falling between 0.07 and 0.41°. Like bottleneck slides, the elongate forms exhibit upslope source areas that are bounded by multiple arcuate scarps and cracks. The channel or chute through which the source material is transported downslope is long and narrow. Sea-floor texture within these features is similar to that of forms previously described. The source area exhibits a blocky or hummocky texture, and individual block size generally decreases down the elongate chutes. Arcuate slump blocks and localized small-scale horsts and grabens are common in the source area.

Chutes are frequently confined by elongate scarps and cracks paralleling the chute's general axial trend. Some chutes are very straight, but the general rule is for these narrow, channel-like features to be sinuous. In a downslope direction the chutes rather impercep-

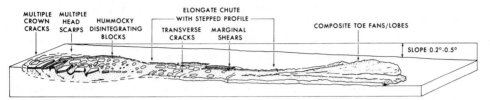

FIGURE 9 *Schematic illustration of an elongate slide and its characteristic morphological components. (Courtesy of David Prior.)*

tibly merge with a series of overlapping depositional lobes, a situation that suggests that these slides are periodically reactivated. Additional evidence for this episode activity is the existence of multiple crown cracks and head scarps in the source area. With each occurrence of slide activity within the feature, localized loading occurs in chute and lobe areas. Eventually, the long profile of the feature is degraded below the angle of stability and well below the surrounding regional slope.

Deformational features occurring on the low-angle slopes of the interdistributary bays are occasionally controlled by local slope and thus display a range of long-axis orientations. Their exact orientations are probably controlled by sea-floor microrelief within the environment. However, as a general rule most forms are oriented with their long dimensions at high angles to the regional bathymetric contours.

DELTA-FRONT ENVIRONMENTS

In the vicinity of distributary mouths are areas characterized by rather smooth bottom topography and shallow water. These areas of coarse-grained sediment deposition constitute river-mouth bars (Fig. 1). Seaward of these areas, in water depths of approximately 10–70 m, is the highly irregular bottom topography of the upper delta-front complex. The sea floor in this zone has an average slope of about 0.7°. Local slopes associated with the complex topography may be somewhat greater than regional slopes. The complexity of the bottom configuration in this zone arises from numerous arcuate slumps and well-defined gullies. These trough-like features occur most frequently at high angles to the regional bathymetric contours. Seaward of the complex upper delta-front topography is an area where the regional slope flattens slightly and bottom topography is less abrupt. These topographic changes represent in part the coalescing of depositional lobes that are discharged from gully systems upslope. Depth limits for this intermediate delta-front complex are between roughly 70 and 120 m. From the seaward limit of this zone to the shelf edge lies the lower delta-front complex, which is traversed by complicated compound gully systems and deposition lobes. Unlike the gullies of the upper delta front, these features occur less frequently and are commonly compound and more complex in outline. Occasionally, deep-seated faults are expressed on the sea floor as scarps.

The aforementioned zones illustrated in Figure 1 occur around the entire periphery of the subaqueous Mississippi River delta. Sediment characteristics and sea-floor topography within each zone are controlled to a large extent by sediment-deforming processes.

Sediment deformation features that occur within the delta-front environments and generally result in downslope movement of sediments consist of a variety of slumps, collapse features, landslides, and depositional lobes, all of which are of various dimensions and shapes. Although differential weighting of the advancing deltaic mass over incompetent prodelta clays has caused injection of mud spines and ridges into overlying sediments, these diapiric structures, or "mudlumps," will not be discussed in this paper. Processes of mudlump generation and resulting response features have been thoroughly addressed by Morgan (1961, 1968), Morgan et al. (1963), Coleman et al. (1974), and, most recently, Coleman and Garrison (1977). The discussion of delta-front deformation presented here will focus on surface slump and landslide features as well as related depositional forms.

Sediments of the delta-front environments vary from fine-grained quartz sands of the river-mouth bars to montmorillonitic clays with essentially no silt content in undisturbed areas near the shelf margin. The sands, however, are generally localized and occupy a relatively small part of the delta-front deposits. Silty clay is the most common sediment type in these environments. However, because of deformational processes sand-rich blocks of bar-front material are displaced into deeper water environments, where only clay-sized particles would be deposited if sedimentation were controlled by hydraulic factors alone.

Sedimentation rates in excess of 1 m/yr have been reported for the rapidly accumulating river-mouth bar areas, while seaward progradation of the bar has been estimated to average 60 m/yr (Coleman et al., 1974). Average accumulation rates in other delta-front environments are extremely variable because processes of sedimentation and accumulation are so closely related to deformational phenomena. Coleman and Garrison (1977) state that "sedimentation rates in delta front water depths of approximately 30 m vary from 0.3 to 0.5 m/year, an extremely high accumulation rate when compared with those areas on the adjacent continental shelf, where in the same water depths accumulation rates are measured at a few millimeters per year."

As is characteristic of sediments in the interdistributary bays, a high sedimentary gas content is present in the rapidly deposited clays of the delta-front environments. Figure 10 illustrates a composite of sedimentological, geochemical, and geotechnical data collected from a sediment boring taken approximately 12 km south of South Pass in 73 m water depth (Roberts et al., 1976). This boring, from the transition area between the upper delta-front deposits and the intermediate delta-front complex (Fig. 1), was collected in an acoustically amorphous region thought to represent a composite deposition lobe or mudflow feature. Sediments throughout the boring display numerous gas expansion and migration features (Fig. 11). Flow structures are common to a depth of approximately −49 m. Sediment structures, as well as anomalous profiles of geochemical and geotechnical data, strongly suggest remolding of the sedimentary materials owing to mass-movement processes. The erratic but generally high sedimentary gas content possibly suggests convective mixing during sediment transport, entrainment of new seawater, and subsequent reactivation of the methane-producing process as discussed by Whelan et al. (1975). Examination of numerous sediment cores with the aid of X-ray radiography has shown that shelf clays slowly deposited outside of areas of mass movement display numerous biogenically related structures, occasional undeformed laminations, and a host of diagenetic features (Fig. 12). Sediments of this description are commonly used to define the lower limit of rapidly deposited and generally deformed prodelta clays.

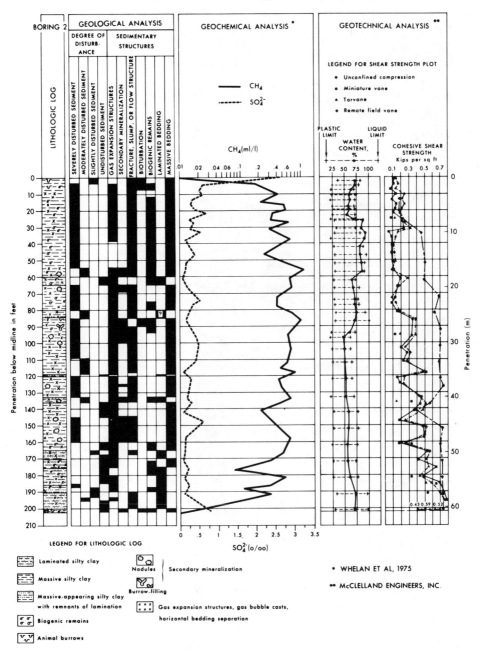

FIGURE 10 *Data summary log for a boring taken from a suspected mudflow feature (composite depositional lobe) located approximately 12 km south of South Pass at a depth of 73 m. Note the erratic but generally high methane content, low shear strengths, and presence of flow structures, coupled with lack of horizontal stratification.*

FIGURE 11 *X-ray radiograph of a typical silty clay from the delta-front depositional lobe or mudflow displaying a lack of internal stratification (remolded sediment) and numerous gas expansion voids and elongate separations. The core diameter is 5.6 cm.*

Surface Deformational Features and Their Morphology

The steepest regional slopes of the modern subaqueous Mississippi River delta are associated with the bars that form adjacent to the main distributary mouths. As previously mentioned, these depositional features are characterized by flat, smooth topography on the bar crest and are composed of relatively coarse sediment (fine sand and coarse silt).

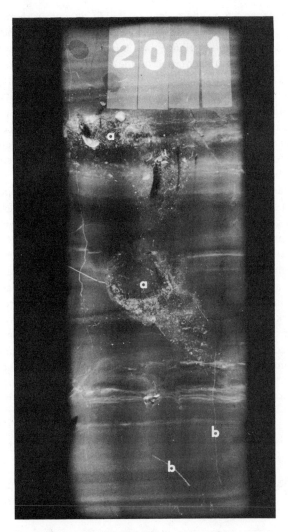

FIGURE 12 *X-ray radiograph of a shelf sediment displaying thin laminations disturbed only by bioturbation. Both large (a) and small (b) burrow structures are present. Finely divided shell debris, plus zone of local cementation, is evident in the large burrows. The absence of contorted bedding and gas-related features, as well as the increase in biogenic debris, clay content, and diagenetic features, strongly suggests an undisturbed shelf sediment that accumulated slowly. The core diameter is 5.6 cm.*

On the bar front, however, where slopes are on the order of 0.2–0.6°, numerous slumps give this environment a distinctly stepped profile. Coleman et al. (1974) have termed these features peripheral rotational slumps and attribute their initiation on the bar front to slope oversteepening resulting from rapid sedimentation on the bar crests during floods and subsequent seaward migration of this material, causing slope instability. The configuration of the bar front is in a continual state of change because of the slumping activity. Figure 13 illustrates the variation in bar-front profiles off South Pass in two successive years. It is clear from this illustration that a large quantity of sediment was displaced downslope. Numerous fathometer profiles and side-scan lines from these delta-front environments illustrate the importance of this process in moving large volumes of sediment from shallow-water areas high on the upper bar front to deeper areas of the shelf.

The slumps are commonly arranged in coalescing arcuate groups that face downslope. Scarp heights vary from 3 to 8 m and display local slopes of 1–4° (Prior and Coleman, 1978a). In some examples scarps can be traced over distances of up to 10 km. However, in most cases the slump features extend over a considerably shorter distance and are generally curved or curvilinear in outline. Side-scan sonar records of these slump features (Fig. 14) illustrate that many of the large downthrown blocks have been rotated toward the initial failure plane, forming a reverse slope. These displaced blocks of material are reflected in sediment cores by inclined bedding. Other forms of complex bottom features, such as horsts, grabens, and blocky bottom texture, are associated with the slump systems. Coleman et al. (1974) have shown from 6.5-kHz seismic data that the shear planes associated with these slumps are concave upward, tend to merge at depth, and penetrate on the average 30–35 m of sediment. Morphology of this description is indicative of rotational sliding accompanied by translational downslope movements. Such morphology suggests that upslope features, starting as arcuate slump systems, evolve into translational

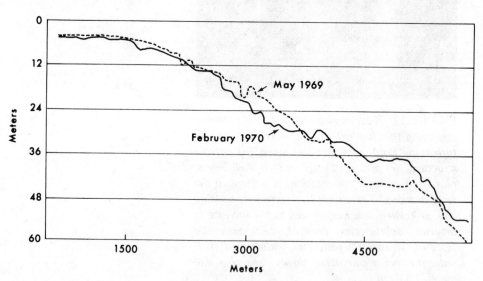

FIGURE 13 *Comparison of fathometer profiles off South Pass. Data were collected in May 1969 and February 1970. Note the loss of material from shallow water and addition of material in deeper water.*

elongate slide features downslope. The gross morphology of the upper delta front confirms this relationship.

The bathymetry of Figure 1 clearly illustrates a complex system of linear depressions or gullies that occur most frequently around the distributary-mouth areas. Systematic study of these features was first undertaken by Shepard (1955), who referred to them as "delta-front valleys." Recently acquired side-scan sonar data from upper delta-front environments have verified the close relationship between the areas of slump topography and upslope ends of the gullies. It is clear that the upslope margin of each gully terminates in a series of step-like slump structures (Fig. 14). Schematically, Figure 15 shows this relationship as well as the salient morphological characteristics of these common delta-front deformational systems. Geomorphic details are shown from the curvilinear slumps at the upslope gully margin to the composite depositional lobes at the deeper distal end of the system.

Delta-front gullies are elongate systems of sediment instability that extend downslope at nearly right angles to the regional bathymetric contours but vary greatly in plan-view outline and dimension. However, they are generally 400–700 m wide, average 8–10 km long, vary up to approximately 25 m in depth (below the surrounding sea floor), and are bounded by linear escarpments that generally parallel the long dimension of the system. The downthrown areas enclosed by the bounding linear escarpments are characterized by irregular, blocky bottom topography (Figs. 15 and 16). Coleman and Garrison

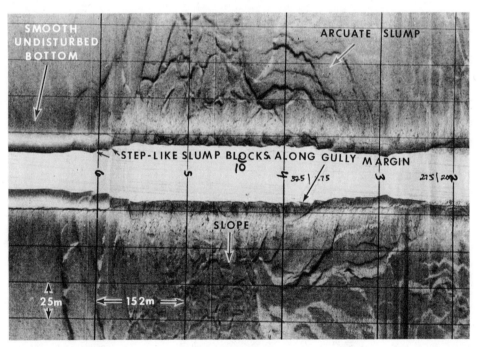

FIGURE 14 *Side-scan sonograph illustrating the step-like profile and curvilinear plan view of arcuate slumps in upper delta-front sediments. Note the complex nature of the bottom topography downslope of this slump system.*

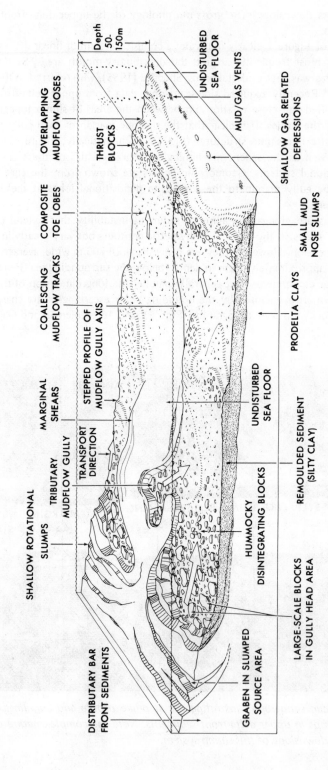

FIGURE 15 *Schematic representation of two delta-front mudflow gully systems from the arcuate slumps at the upslope margin to the composite lobes at the downslope end. The relative length of the system has been greatly shortened for illustrative purposes.*

SHALLOW ROTATIONAL SLUMPS

TRIBUTARY MUDFLOW GULLY

TRANSPORT DIRECTION

MARGINAL SHEARS

STEPPED PROFILE OF MUDFLOW GULLY AXIS

COALESCING MUDFLOW TRACKS

COMPOSITE TOE LOBES

THRUST BLOCKS

OVERLAPPING MUDFLOW NOSES

Depth 50-150m

UNDISTURBED SEA FLOOR

MUD/GAS VENTS

SHALLOW GAS RELATED DEPRESSIONS

SMALL MUD NOSE SLUMPS

PRODELTA CLAYS

UNDISTURBED SEA FLOOR

REMOULDED SEDIMENT (SILTY CLAY)

HUMMOCKY DISINTEGRATING BLOCKS

LARGE-SCALE BLOCKS IN GULLY HEAD AREA

GRABEN IN SLUMPED SOURCE AREA

DISTRIBUTARY BAR FRONT SEDIMENTS

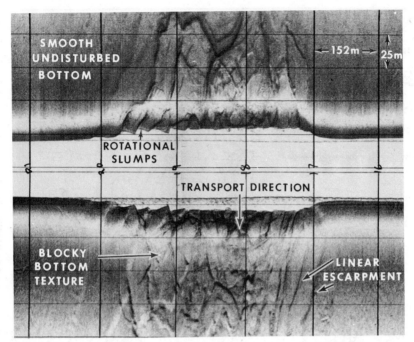

FIGURE 16 *Side-scan sonograph across an elongate mudflow gully showing subparallel linear escarpments along the gully margin and highly chaotic, blocky topography on the floor of the feature.*

(1977) report that gully walls commonly display slopes from 0.6 to 6.0° and occasionally steepen to slopes in excess of 12°.

Most gullies are not straight but assume a rather sinuous shape (Fig. 17). Much of the sinuosity in these systems arises from the numerous "feeder areas" to the gullies en route to the major depositional lobes downslope. The sources or "feeder areas" are generally bulbous compared to the relatively straight and narrow chutes between them that function as channels for the downslope movement of debris. Elongate cracks can be observed paralleling the main chute escarpment in many examples. Occasionally, long, linear block faults can be observed along the chute margins (Fig. 16). The wider, bowl-shaped source areas are bounded by escarpments and multiple curvilinear slumps, suggesting shallow rotational failure. Margins of these downthrown areas display a very coarse and blocky bottom texture. Block size generally decreases downslope and in the chutes. Headward erosion of the source areas occasionally leads to the intersection of two gullies, creating a compound system having tributaries.

An axial profile down one of these gullies is step-like. The narrow chutes generally constitute the steepest parts of the profile, while the wider rough-floored source areas exhibit lower slopes. Wider areas in the system function as accumulation points for minor lobes of material derived from surrounding source areas by headward rotational slumping and side slumping of nearby tributaries. Therefore, the gully systems contain various points and degrees of instability, dependent upon the amount of loading, degree of slope

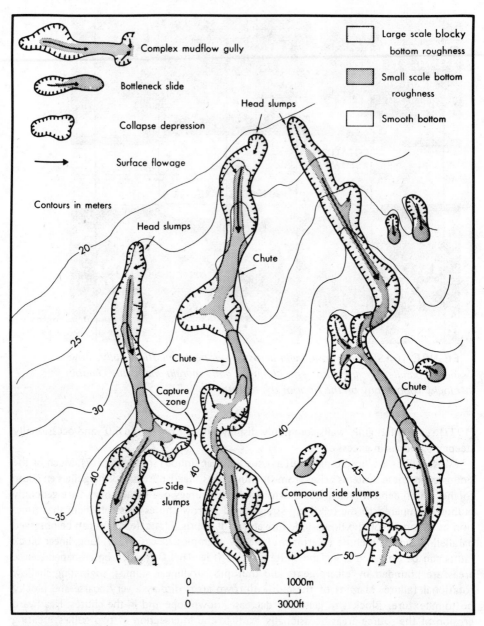

FIGURE 17 *Map of three major delta-front gully systems near South Pass compiled from closely spaced side-scan sonar data. The salient morphological components of each system are indicated.*

modification, and condition of remolded material at localized positions along the gully courses. In comparison with subaerial systems, it has been shown that a significant characteristic of debris flows and mudflows possessing long stepped profiles is their capacity to move episodically in pulses or surges (Johnson, 1970; Hutchinson et al., 1974). From the morphology of these systems it becomes apparent that throughout the entire mudflow gully tract there are various sites of instability, each with its own set of threshold conditions. Further instability is inherent in the system beyond the point at which the gully can be identified as a geomorphic feature (generally in water depths of 50–70 m).

Beyond the downslope distal ends of the upper delta-front gullies the regional slope commonly flattens slightly owing to deposition of remolded debris discharged from the gullies. As shown schematically in Figure 15, sediments discharged from the gullies are deposited as broad lobes that tend to intersect and overlap. Much like the axial trace of the gullies, downslope profiles across these depositional areas also have a step-like appearance, indicating the episodic sediment transport-depositional process (Fig. 18). Slopes on each lobe are very flat to gently inclined (<1°) except for the relatively abrupt terminal scarps, which commonly exhibit slopes of up to 8–10°. Side-scan sonar data have revealed that lobe surfaces are composed of hummocky topography, which generally becomes increasingly subdued in the downslope direction. Irregular and blocky surface roughness similar to that described in association with the floors of gully systems is characteristic of the upslope portions of most depositional areas. Downslope, toward the

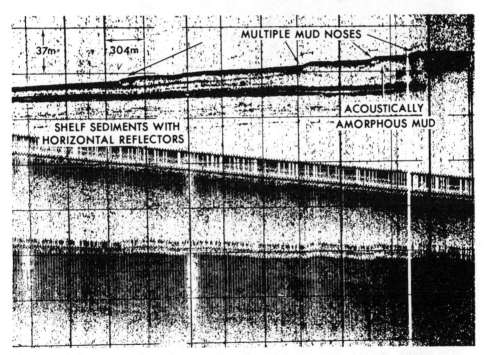

FIGURE 18 *3.5-kHz seismic profile across the depositional lobes at the downslope ends of mudflow gully systems in the upper delta-front complex off South Pass. Note the step-like nature of these deposits. (After Garrison, 1974.)*

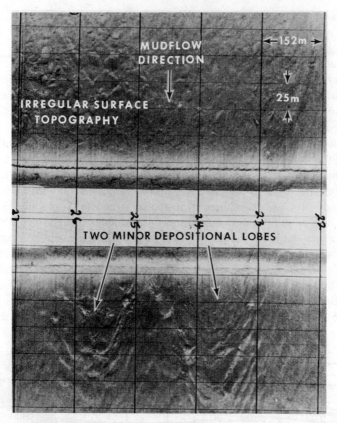

FIGURE 19 *Side-scan sonar record of a major lobe of flowed material with a complex leading edge composed of two minor lobes. Note the rough surface texture of the displaced material.*

margins of older lobes, block size tends to decrease, surface relief of individual blocks becomes less, shallow depressions, vents, and mud volcanoes appear, and low-relief pressure ridges are often present (Fig. 19). Surface texture is generally rough again near the mudflow's terminal scarp.

The relatively abrupt distal ends of the depositional lobes are commonly referred to as "mud waves" or mud "noses." Relief of these distal margins varies greatly. Each overlapping lobe has its own "nose," or terminal scarp. Composite scarps also occur which involve numerous depositional lobes ranging from a few meters to a maximum of about 30 m. Coalescence of adjacent lobes commonly creates a complex sinuous relief feature that can be many kilometers in length.

On seismic profiles (Fig. 20) mudflow lobes are characterized by an acoustically amorphous response. Very few internal reflection events are discernible within these features. Occasionally, curvilinear reflectors near the distal ends of lobes appear to suggest shear planes where failures have occurred and blocks of material have moved downslope.

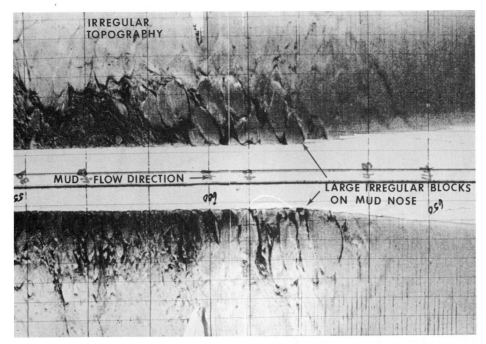

FIGURE 20 *3.5-kHz seismic profile of mudflow depositional lobe ("mud nose"), showing its lack of internal reflection horizons and complex leading edge. (Courtsey of Peter Trabant.)*

Other internal reflection events appear to separate successive lobes of remolded debris in these composite features. However, there are generally few or no continuous internal reflection events, a condition resulting from elimination of internal reflectors by mixing during transport or by high gas content or both. Occasionally, chaotic reflectors are present, a situation that tends to suggest highly disorganized and disrupted bedding. As is illustrated in Figure 20, seismic profiles commonly suggest that the leading edges of these features are complexly sculptured by many small-scale slumps, causing an extremely rough "textured" bottom. Side-scan sonar data confirm the presence of this coarse, blocky topography (Fig. 21). In many cases large blocks derived from the frontal scarp of the mudflow lobe are displaced downslope and tilted back toward the flow feature, indicating shallow rotational slumping.

Downslope from the major mud nose scarp a thin blanket of acoustically amorphous sediment is commonly present (Fig. 20). The top of this unit generally grades seaward into bedded material, as defined by horizontal and continuous reflection events on seismic records. In some cases it appears that acoustically amorphous material has been injected between reflecting horizons in apparently undisturbed sediments seaward of the advancing mud nose. Garrison (1974) described an area of the shelf off South Pass where flat-lying reflectors were elevated 7–8 m above the adjacent unaffected bottom, apparently by injection by material into the subsurface. Coleman and Garrison (1977) suggest that this process could take place only if the injected material were preceded by a rise in pore

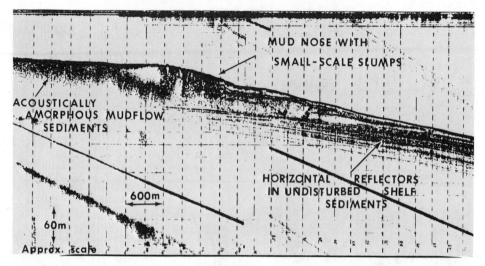

FIGURE 21 *Side-scan sonar record of a "mud nose" showing the extremely complex bottom topography associated with these features.*

pressure, which would provide the hydraulic jacking action necessary to permit injection to take place. As pointed out by Prior and Coleman (1978b), the presence of mud vents on the periphery of the mud noses is in agreement with the principle of undrained loading and the injection concept.

The composite depositional forms, mudflows, are commonly quite extensive. An isopach map of a mudflow off South Pass compiled by Coleman and Garrison (1977) from closely spaced high-resolution seismic profiles (Fig. 22) exhibits a length of over 50 km and covers an area of 770 km^2, while maximum thicknesses are in excess of 36 m. The importance of mudflows as transport mechanisms is reflected in the 11.2 × 10^6 m^3 of sediment estimated to be contained by this single feature (Fig. 22).

It is apparent from inspection of Figure 22 that the leading edge of the flow is highly crenulated, with large-scale lobes and reentrants and occasional bifurcations. This particular feature has moved across the shelf and is now encroaching on the upper continental slope in water depths of over 500 m. Although repetitive data are not sufficient to convincingly quantify rates of movement, existing data suggest sporadic advance of localized areas, with individual lobes prograding up to several hundreds of meters per year.

Mudflows, resulting from the accumulation of sediment discharged at the downslope ends of subaqueous landslides, are not restricted to those associated with the numerous upper delta-front gullies. Intermediate and lower delta-front environments also are traversed by linear landslide tracts that distribute sediments to form complex mudflow features (Fig. 23). Gully systems of these deeper environments occur less frequently and are more complex in general outline than their shallow-water counterparts. Perhaps the decreased slope of the deep delta front, coupled with lower sedimentation rates and greatly decreased physical process intensity (ocean waves and currents), is responsible for differences in shape and frequency of occurrence between the shallow and deep systems.

Because it is difficult to routinely acquire high-quality side-scan sonar records from the deep shelf and upper continental slope, the distal parts of these systems have been

SURFACE MUDFLOW -1975

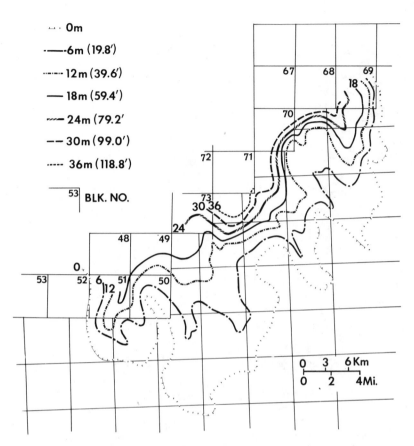

FIGURE 22 *Isopach map of two mudflows off South Pass as compiled from high-resolution seismic data. (Coleman and Garrison, 1977.)*

studied primarily from high-resolution seismic and bathymetric data. The deep shelf gullies are generally longer and broader features that exhibit a more sinuous outline than shallower systems. Scarp heights are commonly larger, some displaying gully walls up to 12 m high. Tributary gullies are more numerous, giving the source areas a complex branching form. Overlapping depositional lobes that form large mudflows which prograde onto the upper continental slope are common. Sediments discharged at the downslope ends of deep gully systems are deposited over shelf clays that have accumulated slowly, contain minimal amounts of sedimentary gas, and exhibit normal consolidation. Therefore, on the deep shelf relatively high relief depositional lobes that are generally acoustically amorphous are separated by low-relief "windows" of normal shelf clays. On the upper delta front the topographic highs represent gas-rich prodelta clays that are often acoustically amorphous, while low-relief gully floors commonly constitute acoustic windows (Fig. 24).

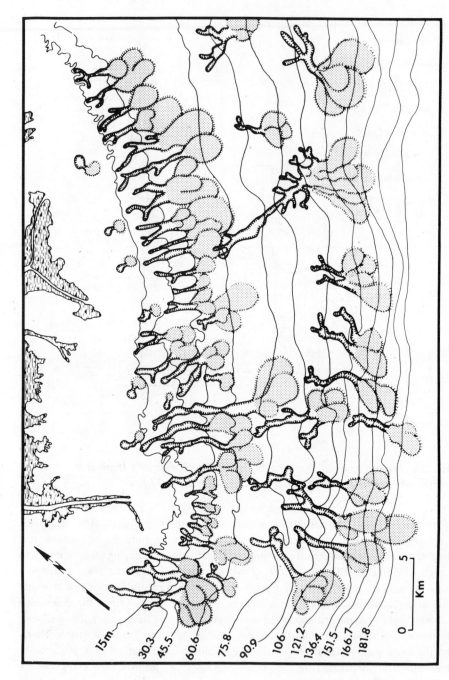

15m
30.3
45.5
60.6
75.8
90.9
106
121.2
136.4
151.5
166.7
181.8

0   5
Km

FIGURE 23 *Schematic representation of the major lineated subaqueous landslide features and associated depositional lobes of the Mississippi River delta-front environments.*

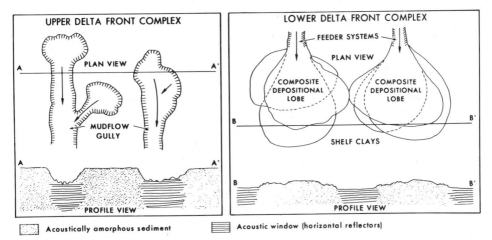

FIGURE 24 *Schematic representation of the relief and acoustic quality of sediments in the shallow versus deep delta-front environments.*

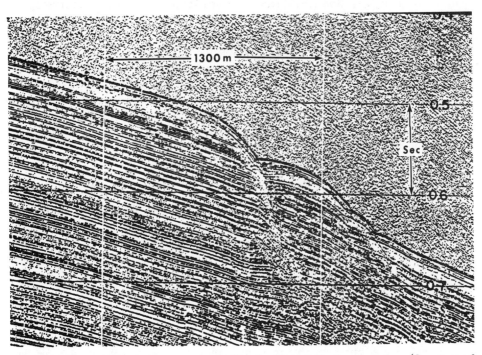

FIGURE 25 *High-resolution seismic profile showing shelf-edge slumping. (Courtesy of Peter Trabant.)*

In addition to mudflows moving over the shelf margin and onto the upper continental slope, a wide variety of slump and fault features are found at the shelf edge (Fig. 2). Probably the most impressive shelf-edge deformational features are associated with large-scale slump systems. These slumps occur in arcuate groups or families that may be as great as 20 km across in plan view. Complex stairstep topography is generally associated with the slump groups (Fig. 25). Sea-floor scarps produced by shelf-edge slumping are often in the range 20–30 m and cut up to 200 m of sediment. Downslope from these major arcuate slump zones, in water depths of 500–1000 m, chaotic bedding, accumulation lobes, and occasional massive blocks of sediment indicate the transport of large volumes of sediment from the shelf edge to the continental slope. The acoustic properties of large relief features found downslope of these systems of en echelon slumps suggest a diapiric origin. Such intrusions appear to be composed of material exhibiting chaotically arranged acoustic reflectors or material that is acoustically amorphous. Upturned reflection horizons commonly occur on their flanks, and horizontal reflectors generally appear beneath the acoustically amorphous material. Relief up to 120 m has been measured on such diapirs.

Other types of common shelf-edge deformational features include a variety of faults, including normal down-to-the-basin faults, some of which cut deeply into the sedimentary column (in excess of 500 m below the sea floor). En echelon contemporaneous faults exhibit the effects of deposition and continuous movement and are frequently observed on high-resolution seismic records. These faults display increased offsets of marker beds with depth, which are related to continued deposition contemporaneous with fault movement. Coleman and Garrison (1977) report that sparker profiles indicate that these faults extend to depths of 700–800 m below the sea floor. Surface mudflows frequently cross faulted areas at the shelf edge, tending to smooth the step-like topography caused by contemporaneous faulting by filling the offset on the downthrown block. The mudflow material then thickens as it crosses the fault, sometimes completely eliminating the surface expression of the fault. As the fault continues to move a new scarp will appear, and, given enough time, another mudflow will obscure this feature and add new sediment on the downthrown side. This interaction between sedimentation by mass movement and contemporaneous faulting may play an important role in controlling offsets and fault movement by preferential loading of the downthrown block.

THRESHOLDS OF SEDIMENT INSTABILITY

All the previously described instability features indicate that critical geomorphic thresholds are regularly exceeded in the Mississippi Delta area. This is the case because the delta sediments are in a condition of incipient failure as a result of high pore-water pressures. External factors such as wave loading, sedimentary loading, or oversteepening can provide a triggering mechanism so that the instability threshold is exceeded. To understand how little external stimulus is needed to cause failure, we must first examine the sediment pore pressures.

Preliminary results of pore-water pressure measurements within the subaqueous sediments on the Mississippi delta-front slope have been reported by Bennett et al. (1976). These were achieved using variable resistance pressure transducer piezometers installed in Block 28, South Pass area, in approximately 19 m water depth. The piezometers were established at depths of 8 and 15 m below the mudline. The pore-water pressure data revealed large excess pressures after 7 hr of stabilization of the system, and values were considerably in excess of hydrostatic pressure. The ratio of the pore-water pressure (u) to the hydrostatic pressure ($\gamma_x z$), $u/\gamma_w z$, had values of 1.40–1.48. Significantly, the ratios of $u/\gamma' z$, where γ' is the unit weight of the sediment, approach the geostatic condition, with values of 0.895–0.936. Even larger pressures have been recorded elsewhere in the off-shore delta region, with $u/\gamma' z$ ratios of 0.986 at a depth of 15 m within the sediment (Bennet, personal communication).

Also, indirect evidence of large pressures within the Mississippi Delta sediments is provided by the presence of widespread mud, water, and biogenic methane gas vents. These are readily identified on side-scan sonar records (Fig. 3) either as small circular holes or as small cones of mud on the subaqueous sediment surface. They are formed by the release, under pressure, of gas, water, and liquid mud from within the sediments. Where only gas and water are discharged, they erode the sediment surface, but when the emissions carry substantial amounts of mud they are accretionary. The vents occur in depositional areas downslope from the gullies, and notably along the lines of tensional cracks that develop across areas of stable sediment immediately adjacent to gullies and slumps (Prior and Coleman, 1978a). These features are remarkably similar to "mud volcanoes" reported in association with subaerial landslides where artesian pore-water pressures are involved in the initiation and maintenance of instability (Hutchinson and Bhandari, 1971; Hutchinson et al., 1974; Hutchinson and Kojan, 1975). The venting of pore fluid was also observed in the laboratory by Doyle (1973) during a study of wave-induced sediment instability.

From the geometry of the features that have been identified, an initial analysis of the mechanisms of these subaqueous slope instabilities can be made in terms of effective stresses if they are considered shallow planar slides on infinite slopes using the equation (Skempton and DeLory, 1957)

$$F = \frac{c' + (\gamma' z \cos^2 \beta - u) \tan \phi'}{\gamma' z \sin \beta \cos \beta} \tag{1}$$

where c' is the cohesion, ϕ' the angle of frictional resistance, γ' the unit weight of the sediment, z the depth of soil above the slip surface, and β the slope/slip surface angle. On the assumption that $F = 1$ (failure condition), it is possible to calculate the pore-water pressure (u) needed to initiate failure from the equation

$$u = \frac{c' - F (\gamma' z \sin \beta \cos \beta)}{\tan \phi'} + \gamma' z \cos^2 \beta \tag{2}$$

The general angle of the delta-front slope and the basal shear surface (β) is taken as $0.5°$, although locally slopes may achieve $1.7°$ (Coleman and Garrison, 1977). The unit weight of the delta-front clays is approximately 16,000 N/m³ (100 lb/ft³) (McClelland, 1967). Calculations of pore-water pressure needed for failure have been made using three sets of c' and ϕ' values (Table 1) and for two different thicknesses of sediment (16 m and 33 m). The latter applies to the position of basal shear planes inferred from seismic data. The shallower depth value is also used because Whelan et al. (1976) reported a typical strength profile with depth from borehole analysis that shows a pronounced "crust" over a zone of substantially weaker sediment. Values of undrained shear strength (cu) from vane tests typically indicate a zone of maximum strength up to 23.9 kPa (500 lb/ft^2) in the upper 14–15 m of sediment overlying materials that have strengths as low as 4.8 kPa (100 lb/ft^2).

Table 1 shows calculated values of pore-water pressure needed for failure. The pressures are expressed in both kilopascals (pounds per square foot) and as pore-water pressure ratios $u/\gamma_w z$ (ratio of hydrostatic pressure) and $u/\gamma' z$ (ratio of geostatic pressure). For all cases, the pore-water pressure needs to be very large for failure to occur. It must be in excess of hydrostatic pressure (ratios of 1.53–1.57), and this constitutes strong artesian pressure. In addition, it is clear that the calculated values approach (or very slightly exceed) geostatic pressure, representing a condition of almost zero effective stress.

TABLE 1 *Strength Properties and Calculated Pore-Water Pressures and Ratios for Failures[a]*

Source	Strength Parameters			16-m Depth (z)			33-m Depth (z)		
	c'	ϕ'	u	$u/\gamma_w z$	$u/\gamma' z$	u	$u/\gamma_x z$	$u/\gamma' z$
Oil Company	9.58	24°	256	1.67	1.070	491	1.60	1.025
Dunlap (personal communication)	3.35	26°	242	1.58	1.011	477	1.58	0.996
Henkel (1970)	0	20°	234	1.53	0.976	468	1.53	0.976

[a]Units are kilopascals.

Both the indirect and the direct evidence indicate large pore pressures within the delta-front sediments. Thus, initial failures (i.e., collapse depressions, bottleneck slides) may be triggered by small increases in pore pressure or small external forces. For example, the $u/\gamma' z$ of 1.54 has to be increased by less than 20.7 kPa (430 lb/ft^2) for failure to occur, using the highest-strength data available for either depth. Fluctuations of 7 to 15 kPa (150 to 300 lb/ft^2) have been recorded owing to passage of storm waves (Bennet, personal communication).

We can also examine the probable stability of sediments after some initial movement has taken place. Consider the chute sediments and those deposited in the toe areas. These sediments have been reworked through movement so that their physical properties are different from the initial state. In the chute areas, the remolded sediments could reasonably have a residual cohesion $c'_r = 0$ and $\phi'_r = 10°$. In this case the chutes would fail under pore pressures that are about 95% of geostatic ($u/\gamma' z = 0.95$), as opposed to about 98% for the unmoved sediments. Thus, given an equal increase in pore pressures (e.g., wave loading), the chute sediments would exceed the threshold for movement first.

The toe-area sediments display characteristics of being potentially more stable than initial deltaic sediments. Having been moved and deposited, the excess pore pressure which initially caused movement may have been dissipated. The strength of the toe sediments could then be larger than the initial condition. The scarp height of the depositional areas can be used to estimate the strength of the sediments. The height of a flow-ing wedge of sediments has been related to the strength of the sediment by Johnson (1970) in the following form:

$$H = \pi \frac{c}{\gamma_1} \left(1 - \frac{\sigma}{90}\right) \tan \left(\frac{\pi}{4} + \frac{\phi}{2}\right) \tag{3}$$

where H is the scarp height, c the cohesion, ϕ the friction angle, σ the slope in degrees, and γ_d the unit immersed weight. In the delta area, scarp heights range from 4 m to as high as 20 m. This implies strengths of 4.8 kPa (100 lb/ft^2) to 24.0 kPa (500 lb/ft^2). These values are consistent with typical near-surface values measured in the toe areas (see Fig. 10). The toe areas would be the least likely to move under a given increase in regional pore pressure. The pore pressure ratios for failure would need to be as large as $u/\gamma'z = 1.17$ for the toe areas.

SUMMATION AND CONCLUSIONS

Systematic evaluation of high-resolution seismic profiles, side-scan sonographs, and bathymetric records has led to the identification of subaqueous landforms whose presence and frequency of occurrence indicate the importance of sediment instability and mass-movement processes in sediment transport on the Mississippi River delta front. Most of these features can be described as submarine landslides that occur on very low angle slopes (generally between 0.2 and 1.7°).

Some of the flattest regional slopes associated with subaqueous deltas are found in the broad, shallow bays between major distributaries. Three main types of instability landforms have been identified from interdistributary-bay environments: collapse depres-sions, bottleneck slides, and elongate slides. The collapse depressions are generally small features, sometimes less than 100 m in diameter, that appear to be related to subsidence accompanied by dewatering and degassing of the sediments. They occur most frequently where regional slopes are extremely low, as in the middle depth to shallow parts of the interdistributary bays. As regional slopes increase, more elongate instability features such as bottleneck slides appear. They clearly exhibit evidence of both subsidence and trans-lational movements. Elongate slides are more complex in their development than collapse depressions and bottleneck slides because they involve both downslope and upslope ex-tension and possibly localized loading within the system. It is clear that the sediment-deforming features identified from interdistributary bays occur in a spectrum of shapes, from small, nearly circular forms to sinuous elongate ones, some of which are compound features.

Within the upper delta-front deposits rapid deposition near the distributary mouths promotes rotational sliding in the form of large slumps that have concave-upward shear planes. The slumping process is combined with translational movements downslope. In these upper delta-front environments instabilities generally are initiated as shallow rota-tional slumps, but as movements become more translatory many slump systems "feed" elongate mudflow gullies. The delta-front gullies clearly display features that indicate that numerous source areas, as well as sites of deposition, are built into the system. Numerous branching tributaries commonly discharge sediment into a main channel whose step-like profile indicates the presence of local failures as well as sites of deposition and loading within the system. Downslope from the distal ends of mudflow gullies are areas of irregular bottom topography representing overlapping mudflow lobes. Step-like profiles across these depositional zones indicate the positions of relatively abrupt distal scarps or noses of individual mudflow lobes. Intersecting lobes fed from numerous upslope gully systems produce compound mudflow features that are tens of kilometers in length and in excess of 30 m thick. Mudflow lobes and noses are capable of advancing downslope at exceptionally high rates. From the limited data available from repeated surveys, movements appear to be sporadic and to involve localized lobes of the compound mudflow feature. Rates as well as other characteristics of mudflow movement are not well known because of the lack of repetitive data sets from critical areas.

In the deeper delta-front environments gully systems are somewhat larger, spaced farther apart, and generally more complex in plan view because of the sometimes numerous tributaries. Sediments discharged from these systems commonly feed mudflows that have progressed over the shelf edge onto the upper continental slope, occasionally obscuring scarps from a number of shelf-edge faults and slumps.

The shelf edge is frequently broken by large slump systems, deep-seated contemporaneous faults, and normal down-to-the-basin faults. Of these deformational features the large, arcuate-shaped families of shelf-edge slumps are the most impressive. They are similar in form to the small-scale slump features associated with shallow delta-front deposits. At this time, data on the upper continental slope are insufficient to map downslope surface deformational systems comparable to the delta-front mudflow gullies. However, rotational slumping and translation of large blocks downslope have been confirmed by high-resolution seismic profiles.

The existence of numerous sediment deformational features from the shallow water of interdistributary bays to the shelf edge indicates that critical geomorphic thresholds are being consistently exceeded. It seems apparent that slope and sediment instabilities associated with these various features are not the result of a single causal mechanism, but more commonly represent very complex process interactions. At present many of these conditions that result in initiating failure can be only qualitatively discussed. Prior and Coleman (1978b) emphasize that individual thresholds, when stresses exceed the strength of the material and instability occurs, are likely to be achieved by quite different combinations of the same basic factors over time and space. For example, large wave-generated bottom pressures associated with a hurricane may have a potentially greater effect if reduction in strength is well advanced, either by cohesive strength changes or gas- and water-pressure generation, than if strength characteristics have not been altered by these processes. Alternatively, rapid generation of in situ methane gas, or its mobility from one zone to another, or in combination with excess pore-water pressure caused by sedimentary loading, may result in failure without any external changes in stress conditions.

Using the geometry of deformational features already identified, initial evaluations of slope instability mechanisms have been made in terms of effective stresses using infinite slope analysis. Since much deltaic sediment is in a state of near-failure because of rapid deposition and subsequent production of excess pore-water pressures, very little external stress is needed to surpass the thresholds of sediment strength and cause failure. Excess sedimentary gas pressures can also create a condition of instability and in some cases initiate failure. Several lines of evidence, including direct field measurements, confirm the existence of large excess pore-water pressures in the dominantly fine-grained sediments of the interdistributary-bay and delta-front environments. Therefore, initial failures may result from small increases in pore pressures or small external forces.

Analysis of the stability of sediments that have already moved shows that chute areas containing remolded sediment would fail under pore pressures that are about 95% of geostatic compared to about 98% for undeformed sediment, which would cause the chute sediments to exceed their stability threshold first, resulting in failure, given an equal increase in pore pressures.

Sediments in the toe lobes of deformational systems such as delta-front gullies have been transported, remolded, and redeposited. During this process, it is thought, excess pore-water pressures are somewhat dissipated, leaving the toe lobe sediments with higher strength properties or higher thresholds of instability than initially deposited deltaic sediments. Using the scarp height as an indicator of sediment strength on mudflow lobes, it was found that toe areas are less likely to fail given an increase in regional pore pressure than undisturbed sediments.

In conclusion, it is apparent that thresholds of sediment stability are being surpassed in all subaqueous environments of the Mississippi River delta, from the interdistributary bays to the shelf edge. The factors that contribute to stress increase or strength reduction, leading to slope failure, are complex and clearly show that there are several different ways that critical strength/stress disequilibrium can be achieved. One characteristic of this very important method of sediment transport is that movements are sporadic or episodic. A stair-step geomorphic response generally results from these movements. Therefore, within any given deformational system such as a mudflow gully there is a nested array of geomorphic features, each with its own characteristics and stability thresholds. Only in recent years have the aforementioned processes of sediment deformation been fully appreciated as important sediment transport mechanisms on such low-angle slopes.

ACKNOWLEDGMENTS

This paper is largely a compilation of results of research supported by two primary agencies: the Geography Programs, Office of Naval Research, Arlington, Virginia 22217, and the U.S. Geological Survey. Cooperative research efforts between personnel at Coastal Studies Institute, the U.S. Geological Survey, and the Departments of Oceanography and Civil Engineering, Texas A and M University, have input greatly to this summary. In addition, the petroleum industry and various consulting firms are acknowledged for providing access to valuable data which helped formulate concepts concerning deformational processes in the subaqueous delta. Peter Trabant of Texas A and M University is gratefully acknowledged for acquainting the authors with high-resolution geophysical data of the Mississippi delta-front area taken on Texas A and M Cruise 72-A-19.

REFERENCES CITED

Bea, R. G., Bernard, H. A., Arnold, P., and Doyle, E. H., 1975, Soil movements and forces developed by wave-induced slides in the Mississippi Delta: J. Petrol. Technol., p. 500–514.

Belderson, R. H., Kenyon, N. H., Stride, A. H., and Stubbs, A. R., 1972, Sonographs of the sea floor: Amsterdam, Elsevier, 185 p.

Bennet, R. H., Bryant, W. R., Dunlap, W. A., and Keller, G. H., 1976, Initial results and progress of the Mississippi delta sediment pore water pressure experiment: Marine Geotechnol., v. 1, p. 327–335.

Coleman, J. M. and Garrison, L. E., 1977, Geological aspects of marine slope stability, northwestern Gulf of Mexico: Marine Geotechnol., v. 2, p. 9–44.

——, Suhayda, J. N., Whelan, T., and Wright, L. D., 1974, Mass movements of Mississippi River deltas: Proc. 24th Conf. Gulf Coast Assoc. Geol. Soc., Lafayette, La., p. 49–68.

Doyle, E. H., 1973, Soil wave tank studies of marine soil instability: Fifth Offshore Technol. Conf., Houston, Tex., n. 1901.

Flemming, B. W., 1976, Side-scan sonar: a practical guide: Int. Hydrographic Rev., v. 53, n. 1.

Frazier, D., 1967, Recent deltaic deposits of the Mississippi River, their development and chronology: Trans. Gulf Coast Assoc. Geol. Soc., v. 17, p. 187–315.

Garrison, L. E., 1974, The instability of surface sediments on parts of the Mississippi delta front: U.S. Geol. Survey, open-file report, Corpus Christi, Tex., 18 p.

Hutchinson, J. N., and Bhandari, R. K., 1971, Undrained loading, a fundamental mechanism of mudflows and other mass movements: Geotechnique, v. 21, p. 353–358.

——, Prior, D. B., and Stephens, N., 1974, Potentially dangerous surges in the Antrim mudslide: Quart. J. Engr. Geol., v. 7, p. 363–376.

——, and Kojan, E., 1975, The Mayunmarca landslide of 25 April 1974: UNESCO Reports 3124/Rmo:RD/SCE, Paris, 24 p.

Johnson, A. M., 1970, Physical processes in geology: San Francisco, W. H. Freeman, p. 433–459.

Kolb, C., and Van Lopik, J., 1958, Geology of the Mississippi River deltaic plain, southeastern Louisiana: U.S. Army Corps Engr. Waterways Experiment Station, Vicksburg, Miss., Tech. Rept. 3–483.

——, and Van Lopik, J., 1966, Depositional environments of the Mississippi River deltaic plain, southeastern Louisiana: in Shirley, M., Deltas in their geologic framework: Houston, Tex., Geol. Soc., p. 17–61.

McClelland, B., 1967, Progress of consolidation in delta front and pro-delta clays of the Mississippi River: Marine Geotechnique, Urbana, Ill., University of Illinois Press, p. 22–33.

Morgan, J. P., 1961, Mudlumps at the mouths of the Mississippi River: in Genesis and paleontology of the Mississippi River mudlumps: Louisiana Dept. Conservation Geol. Bull. 35, p. 1–116.

——, 1968, Mudlumps: diapiric structures in Mississippi delta sediments: in Diapirism and diapirs: Amer. Assoc. Petrol. Geol, Mem. 8, p. 145–161.

——, Coleman, J. M., and Gagliano, S. M., 1963, Mudlumps at the mouth of South Pass, Mississippi River: sedimentology, paleontology, structure, origin, and relation to deltaic processes: Coastal Studies Inst., Louisiana State Univ., Baton Rouge, Coastal Studies Ser., 10, 190 p.

Newton, R. S., Seibold, E., and Werner, F., 1973, Facies distribution patterns on the Spanish Sahara continental shelf mapped with side-scan sonar: "Meteor" Forsch-Ergebnisse, Reihe C, n. 15, p. 55–77.

Prior, D. B., and Coleman, J. M., 1978a, Submarine landslides on the Mississippi delta front slope: Geoscience and Man, v. 19, p. 41–53 (School of Geosciences, Louisiana State University, Baton Rouge).

——, and Coleman, J. M., 1978b, Disintegrating, retrogressive landslides on very low angle subaqueous slopes, Mississippi delta: Marine Geotechnol., v. 3, p. 37–60.

Roberts, H. H., Cratsley, D., and Whelan, T., III, 1976, Stability of Mississippi delta sediments as evaluated by analysis of structural features in sediment borings: Eighth Offshore Technol. Conf., Houston, Tex., p. 9–28.

Shepard, F. P., 1955, Delta-front valleys bordering the Mississippi distributaries: Geol. Soc. Amer. Bull., v. 66, p. 1489–1498.

Skempton, A. W., and DeLory, F. A., 1957, Stability of natural slopes in London clay: Proc. Fourth Int. Conf. Soil Mech., London, v. 2, p. 378–381.

Sterling, G. H., and Strohbeck, E. E., 1973, The failure of the South Pass 70 "B" platform in Hurricane Camille: Fifth Offshore Technol. Conf., Houston, Tex.

Suhayda, J. N., Whelan, T., III, Coleman, J. M., Booth, J. S., and Garrison, L. E., 1976, Marine sediment instability: interaction of hydrodynamic forces and bottom sediments: Eighth Offshore Technol. Conf., Houston, Tex., n. 2426, p. 29–40.

Whelan, Thomas, III, Coleman, J. M., Suhayda, J. N., and Garrison, L. E., 1975, The geochemistry of recent Mississippi River delta sediments: gas concentration and sediment stability: Seventh Offshore Technol. Conf., Houston, Tex., p. 71–84.

——, Coleman, J. M., Roberts, H. H., and Suhayda, J. N., 1976, The occurrence of methane in recent deltaic sediments and its effects on soil stability: Bull. Int. Assoc. Engr. Geol., v. 14, p. 55–64.

III

HYDROGEOLOGIC REGIMES

III

8

THRESHOLDS IN DETERMINISTIC MODELS OF THE RAINFALL–RUNOFF PROCESS

A. I. McKerchar

INTRODUCTION

Thresholds are encountered when a deterministic approach is adopted to hydrological problems. The general definition of a threshold—the point at which a stimulus begins to produce a response—is applicable to these problems. A stochastic analysis may proceed without considering the causal mechanism of thresholds, whereas in the deterministic approach the physical processes causing thresholds are specified. The two approaches are in many instances complementary. In hydrology the mechanisms involved in the catchment process of translating precipitation input into runoff output are not well defined and modeling of this process is therefore difficult. Nevertheless, much has been achieved in the deterministic modeling of catchment response. For such models to accurately represent reality, thresholds encountered in the catchment process must be identified and correctly incorporated within a catchment model.

In the catchment process some thresholds, such as the time for initiation of runoff, are immediately obvious and can be simply measured. Other thresholds, associated with individual processes operating within the catchment, are readily understood at individual points but may not be amenable to field observation and measurement over a catchment. The former thresholds are associated with integrated quantities, such as discharge, whereas the latter are parts of hydrological processes operating over the catchment area. Those associated with integrated quantities might also be termed gross thresholds, since in a deterministic framework these would be by-products of a complete understanding and integration of the individual physical processes.

The integrating effect can be reproduced by a deterministic model which is a set of mathematical expressions representing the physical processes knitted together by a logic dictated by individual threshold levels. Such hydrological models are termed "conceptual" since they represent physical concepts of the processes operating on precipitation moving through the catchment.

In comparison with stochastic modeling which has been undertaken by statisticians and others over many years, the history of conceptual modeling in hydrology is short, dating from the availability of large digital computers. Nevertheless, this short interval has seen considerable progress in developing models and in applying them to different catchments. Fleming (1975) for example, reviews 19 such models, and the World Meteorological Organization (WMO) recently reported (WMO, 1975) on a comparison of mathematical catchment models, nine of which were conceptual models.

Despite this apparent progress, conceptual modeling now seems to have reached what Chapman (1975) terms a "plateau of achievement," beyond which it appears that progress will not be possible until certain issues are resolved. At the center of these issues is the estimation and interpretation of the parameters of the models, which can be achieved only when threshold levels for individual processes are correctly characterized and accurately estimated.

The purposes of this chapter are to (1) indicate where thresholds usually occur in catchment processes, (2) demonstrate how thresholds are typically built into conceptual models, and (3) outline the parameter estimation problems raised by their presence. This chapter is structured as follows: first conceptual models and thresholds are discussed with reference to a typical example; this is followed by a brief review of the parameter estimation problem; particular difficulties attributable to thresholds are then described; and finally some directions for future work in the area are indicated.

CONCEPTUAL MODELS AND THRESHOLDS

The discharge from a catchment is the integrated response to precipitation input. This integration extends in time and space over the catchment and over the individual processes operating on and below the catchment surface, and in the stream channel. The processes include interception, ponding, evaporation, snowmelt, infiltration, transpiration, soil moisture storage, groundwater storage, capillary rise, and channel routing.

Thresholds are manifest in these processes in several ways. Where catchment processes incorporate storages, a new mode of operation commences, or a new process is included when a storage is filled. An example is the delivery of net precipitation onto the ground after the interception storage in the vegetation canopy is filled. Thresholds are met when rates of operation of processes reach critical levels; for example, when the net precipitation rate exceeds the infiltration rate, overland flow can commence.

For single points on a natural catchment the physics of these processes are reasonably well understood and can be expressed by mathematical equations and logical expressions. The fundamental problem in catchment modeling is to apply this point knowledge to a heterogeneous catchment when processes vary in time and space and where boundary conditions are known only approximately (Philip, 1975). In conceptual modeling, small catchments, or subareas of large catchments, are most often considered to be homogeneous; a single average precipitation input is applied over the catchment and individual processes, which in fact can vary widely over the area, are represented by lumped single indices. Incorporation of naturally distributed areal variability into models leads to considerable complexities in models, often requiring data that are unavailable. A conceptual model, be it lumped or distributed, is calibrated by assigning values to the parameters of the individual processes and to the parameters that specify thresholds.

Ideally, these parameters should (a) have physical interpretation, and (b) be measurable in the field. Property (a) would enable the model to be used in assessing the influence of proposed changes in land use; property (b) would allow the model to be applied to ungaged catchments. In practice these ideals are rarely achieved: all parameters can rarely be interpreted physically in conceptual models, and attempts to utilize field estimates of parameters have met with little success. Most conceptual modeling relies upon numerical search procedures to define at least some of the model parameters.

Despite the estimation difficulties, conceptual models have been extensively used in hydrology. Clarke (1973) suggests three principal uses for conceptual models:

1. To forecast river flows, both in operational or "real-time" situations, and in hypothetical or design situations.

2. To estimate records of flows that correspond to long precipitation records.

3. To predict possible effects of proposed changes in the catchment on river flow characteristics. Such changes include urbanization, afforestation, and logging.

The success of a model in these uses depends in many instances on the ease of parameter estimation, including those parameters which involve thresholds.

THE STRUCTURE OF CONCEPTUAL MODELS

The conceptual models of particular hydrological interest are mathematical representations of catchment response to precipitation based on understandings of the operative physical processes. These models are normally implemented with the aid of digital computers to estimate sequences of flow data corresponding to observed precipitation records. As an example, the conceptual model of Dawdy and O'Donnell (1965) is considered. This model has been used extensively in experiments for the numerical estimation of parameters (Ibbitt and O'Donnell, 1971). It is designed to estimate a sequence of flow data corresponding to a precipitation record and is typical of a number of conceptual models reported in the literature. It operates by the successive routing of the precipitation and evaporation inputs for each time interval through four storages (Fig. 1). A hierarchy of relationships and logical instructions dictate the functioning of each component. Three storages incorporate thresholds at which new modes of operation become effective.

The storage elements are as follows: Surface and interception storage R is increased by precipitation and is depleted by evaporation E_R and by infiltration F. When R exceeds its threshold R^*, it is depleted by channel inflow Q_1. The channel storage S is increased by discharge Q_1 from interception and surface storage, and depleted by discharge from channel storage Q_S. The soil moisture storage, M, is increased by infiltration F and by capillary rise C, and is depleted by transpiration E_M. When M exceeds a threshold M^* it is also depleted by deep percolation D. Finally, groundwater storage G is increased by

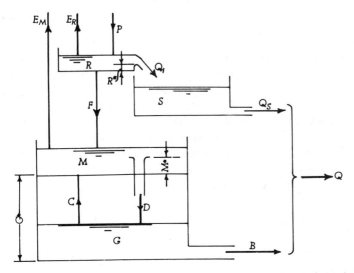

FIGURE 1 *Conceptual models of Dawdy and O'Donnell (1965).*

deep percolation D and depleted by capillary rise C and baseflow B. For intervals when G exceeds the threshold $G*$, the soil moisture and groundwater storages are merged and the capillary rise and deep percolation no longer operate.

There are nine parameters controlling the operation of the model: (1) the threshold values $R*$, $M*$, $G*$; (2) channel and groundwater reservoir routing coefficients K_S and K_G, respectively; (3) the maximum rate of capillary rise C_{max}; and (4) the three infiltration parameters in the Horton infiltration equation,

$$f_t = f_{min} + (f_{max} - f_{min}) \exp (-kt) \tag{1}$$

where f_t is the potential infiltration rate at time t, f_{max} and f_{min} the maximum and minimum potential rates, and k a decay constant. These nine parameters must be estimated to operate the model.

Although not included in this model, snowpack can be an important component in some hydrological models. A complete statement of the dynamics of the accumulation and ablation of snow requires budgeting of the energy balance by taking account of the energy transfer by radiation, convection, and conduction, and phase changes. These are subject to large diurnal fluctuations and considerable areal variability within a natural catchment. Embedded in this variability are thresholds defining levels above which the snowpack melts and levels at which runoff occurs. This distinction is necessary because meltwater may be held by capillary action within the snowpack (termed snow "ripening"). The sophistication of snowmelt models is constrained by the availability of suitable data. For example, the British Institute of Hydrology catchment model (Douglas, 1974), lacking accurate precipitation measurements during periods of snowfall and detailed temperature records, uses a very simple snowmelt component with a "degree-day" approach to estimating melt. Melt in any one period is taken as a linear function of temperature above a threshold level. In situations with sparse data, greater detail may be unwarranted. In continental climates where large proportions of runoff originates as snow, a more thorough approach to the modeling of snowmelt is obviously necessary.

ESTIMATION OF PARAMETERS

Because field estimation of model parameters has not been very successful, procedures for numerical parameter estimation have been used. These rely on using an initial set of parameters to estimate a flow hydrograph by routing precipitation and other inputs through the catchment model. The calculated hydrograph is compared with the observed hydrograph for the same period and the model parameters are then adjusted to improve the fit of the computed hydrograph. The goodness of fit may be subjective, or may be measured objectively by using a criterion function. One criterion function frequently used for its attractive analytical properties is the sum of squares of differences between observed and computed hydrographs. That is,

$$F = \sum_{t=1}^{N} (q_t - \hat{q}_t)^2 \tag{2}$$

where N = length of record under consideration

$\qquad q_t$ = observed discharge at time t

$\qquad \hat{q}_t$ = computed discharge at time t

and the set of values for model parameters is sought that will minimize F. Such values are "optimal" with respect to F. In theory a global minimum for an objective function such as equation 2 is sought but may not be achieved.

The method of finding a "best-fit" parameter set may be by trial and error, relying on the modeler to adjust parameter values to achieve a satisfactory fit between the observed and computed hydrograph for the portions of the hydrograph of interest. Alternatively, an automatic numerical search procedure (Johnston and Pilgrim, 1973) can be employed to determine a parameter set corresponding to an optimal value of the criterion function. Ibbitt and O'Donnell (1971) describe problems that can be encountered in applying automatic search procedures, some being met in numerical optimisation problems in all fields, and some appearing to be confined to hydrological modeling. The former include the presence of local optima, saddle points, and valleys on the multidimensional objective function surface; the latter include the presence of plateaus and depressions (really a special sort of local optima). The problems caused by these features are such that models constructed without regard for the fitting procedure can lead to intractable situations.

For example, parameter interdependency in the case of a two-parameter model is manifest as valleys on the objective function surface. Here the minimum is ill-defined, many pairs of parameter values giving almost identical values to the true minimum of the objective. In models having several parameters to be determined by search of the multidimensional objective function surface, potential interaction may not be obvious. Although a point near the minimum of the objective function can be reached, and the model may function satisfactorily, the situation negates the whole philosophy of conceptual modeling. Since many parameter sets are possible, any one set can have no physical meaning, nor bear any relationship to field observations. Therefore, models have to be designed to facilitate fitting, a task that is not straightforward in the presence of thresholds.

ESTIMATION PROBLEMS DUE TO THRESHOLDS

Storage processes are basic parts of many conceptual models. Figure 1, for example, which illustrates the model of Dawdy and O'Donnell (1965), contains four elements, to represent the dynamics of interception and surface storage, channel storage, soil moisture storage, and groundwater storage. Three of these storages have associated threshold values, indicating capacity; channel storage is taken to be semiinfinite. Changes in modes of operation when these thresholds are reached are indicated in Figure 1. Full interception and surface storage leads to direct runoff into channels; filling the groundwater storage merges it with the soil moisture storage and causes waterlogging of the catchment.

If, in automatic fitting, the range of possible values assigned to a threshold is set too high, such that the threshold level is never reached, the change of mode can never occur, and both the parameter defining the threshold, and the parameters associated with the processes operative in the new mode, are not considered; all have zero influence on the calculation of the objective function. Ibbitt and O'Donnell (1971) give a particular example; if for the whole of a record the quantities in interception and surface storage never exceed the threshold R^*, no channel inflow Q_1 can occur. Thus, if the channel storage S is zero at the start of the record, it will remain zero, and no channel discharge Q_S

occurs. Hence, the channel storage constant K_S is never tested and all values have zero effect on the objective function. This is termed a "plateau" situation, because for any value taken by K_S there will be no change in the objective function. K_S can thus be considered as a "nuisance" parameter. In this plateau situation any increase in the threshold R^* will similarly have no effect. However, unlike K_S, a decrease in R^* will eventually result in channel inflow Q_1, allowing K_S to become active, leading to a change in objective function F. Similar fitting problems for thresholds are encountered for the soil moisture storage M^* and the groundwater storage G^*; here the carryover effect applies to maximum rate of capillary rise C_{max} and the groundwater reservoir routing constant K_G.

Deceptively good results can be obtained by fitting a model to a set of data that do not sufficiently exercise all the components of the model, leaving some parameter values either undetermined, or determined only by an inadequate sample of data. The inadequacy will be evident only when the model provides poor forecasts of flows for periods when the components are exercised outside the fitting range, despite the fact that good reproduction of the data for the calibration interval may have been achieved. It is therefore essential that models be well exercised in fitting. This generally means using a sequence of data for calibration covering an interval of several years, a task that can be very demanding on computing facilities if several parameters are to be estimated jointly, and several time intervals for each day are used.

These difficulties have meant that estimates of physically meaningful parameters have rarely been obtained, and may be the reason why conceptual models have not, in general, fulfilled the early expectations held for them in predicting the influence of changes in land use on flow characteristics. The change in the flow characteristics of a catchment caused by a conversion from grassland to forest, for example, is an outstanding hydrological issue in the United Kingdom, Australia, and New Zealand to which conceptual modeling has contributed very little understanding. Nevertheless, conceptual models have found wide applications in situations where adequate concurrent precipitation, evaporation, and flow records are available for satisfactory model calibration (Fleming, 1975). Also, some progress has been made in studying hydrological effects of urbanization; here the treatment, namely more impervious areas, is readily quantifiable.

It seems clear that a wider use of conceptual models will be possible only when thresholds can be incorporated in ways more amenable to numerical optimization of parameters. Ibbitt (1972), for example, proposes a physically interpretable scheme for representing the interception process on a daily basis which is amenable to numerical optimization. These ideas await wider testing and application.

CONCLUSIONS

Usually, a few variables and thresholds are considered in stochastic models which are relatively simple in their structure. In contrast, conceptual hydrological models tend to be more complex and represent a variety of processes and several thresholds.

If a catchment is considered homogeneous, hydrological processes such as interception, soil moisture storage and groundwater storage may be represented as storages with thresholds being equivalent to full storages. As these processes may vary widely within a catchment, field estimation of the storage parameters is complicated. Consequently, numerical search procedures have been widely used to estimate them. There appears to have been little progress in comparing numerically derived estimates with field observations and examining reasons for discrepancies. Wider use of conceptual models continues to be hampered by this difficulty.

Complications in search procedures caused by thresholds have not been recognized widely. Also, thresholds in conceptual hydrological models do not appear to have received the attention they deserve. There is, for example, little information available on the sensitivity of models to changes in these thresholds.

ACKNOWLEDGMENTS

Support of the Water and Soil Division, Ministry of Works and Development, is acknowledged. Helpful discussion with A. R. Rao, R. P. Ibbitt, and M. J. McSaveney is appreciated.

REFERENCES CITED

Chapman, T. C., 1975, Trends in catchment modelling: *in* Chapman, T. G., and Dunin, F. X., eds., Prediction in catchment hydrology: Canberra, Aust. Acad. Sci.

Clarke, R. T., 1973, A review of some mathematical models used in hydrology, with observations on their calibration and use: J. Hydrol., v. 19, p. 1—20.

Dawdy, D. R., and O'Donnell, T., 1965, Mathematical models of catchment behaviour: Proc. Amer. Soc. Civic Engineers, v. 91, n. Hy 4, p. 123—137.

Douglas, J. R., 1974, Conceptual modelling in hydrology: Institute of Hydrology, Rept, 24, Wallingford, Oxon.

Fleming, G., 1975, Computer simulation techniques in hydrology: Amsterdam, Elsevier.

Ibbitt, R. P., 1972, Development of a conceptual model of interception, part I: Hydrol. Progr. Rept. 5, Ministry of Works, N.Z.

——, and O'Donnell, T., 1971, Designing conceptual catchment models for automatic fitting methods: Mathematical models in hydrology, Proc. Warsaw Symp., v. 2, IAHS, Publ. 101, p. 461—475.

Johnston, R. P., and Pilgrim, D. H., 1973, A study of parameter optimization for a rainfall-runoff model: Water Res. Lab., Rept. 131, Univ. N.S.W.

Philip, J. R., 1975, Some remarks on science and catchment prediction: *in* Chapman, T. G., and Dunin, F. X., eds., Prediction in catchment hydrology: Canberra, Aust. Acad. Sci.

World Meteorological Organization, 1975, Intercomparison of conceptual models used in operational hydrological forecasting: Operational Hydrol. Rept. 7, WMO 429, Geneva, World Meteorological Organization.

9

STOCHASTIC ANALYSIS OF THRESHOLDS IN HYDROLOGIC TIME SERIES

A. Ramachandra Rao

INTRODUCTION

The point at which a stimulus initiates a response is commonly understood as a threshold. Both the magnitude of the response and the time of its occurrence are important in defining a threshold. If both stimulus and response can be measured, the characterization of thresholds is relatively easy.

In many of the common situations in hydrology only one variable (either a stimulus or a response), such as rainfall, runoff, or temperature, can be measured. In such cases the definitions of stimulus and response become vague. For instance, although rainfall is a stimulus for runoff, it is also a result of atmospheric conditions and hence may be considered a response to these conditions. The behavior of these hydrologic variables show remarkable variations. Their changes may be broadly classified as in Figure 1. Changes such as step, ramp, and transient changes may be distinguished from "regular" behavior of sequences. Sequences of hydrologic variables such as rainfall and runoff typically involve combinations of steps and ramps (Fig. 2).

The variability in the modular coefficients ($x_t/\bar{x}$, where $\bar{x}$ is the arithmetic average of x_t values) of several annual hydrologic series are shown in Figure 2. The Nile River data

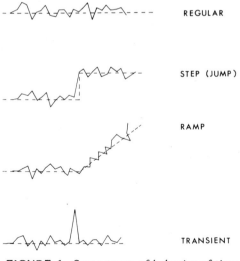

REGULAR

STEP (JUMP)

RAMP

TRANSIENT

FIGURE 1 *Some types of behavior of time series.*

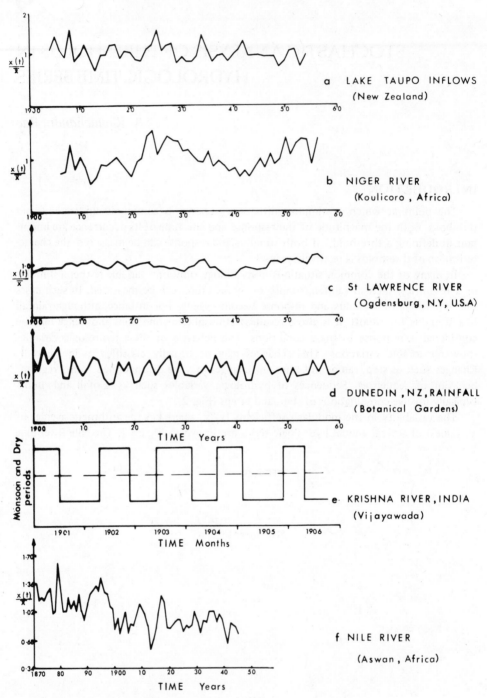

FIGURE 2 *Some hydrologic time series.*

(Fig. 2f) show two regions, one prior to construction of the Aswan Dam (1902) and another after completion. A trend of decreasing flows is also discernible in the period 1871–1902. The flows measured from 1902, however, are more accurate than those prior to 1902 (Hurst et al., 1965). Several reasons for decreased flows in the Nile River since 1902 have been discussed by Yevjevich and Jeng (1969). We may consider the Nile River flows as a decreasing ramp function (up to 1902) followed by "regular" behavior from 1902 onward. A similar situation, although not as striking, exists for annual rainfall data at Dunedin, New Zealand (Fig. 2d). The Niger River flows (Fig. 2b) are quite variable, comprised of rising and falling ramps.

Inflows to Lake Taupo (Fig. 2a) and St. Lawrence River flows (lake outflow) (Fig. 2c) do not exhibit the pronounced variability of the other examples. The St. Lawrence River flows are stabilized by storage in the Great Lakes and hence remain above or below the mean for long periods. Large storage is absent from the Lake Taupo inflows. Yet both series are more "regular" than the others shown (Fig.2).

Another important type of hydrologic behavior is the seasonal variation in flows of many rivers. The Krishna River (Fig. 2e) alternates between low flow (flows less than 500 m³/sec) and monsoon runoff (flows greater than 500 m³/sec). Although the

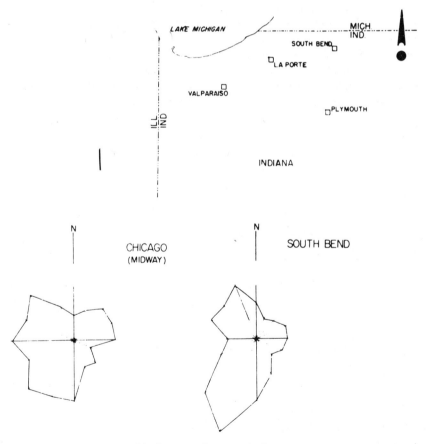

FIGURE 3   *Location of LaPorte and some wind roses.*

duration of high monsoon flows varies, the river may be visualized as having two distinct regimes, with thresholds at the onset and end of the monsoon.

One of the hydrologic series that has received considerable attention is precipitation at LaPorte, Indiana. LaPorte is east of the Chicago–Gary complex (Fig. 3) and the predominant wind direction at Chicago is westerly. As the Chicago–Gary area became heavily industrialized atmospheric particulate pollution increased, as did the heat island effect. A present hypothesis is that these pollutants and increased convection in the Chicago–Gary area have caused increased precipitation in the LaPorte area. These increased precipitation and rain-related activities have been referred to as the "LaPorte Anomaly" (Changnon, 1968, 1969).

The evidence suggests the plausibility of the hypothesis. For example, the monthly rainfall between 1898 and 1928 was 3.03 inches and it increased to 3.72 inches between 1929 and 1968, an increase of 23%. The mean annual rainfall in the corresponding periods of 1915–28, and 1929–68 changed from 36.40 inches to 44.59 inches. Mass curve, CUSUM (Woodward and Goldsmith, 1964), and the annual data were used to select 1929 as the year in which the rainfall increased (Fig. 4). Mean rainfall during different months and corresponding standard deviations (Fig. 5) also indicate considerable differences in these characteristics.

In general, a hydrologic process fluctuates around an unknown mean value, only an estimate of which is available. The second- and higher-order properties of the data are estimated by assuming a constant mean. If the mean value of the process changes, due to man as in the case of Nile River flows and perhaps LaPorte rainfall, or naturally, as in the Niger River flows, then the hydrologic series is nonstationary.

Not all fluctuations in the means of hydrologic time series are statistically significant, however. If the mean value of a time series is thought to have changed, the statistical

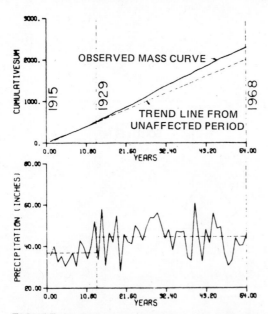

FIGURE 4 *Annual precipitation at LaPorte and the mass curve.*

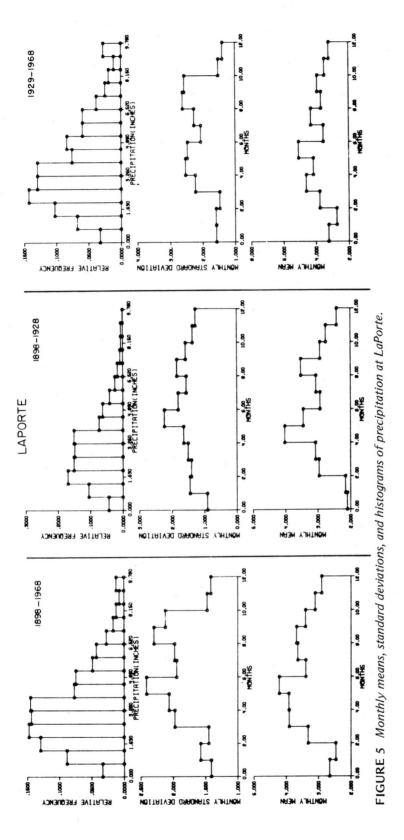

FIGURE 5 *Monthly means, standard deviations, and histograms of precipitation at LaPorte.*

significance of the change must be tested. If the change is significant, the hypothesis of change in the mean may be accepted. A similar analysis must be applied to other changes also. The times when significant changes occur in a time series may then be considered as points at which the process has entered a new "regime" through crossing a threshold. If the change in the mean or the trend is not statistically significant, the process is assumed to have remained as before. Consequently, in defining thresholds in hydrologic processes, where only a single variable may have been observed, the statistical significance of changes assumes a dominant role.

The significance of changes observed in hydrologic sequences, such as LaPorte rainfall or Nile River flow, cannot be tested by using statistical methods based on the assumption of independence of these sequences. Both monthly and annual rainfall values are usually autocorrelated, although the correlation in monthly data is stronger than that in annual data. In these cases methods and models of time-series analysis are the basic tools that are necessary. Basic techniques of stationary time-series analysis such as the correlogram

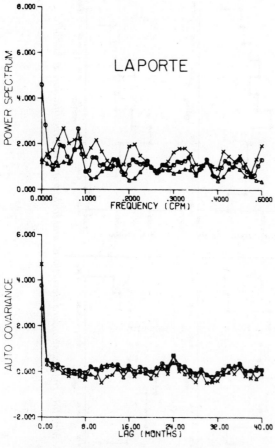

FIGURE 6 *Correlograms and power spectra of monthly precipitation at LaPorte: (X : 1898–1968, Δ : 1898–1928, O : 1929–68).*

and power spectral analysis cannot bring out the differences in characteristics of changing processes. For instance, correlograms and power spectra of LaPorte rainfall in the periods 1898–1928, 1929–68, and the entire period 1898–1968 (Fig. 6) show only a slight predominance in power in the spectra of the first (1898-1928) and the entire period (1898-1968) which is expected. Otherwise, the results do not give insight into the nature of change. Models of nonstationary time series are basic for the definition of thresholds.

For the present discussion "hydrologic thresholds" are defined to include two factors. The most important factor is the occurrence of significant change in the series whether natural or due to intervention of man. The most important characteristic of this factor is the mean. The second factor is the time at which the change occurs.

Because of the concern about changes in hydrologic variables, many of these have been observed over centuries at several sites. Large observation networks have only recently been established, however. In the absence of direct observation, indirect evidence such as the growth of glaciers and tree rings have been used to investigate climatic changes (National Academy of Sciences, 1975). The investigation of thresholds is further complicated by the intervention of man in the hydrologic cycle, which causes local and possibly global changes in these variables.

Although significant changes in hydrologic variables are related to climatic change, the focus of this chapter is determining thresholds in single hydrologic series. As changes in each of the series in an area provide the evidence for climatic change, the present approach is not as limited as it may appear to be.

The methods of analyses, development of models, and inferences from these models for processes whose characteristics change with time are considered in the remainder of this chapter. The emphasis is on the problems of characterization and modeling of data in which the mean changes. As discussed above, the inferences about changes in mean lead to the characterization of thresholds. Effects of a changing mean on the characteristics of independent random variables are discussed in the next section. The problem of characterizing time series with changes in mean values is discussed in the third section. Development of models of seasonal time series with changes in mean and inferences from these are considered in the fourth section. A general discussion and a set of conclusions are presented in the last section.

STATISTICAL CHARACTERISTICS OF NONSTATIONARY DATA

Thresholds in hydrologic data may be approximated by jumps or steps in the data sequences. These step changes and times of their occurrence can have considerable impact on the statistics of the data, and on the development of hydrologic models. Moreover, mean values that vary over time can affect not only the basic statistics, such as the variance, but also the correlation coefficients. The effect of step changes in the mean will be considered on basic statistics of the data, on probability distributions, and on serial correlation coefficients. The data are independent except for the presence of jumps. Many of the details in the following discussion are based on an investigation by Yevjevich and Jeng (1969). The occurrence of a single jump in the mean is discussed in some detail, while the results for the case of multiple steps are only summarized.

Let us assume that N observations of the observed series x_i are available, of which the first m values have the mean L and the second n values have the mean $L + \delta$. Let

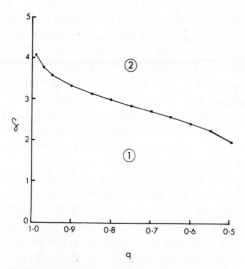

FIGURE 7 *Regions indicating the occurrence of one or two peaks in the probability density function: (1) one-peak region; (2) two-peak region. (After Yevjevich and Jeng, 1969.)*

$m/N = p$ and $n/N = q$, so that $p + q = 1$. The data sequence for this case of a single jump may be written in terms of random variable ϵ_i as in

$$X_i = \begin{cases} \epsilon_i + L & i = 1, 2, \ldots, m \\ \epsilon_i + \delta + L & i = m + 1, m + 2, \ldots N \end{cases} \tag{1}$$

The random variable ϵ_i is assumed to be Gaussian with zero mean ($\bar{\epsilon} = 0$), variance σ_ϵ^2,

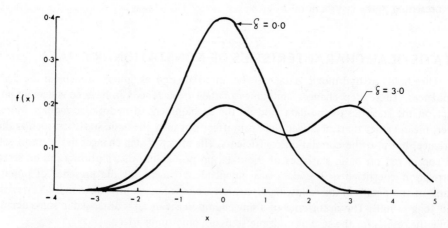

FIGURE 8 *Probability density function for $\delta = 0.0$ and $\delta = -3.0$, $\bar{\epsilon} = 0.0$, $\sigma_\epsilon = 1.0$, and q = 0.5. (After Yevjevich and Jeng, 1969.)*

skewness coefficient β_ϵ, and kurtosis γ_ϵ. The corresponding quantities for the X_i series are denoted by the subscript X. Without loss of generality, L is assumed to be zero and the variance of X_i to be unity. Under these assumptions the effects of step changes on the properties of the data are discussed.

Probability Distribution

The probability density function $f(X)$ of the series may be written in terms of the density function $g(X)$ as in

$$f(X) = pg(X) + qg(X - \delta)$$

$$= \frac{p}{\sqrt{2\pi}} e^{-x^2/2} + \frac{p}{\sqrt{2\pi}} e^{-(x - \delta)^2/2} \tag{2}$$

The step in the data produces a distribution that may have more than one peak. To determine the nature of this distribution, we may equate $f'(X)$ to zero to obtain

$$qe^{(X-\delta/2)}(\delta - X) = px \tag{3}$$

Solution of equation 3 yields one or three roots, indicating a single distribution or a double-peaked distribution with an intervening trough. The regions where one or two peaks occur are indicated in Figure 7, and typical examples of the probability distributions for different δ values given by Yevjevich and Jeng (1969) are shown in Figure 8. The significant point in these results is that the combination of the change in the mean and the duration of the change dictate the presence of single or double peaks, which is physically meaningful also.

If a number of step changes of magnitude δ_i occur at various positions i, then we can define q_i as being equal to n_i/N, where n_i is the length of the series after change δ_i and before change $i + 1$ so that $\sum_{i=1}^{h} q_i = 1$ and $\sum_{i=1}^{h} n_i = N$, where h is the total number of subseries, including the first. An example of a natural series with a sequence of step changes is shown in Figure 9. In such a case we may write the expression for the probability density as

$$f(X) = \sum_{i=1}^{h} \frac{q_i}{\sqrt{2\pi}} \exp \left[-\frac{1}{2\sigma_\epsilon^2} (X - L - \delta_i)^2 \right] \tag{4}$$

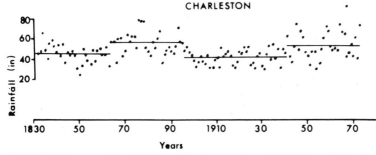

CHARLESTON

FIGURE 9 *Annual rainfall at Charleston, South Carolina. (After Potter, 1976.)*

TABLE 1 *Changes in Statistics Brought About by Jumps (After Yevjevich and Jeng, 1969)*

Function or statistics	No jumps	Single jump	Multiple jumps
Probability density function	$f(X) = \dfrac{1}{\sigma_\epsilon\sqrt{2\pi}}\exp\left[\dfrac{-(X-L)^2}{2\sigma_\epsilon^2}\right]$	$f(X) = \dfrac{p}{\sigma_\epsilon\sqrt{2\pi}}\exp\left[\dfrac{-(X-L)^2}{2\sigma_\epsilon^2}\right] + \dfrac{q}{\sigma_\epsilon\sqrt{2\pi}}\exp\left[\dfrac{-(X-\delta-L)^2}{2\sigma_\epsilon^2}\right]$	$f(X) = \sum \dfrac{q_i}{\sigma_\epsilon\sqrt{2\pi}}\exp\left[-\dfrac{1}{2\sigma_\epsilon^2}(X-L-\delta_i)^2\right]$
Mean	L	$\bar{x} = L + q\delta$	$\bar{x} = L + \sum q_i\delta_i$
Variance	σ_ϵ^2	$\sigma_x^2 = \sigma_\epsilon^2 + (1-q)q\delta^2$	$\sigma_x^2 = \sigma_\epsilon^2 + [\sum q_i\delta_i^2 - (\sum q_i\delta_i)^2]$
Skewness coefficient	β_ϵ	$\beta = \left(\dfrac{\sigma_\epsilon}{\sigma_x}\right)^3 \beta_\epsilon + (2q-1)(q-1)\,q\alpha^3$	$\beta = \dfrac{1}{\sigma_x^3}\left\{\beta_\epsilon\sigma_\epsilon^3 + (\sum q_i\delta_i)\,[2(\sum q_i\delta_i)^2 - 3\sum q_i\delta_i^2] + \sum q_i\delta_i^3\right\}$
Kurtosis	γ_ϵ	$\gamma = \dfrac{1}{\sigma_x^4}[\sigma_\epsilon^4\gamma_\epsilon + 6(1-q)q\delta^2\sigma_\epsilon^2 - q\delta^4(q-1)(3q^2-3q+1)]$	$\gamma = \dfrac{1}{4}\Big\{\gamma_\epsilon\sigma_\epsilon^4 + 6\sigma_\epsilon^2\,[\sum q_i\delta_i^2 - (\sum q_i\delta_i)^2]$ $-\,3(\sum q_i\delta_i)^4 + \sum q_i\delta_i^4$ $+\,(\sum \delta_i q_i)\,[6(\sum q_i\delta_i^2)(\sum q_i\delta_i)$ $-\,4(\sum q_i\delta_i^3)]\Big\}$

All the summations are from 1 to h.

Elementary Statistics

Single or multiple steps alter the elementary statistics, such as the mean, variance, skewness coefficients, and kurtosis. The changes in these statistics are listed in Table 1.

The mean values may be changed by the steps by $q_i\delta_i$. Depending upon the sign of δ_i, the mean may be unaltered from the level L. The variance, however, will always be increased by step changes. Skewness and kurtosis change, or may remain the same, depending upon the magnitude of $q_i\delta_i$, and other variables. No general comments about increase or decrease in statistics (except variance) can be made, except that they may change with step changes.

Serial Correlation Coefficients

The serial correlation structure of time-series data plays an important fundamental role in analysis and development of stochastic models. The estimates of correlation coefficients are given by equation 5 in which σ_X^2 is assumed to be constant. The estimates

$$r(k) = \frac{\frac{1}{N-k}\sum_{i=1}^{N-k}(X_i - \bar{X})(X_{i+k} - \bar{X})}{\sigma_X^2} \tag{5}$$

of serial correlation coefficients $r(k)$ are affected by the presence of step changes. For the case of a single step in data, the correlation coefficients $r(k)$ are given by (Yevjevich and Jeng, 1969)

$$r(k) = \begin{cases} \alpha^2\left[q^2 + \dfrac{q(1 - 2q) - p\lambda}{1 - \lambda}\right] & \dfrac{1}{N} \leq \lambda \leq \min(p,q) \\[2ex] p\alpha^2\left(\dfrac{q - \lambda}{1 - \lambda} - q\right) & p \leq \lambda < q \\[2ex] \alpha^2 q^2\left(\dfrac{\lambda}{\lambda - 1}\right) & q \leq \lambda < p \\[2ex] pq\alpha^2 & \max(p,q) < \lambda < 1 \end{cases} \tag{6}$$

$$\lambda = \frac{k}{N} \qquad \alpha = \frac{\delta}{\sigma_\epsilon}$$

For the case of multiple steps the estimate of the serial correlation coefficient $r(k)$ for $k < \min(n_1, n_2, \ldots, n_h)$ is given by (Yevjevich and Jeng, 1969)

$$r(k) = \frac{1}{N-k}\left\{k\left[\sum_{j=1}^{h-1}(\delta_j - \sum_{i=1}^{h}q_i\delta_i)(\delta_{j+1} - \sum_{i=1}^{h}q_i\delta_i)\right]\right. \\ \left. + \sum_{j=1}^{h}(n_j - k)(\delta_j - \sum_{i=1}^{h}q_i\delta_i)^2\right\} \tag{7}$$

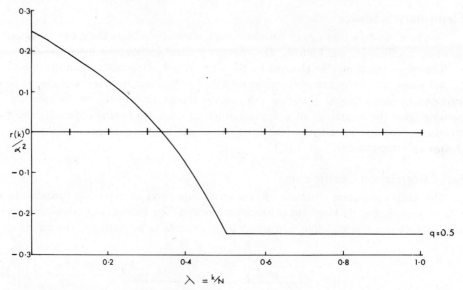

FIGURE 10 *Relationship between $r(k)/\alpha^2$ and $\lambda = k/N$ for $q = 0.5$, for a single constant step. (After Yevjevich and Jeng, 1969.)*

In general, steps in hydrologic data increase the correlation that might exist in the data. Increased correlation may not be evident with a single sample but becomes apparent with several samples.

The correlation given by equation 6 for $q = 0.5$ is shown in Figure 10. A feature of this correlation is a constant negative value attained by the correlation coefficient for higher lags. The data for the Niger River exhibit this behavior (Fig. 11). The existence of a constant negative correlation coefficient at higher lags may be used to detect the

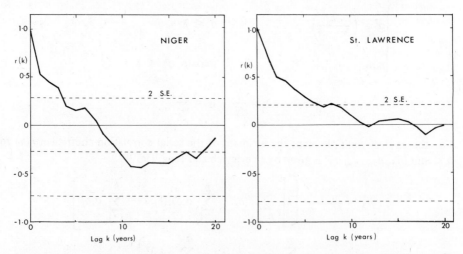

FIGURE 11 *Correlograms of annual flows. Niger at Koulicoro, Africa (1906–57). St. Lawrence (main stem) near Ogdensburg, N.Y. (1860–1957).*

presence of nonstationarity in the data as suggested by Yevjevich and Jeng (1969). An inspection of the Niger data and its correlogram substantiates this suggestion. In comparison, the correlogram of the St. Lawrence River data do not show this effect (Fig. 11).

Some Results and Discussion

In an investigation to explain the Hurst phenomenon (Hurst et al., 1965), Potter (1976) analyzed long-term rainfall records in the United States. Since Potter's results support the general trend of discussion in this section, they are discussed below in some detail by using the results for annual rainfall at Charleston, South Carolina (Fig. 9, Table 2).

TABLE 2 *Statistics of Annual Rainfall at Charleston, South Carolina (After Potter, 1976)*

Years	L_i	δ_i	q_i	$(\sigma_\epsilon)_i$	Weighted average of $(\sigma_\epsilon)_i$	β_i	Weighted average of β_i	$r_i(1)$	Weighted average of $r_i(1)$
1823–63	45.30		0.23	9.05		−0.02		0.34	
1864–95	55.70	+10.40	0.23	10.98		0.40		0.35	
1896–1940	41.12	−14.58	0.31	7.12		−0.06		0.21	
1941–73	52.77	+11.65	0.23	12.40		0.21		0.31	
1832–1973	48.06			11.43	9.65	0.56	0.12	0.45	0.29

Changes in mean δ_i are, typically, quite large, and annual rainfall values remain at a level for as long as 30 yr before crossing a threshold to another level. These changes in mean render the series nonstationary. In fact, changes in mean values yield poor forecasts of annual data when constant coefficient stochastic difference equation models are used to represent the process (Rao et al., 1975). Standard deviations also fluctuate considerably from one period to the next. As discussed above (Table 1), the standard deviation of the entire series is larger than the standard deviation of three out of four periods, which can be attributed to the step changes in the series.

The skewness coefficient values also fluctuate widely, from almost zero in the period 1823–1963, to 0.4 in the period 1864–95. The skewness coefficient of the entire series is considerably higher than that in any individual period.

The serial correlation coefficient of a data series with jumps, according to the previous discussion, should be higher than the correlation coefficient in any of the individual sections that do not have jumps. Potter's results clearly support this analytical result. The lag-one correlation coefficient of the entire data (0.45) is considerably higher than the corresponding values of the individual segments. This result is common to other data considered by Potter.

Modeling nonstationary series by constant coefficient stochastic difference equation models may thus lead to inadequate representations. Alternatively, the assumption of stationarity for hydrologic series is questionable. This aspect has been discussed at length by Klemeš (1976), who has suggested the nonstationarity of the hydrologic series as an explanation for the Hurst phenomenon. In this context it may well be, to quote Philip (1975), that "the Hurst effect has provoked lengthy sophisticated and intellectually important forays into this same field (the theory of stochastic processes), though nobody seems particularly interested in the unexciting explanation of the Hurst effect as an elementary consequence of nonstationarity." Furthermore, in view of the fact that jumps create such large variations in the statistics of the data, "preserving" these in generated data may not be very meaningful. These changes in statistics brought about by jumps and the bias in their estimation (Wallis and Matalas, 1971) create considerable difficulties in hydrologic modeling.

However, before considering hydrologic time series as being nonstationary, which brings with it the concept of thresholds, testing changes in mean values for statistical significance and characterizing them becomes important. This problem and some of the proposed solutions are considered next.

THRESHOLDS IN NONPERIODIC CORRELATED DATA

Many hydrologic series, such as annual rainfall or runoff data, are nonperiodic but correlated, and shift considerably from one level to another as discussed earlier, so that testing the significance of the changes is important. As these series are correlated, the methods such as those discussed in the previous section are not appropriate. In this section methods of analysis are outlined to test statistical significance, and to estimate magnitudes of changes observed in correlated nonperiodic data. Analysis of periodic data such as monthly or weekly means is dealt with in the next section. The use of the techniques discussed in this section is illustrated by using annual rainfall data from LaPorte, Indiana.

The Model

One of the basic models which may be considered for modeling nonstationary data is the so-called ARIMA $(0, 2, 2)$ model,

$$\nabla^2 X_i = (1 - \theta_1 B - \theta_2 B^2) a_i \tag{8}$$

in which a_i is a white-noise sequence with mean μ and variance σ_a^2, B the unit delay operator $(Ba_i = a_{i-1})$, and ∇ is equal to $(1 - B)$. An alternative form of equation 8 is given in equation 9 (Box and Jenkins, 1970), in which $b_i = \mu + a_i$. In the interval $(0, m)$, the first period, the X_i sequence will have shifted or drifted from level L to another level, L', given in equation 11 and the difference in levels $(L' - L)$ is called the drift in the series, which is controlled by the parameter μ. The parameter μ is therefore called the drift parameter and L the initial level of the series (Glass, 1972).

$$X_i = L + \alpha_1 \sum_{j=1}^{i-1} b_j + \alpha_2 \sum_{j=1}^{i-1} (i - j) b_j + b_i \quad i = 0, 1, 2, \ldots, m; \tag{9}$$

$$\alpha_1 = 1 + \theta_2; \quad \alpha_2 = 1 - \theta_1 - \theta_2$$

$$X_m = L' + \alpha_1 \sum_{j=1}^{m-1} a_j + \alpha_2 \sum_{j=1}^{m-1} (i - j) a_j + a_i \tag{10}$$

$$L' = L + \alpha_1 \mu (m - 1) + \alpha_2 \mu m \frac{m - 1}{2} \tag{11}$$

The invertibility conditions on equation 9 are

$$0 < 2\alpha_1 + \alpha_2 < 4, \quad \alpha_1 > 0, \quad \alpha_2 > 0$$

In the second period the process would be at a level $L + \delta$ and may have a different drift parameter $\mu + \eta$. The models for the first and second periods are

$$X_i = L + \alpha_1 \sum_{j=1}^{i-1} (\mu + a_j) + \alpha_2 \sum_{j=1}^{i-1} (i - j)(\mu + a_j) + (\mu + a_i) \quad i = 1, 2, \ldots, m \tag{12}$$

$$X_i = (L + \delta) + \alpha_1 \sum_{j=1}^{i-1} (\mu + \eta + a_j) + \alpha_2 \sum_{j=1}^{i-1} (i - j)(\mu + \eta + a_j)$$

$$+ (\mu + \eta + a_i) \quad i = m + 1, m + 2, \ldots, N \tag{13}$$

Assuming that α_1 and α_2 are known, we can estimate the parameter δ, which may be interpreted as representing the change in the level from the first to the second period. The estimates of μ and $\mu + \eta$ indicate the "drift" of the series (i.e., whether the X_i values are generally increasing or decreasing in the first and second periods). The parameter L represents the initial level of the series. These parameters L, δ, μ, and η are estimated along with their standard errors. The ARIMA (0, 1, 1) model used by Box and Tiao (1965) is obtained by setting μ, η, and α_2 in equations 12 and 13 equal to zero. The ARIMA (0, 1, 1) models for the two periods are given as equations 14 and 15. By using the ARIMA (0, 1, 1) model information about the change in the level of the series may be obtained, whereas the ARIMA (0, 2, 2) model yields information not only about the change in the level of the series but also about the trends in the data. Hence, it is a more general model.

$$X_i = L + \alpha_1 \sum_{j=1}^{i-1} a_j + a_i \quad i = 1, 2, \ldots, m \tag{14}$$

$$X_i = L + \delta + \alpha_1 \sum_{j=1}^{i-1} a_j + a_i \quad i = m + 1, m + 2, \ldots, N \tag{15}$$

The methds of estimation of parameters and their standard errors of these models are discussed in detail in Glass (1972) and Box and Jenkins (1970) and are not repeated here. However, for the ARIMA (0, 1, 1) model, the parameters L and δ may be estimated by equations 16 and 17 (Box and Tiao, 1965), and these are given next as they yield some additional insight into the nature of the parameters, especially the parameter representing the change in mean, δ.

$$\hat{L} = \frac{\alpha_1}{1 - (1 - \alpha_1)^{2m}} \sum_{i-1}^{m}(1 - \alpha_1)^{i-1}X_i + (1 - \alpha_1)^m \sum_{i=1}^{m}(1 - \alpha_1)^{m-i}X_i \qquad (16)$$

$$\hat{\delta} = f_1 - f_2 \qquad (17)$$

$$f_1 = \frac{\alpha_1}{1 - (1 - \alpha_1)^{2n}} \sum_{i=1}^{n}(1 - \alpha_1)^{i-1}X_{m+i} + (1 - \alpha_1)^m \sum_{i=1}^{m}(1 - \alpha_1)^{m-i}X_{m+i} \qquad (17a)$$

$$f_2 = \frac{\alpha_1}{1 - (1 - \alpha_1)^{2m}} \sum_{i=1}^{m}(1 - \alpha_1)^{m-i}X_i + (1 - \alpha_1)^m \sum_{i=1}^{m}(1 - \alpha_1)^{i-1}X_i \qquad (17b)$$

The estimate of $\hat{\delta}$ is thus the difference between two exponentially weighted averages of observations with maximum weight at the point at which the second period commences. The X_i values that are closest to the start of the second period have maximum effect on δ, and this effect decreases exponentially as we move further away from the beginning of the second period.

Inferences from the Models

After estimating the parameters (and their standard errors) of the models, and verifying that the residual sequences a_i are white and without periodicities, the statistical significance of the parameters may be evaluated. If the magnitude of a parameter estimate is less than its standard error, that parameter is concluded to be insignificantly different from zero. For example, if the value of δ is less than its standard error, we conclude that the increase in mean value of the process is statistically insignificant; otherwise, the observed change is significant and we may consider the process to have crossed a threshold at the start of the second period.

Confidence intervals for the parameters may also be estimated and used for inferences. Assuming the residuals a_i to be normally distributed, it can be shown that the difference between the parameter estimate $\hat{\theta}_i$ and its true value θ_i is t-distributed, as in

$$\hat{\theta}_i - \theta_i \sim \sigma_a k_i t(N_e) \qquad (18)$$

where σ_a is the estimate of the variance of the residuals a_i, k_i a function of the data, and N_e the degrees of freedom, which are equal to $N - 2$ for ARIMA $(0, 1, 1)$ and $N - 4$ for ARIMA $(0, 2, 2)$ models. The expression is valid for the parameters L, δ, μ, and η with the corresponding k_i values. The confidence limits at level α for θ_i are thus given by

$$\hat{\theta}_i \pm t_\alpha(N_e)k_i \sigma_a \qquad (18a)$$

Finally, the signs ($\pm$) of the significant parameter estimates indicate whether the process has increased or decreased levels, increasing or decreasing trends. Some other inference procedures are found in Rao and Rao (1974).

An Example

The models fitted to the LaPorte data are considered to illustrate the above theory. The ARIMA (0, 1, 1) and (0, 2, 2) models corresponding to equations 14 and 15 and 12 and 13 are given below, along with the 95% confidence limits, the t-statistics, and standard errors of the parameters, which are shown in parentheses below the parameters.

$$
X_i = \begin{cases}
\underset{(22.79)}{35.33} + 0.01 \sum_{j=1}^{i-1} a_j + a_i & i = 1, 2, \ldots, m & (19) \\[2em]
\underset{(22.79)}{35.33} + \underset{(1.82)}{9.28} + 0.01 \sum_{j=1}^{i-1} a_j + a_i & i = m + 1, \ldots, N & (20)
\end{cases}
$$

95% confidence limits
$$
\begin{bmatrix}
\hat{L}: (38.43, 32.23) & t_{\hat{L}} = 22.79 \\
\hat{\delta}: (12.91, \ 5.64) & t_{\hat{\delta}} = 5.10
\end{bmatrix}
$$

$$
X_i = \begin{cases}
\underset{(2.13)}{34.61} + 0.01 \sum_{j=1}^{i-1} \underset{(1.39)}{(0.18} + a_j) + 0.01 \sum_{j=1}^{i-1} (i - j) \underset{(1.39)}{(0.18 + a_j)} \\
\qquad + (0.18 + a_j) \quad i = 1, 2, \ldots, m & (21) \\[2em]
\underset{(2.13)}{34.61} + \underset{(2.62)}{12.71} + 0.01 \sum_{j=1}^{i-1} \underset{(1.39)}{(0.18} - \underset{(1.88)}{1.21} + a_j) \\[1em]
\qquad + 0.01 \sum_{j=1}^{i-1} (i - j)(0.18 - 1.21 + a_j) + (0.18 - 1.21 + a_i) \\[1em]
\qquad\qquad\qquad\qquad i = m + 1, m + 2, \ldots, N & (22)
\end{cases}
$$

95% confidence limits
$$
\begin{bmatrix}
\hat{L}: (38.87, 30.35) & t_{\hat{L}} = 16.32 \\
\hat{\delta}: (17.95, -7.47) & t_{\hat{\delta}} = 4.86 \\
\hat{\mu}: \ (2.98, -2.62) & t_{\hat{\mu}} = 0.131 \\
\hat{\eta}: \ (2.55, -4.97) & t_{\hat{\eta}} = -0.646
\end{bmatrix}
$$

In the ARIMA (0, 1, 1) model the parameter estimates $\hat{L}$ and $\hat{\delta}$ are greater than their standard errors and their critical t values, indicating their significance. The increase in mean annual rainfall in the second period given by the model is approximately 9.28 inches, which compares well with the observed increase of 7.69 inches. It is interesting to note that the α_1 value remains constant in the two periods, indicating that the correlation structure of the process has remained essentially constant.

Although the initial level of the series given by the ARIMA (0, 2, 2) and ARIMA (0, 1, 1) values are very close to each other (34.61 and 35.33, respectively) and compare well with the observed mean value (36.90 inches) in the first period, the δ values given by the two models are considerably different from each other. The increase in mean annual rainfall estimated by ARIMA (0, 2, 2) model is 12.71 inches whereas that by ARIMA (0, 1, 1) model is 9.28 inches, and both of these are greater than the observed increase in mean annual rainfall, which is 7.69 inches. The drift parameter estimates $\hat{\mu}$ and $\hat{\eta}$ are smaller than their standard errors but are larger than their critical $t_{95\%}$ values, which casts doubts on the existence of trends. Inspection of the data does not indicate the presence of trends. The α_1 values in the ARIMA (0, 2, 2) model are also small, and they remain constant in the first and second periods. Finally, in both models parameter estimates are within their confidence limits, indicating plausibility of their occurrence. Thus, we can conclude that there is indeed evidence to suggest that the rainfall at LaPorte crossed a threshold around 1929.

Variation of the Estimate $\hat{\delta}$ in ARIMA (0, 1, 1) Model with $\hat{\alpha}_1$

The parameter estimate $\hat{\alpha}_1$ affects the value of $\hat{\delta}$ in the ARIMA (0, 1, 1) model. In order to investigate the nature of this effect and to obtain the probability density of the parameter $\hat{\delta}$ we may proceed along the lines developed by Box and Tiao (1965). The probability density of δ is of particular interest in threshold analysis. If this probability density is sharp-peaked, the limits of variation of δ can be established with precision. The method is Bayesian with locally uniform and independent prior distribution of the parameters L, δ, and $\log \sigma_a$. Under these assumptions the joint probability density of δ and α_1 is given by equation 23, in which $p(\cdot)$ indicates the probability density function and X the vector of X_i values. From equation 23 we can derive the probability density function of δ by equation 25. The posterior probability density function of α_1 is given by equation 26, where k_1 is a normalizing constant.

$$p(\delta, \alpha_1 | X) \propto p(\alpha_1) \left[\frac{\alpha_1 (2 - \alpha_1)}{1 - (1 - \alpha_1)^{2N}} \right]^{\frac{1}{2}} (\sigma_a)^{-(N-L)/2} \left[1 + \frac{t^2}{N - 2} \right]^{-(N-L)/2} \tag{23}$$

$$t = (\alpha_1 - \hat{\alpha}_1) \left[\frac{[1 - (1 - \alpha_1)^{2m}][1 - (1 - \alpha_1)^{2n}]}{[1 - (1 - \alpha_1)^{2n}][\alpha_1 (2 - \alpha_1)] \sigma_a^2} \right]^{\frac{1}{2}} \tag{24}$$

$$p(\delta | X) = \int p(\delta, \alpha_1 | X) \cdot d\alpha_1 \tag{25}$$

$$h(\alpha_1 | X) = k_1 \frac{\alpha_1 (2 - \alpha_1)}{\{[1 - (1 - \alpha_1)^{2m}][1 - (1 - \alpha_1)^{2n}]\}^{\frac{1}{2}}} (\sigma_a^2)^{-(N-2)/2} \tag{26}$$

The conditional mean and the standard errors of δ are given by equations 27 and 28. The significance of $\hat{\alpha}_1$ itself can be tested by using the $t(\alpha_1)$ ratio in equation 29 along with $h(\alpha_1 | X)$.

$$E(\delta | \alpha_1) = \hat{\delta} \tag{27}$$

$$\sigma(\delta | \alpha_1) = \left[\frac{\alpha_1 \sigma_a^2 (2 - \alpha_1)[1 - (1 - \alpha_1)^{2N}](N - 2)}{[1 - (1 - \alpha_1)^{2m}][1 - (1 - \alpha_1)^{2n}](N - 4)} \right]^{\frac{1}{2}} \tag{28}$$

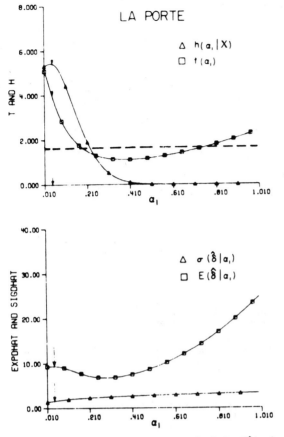

FIGURE 12 *Estimates of* $t(\alpha_1)$, $h(\alpha_1|X)$, $E(\hat{\delta}|\alpha_1)$, *and* $\sigma(\hat{\delta}|\alpha_1)$ *for annual precipitation at LaPorte.*

$$t(\alpha_1) = \hat{\delta}\left\{\frac{[1 - (1 - \alpha_1)^{2m}][1 - (1 - \alpha_1)^{2n}]}{[1 - (1 - \alpha_1)^{2N}]\alpha_1(2 - \alpha_1)\sigma_a^2}\right\}^{1/2} \tag{29}$$

The estimates of $t(\alpha_1)$, $h(\alpha_1|X)$, $E(\delta|\alpha_1)$, and $\sigma(\delta|\alpha_1)$ are shown in Figure 12 for the LaPorte data, along with the critical t-statistic (95%) in dashed lines. The $t(\alpha_1)$ value is greater than the critical value of the t-variate, indicating that the change in annual rainfall is significantly different from zero. The probability density function $h(\delta|X)$ has a maximum around $\alpha = 0.01$, confirming the accuracy of the estimate of α. The value of $\sigma(\delta|\alpha_1)$, the standard error of $\hat{\delta}$, is also quite small.

The probability density function $p(\hat{\delta})$, and the weight functions computed from equations 17a and 17b, are shown in Figures 13 and 14. The probability density function is symmetric and concentrated around the estimated $\hat{\delta}$ value and can be approximated by the normal distribution. The weight functions decay very slowly, indicating that the in-

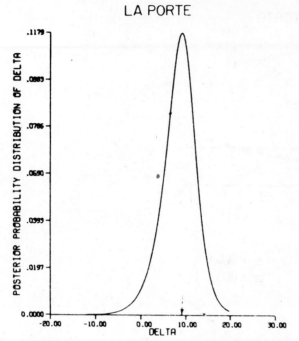

FIGURE 13 *The posterior probability function of* δ, $p(\delta|X)$.

crease in rainfall in the second period is sustained over a long period and is not a transient effect.

Conclusions

The methods of analysis to estimate and test the statistical significance of changes in mean, trends in the data, and probability distributions of some of the parameters have been outlined and illustrated. Based on the results of these tests, one can establish, at prescribed confidence levels, existence of thresholds in the process and establish the magnitude of change in the mean.

CHARACTERIZATION OF THRESHOLDS IN SEASONAL DATA

As defined earlier, X_i, $i = 1, 2, \ldots, m$, represents the observations in the first period and $X_{m+1}, X_{m+2}, \ldots, X_{m+n}$ those in the second period, where $N(= m + n)$ is the total number of observations available. In this section we consider X_i to be seasonal data such as monthly flows. In this section some methods to estimate the change in the mean value of the process in the two periods and to test its significance are discussed.

The Model

The series X_i is represented as in equation 30, which is the sum of the noise N_i and a component D_i which may be a dynamic model representing the exogenous variables.

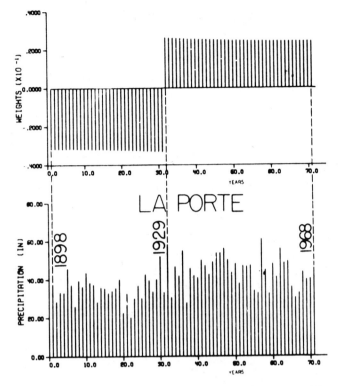

FIGURE 14 *Weight functions f_1 and f_2 (equations 17a and 17b) and annual precipitation at LaPorte.*

The form taken by D_i should be specified, and some of these are discussed below. The noise process N_i may be modeled as a

$$X_i = D_i (\cdot) + N_i \tag{30}$$

mixed ARIMA process as in equation 31. The polynomials $\Phi(B)$ and $\theta(B)$ are

$$\Phi(B)N_i = \theta(B)a_i \tag{31}$$

defined below. The polynomials $\Phi(B)$ and $\theta(B)$ obey stationarity and

$$\Phi(B) = 1 - \phi_1 B - \phi_2 B^2 - \cdots - \phi_p B^p$$
$$\theta(B) = 1 - \theta_1 B - \theta_2 B^2 - \cdots - \theta_q B^q \tag{32}$$

invertibility conditions and may be further decomposed as in equation 33 to represent seasonal nonstationary processes. In equation 33, s stands for seasonal differencing and the superscripts D and d stand for the Dth (seasonal) and dth (nonseasonal) differences.

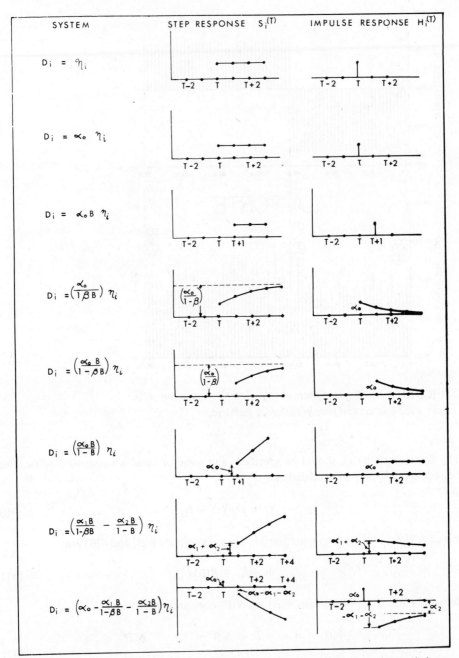

FIGURE 15 *Step and impulse response functions of some commonly used dynamic models.*

$$\Phi(B) = \Phi_1(B)\Phi_2(B^s)$$

$$\Phi_1(B) = (1 - B)^d \Phi_1(B)$$

$$\Phi_2(B) = (1 - B^s)^D \Phi_2(B^s)$$

(33)

$$\theta(B) = \theta_1(B)\theta_2(B^s)$$

The polynomials $\Phi_1(B)$, $\Phi_2(B^s)$, $\theta_1(B)$, and $\theta_2(B^s)$ are, respectively, of orders p_1, p_2, q_1, and q_2, and the roots of $\Phi_1(B)$ lie outside the unit circle. Further discussion of the model is found in Box and Taio (1975).

The form of the dynamic model $D_i(\cdot)$ must be specified after considering the nature of changes exhibited by the data. A number of commonly used dynamic models are given in Figure 15 along with their impulse and step responses. A discussion of these models is found in Box and Jenkins (1970). The step and the impulse responses initiated at time T are, respectively, indicated by $S_i^{(T)}$ and $H_i^{(T)}$. The step and impulse responses are estimated by considering models for D_i such as those shown in equation 34. The polynomials $\alpha(B)$ and $\beta(B)$ are of orders r and s, respectively.

$$D_i(\alpha, \beta, \eta) = \frac{\alpha(B)}{\beta(B)} B^b \eta_i$$

$$S_i^{(T)} = \eta_i = \begin{cases} 0 & t \leqslant T \\ 1 & t > T \end{cases}$$

(34)

$$H_i^{(T)} = \eta_i = \begin{cases} 0 & t \neq T \\ 1 & t = T \end{cases}$$

The roots of $\alpha(B)$ should be outside and those of $\beta(B)$ should be on or outside the unit circle. The series η_i can be an exogenous time series whose effect on X_i is to be investigated. The η_i sequence may be a binary indicator series taking values of zero and 1. The general model can be written as

$$X_i = \frac{\alpha(B)}{\beta(B)} \eta_i + \frac{\theta(B)}{\Phi(B)} a_i$$

(35)

Once a model is specified, parameters of the model and their standard errors may be estimated by using methods given in Box and Jenkins (1970) or Kashyap and Rao (1976). The residual series a_i are checked for whiteness by using the methods given in Box and Jenkins (1970) or Kashyap and Rao (1976). When the residuals are found white, the model may be used to verify the hypothesis that the process has changed or that a "threshold" has been crossed. We will presently consider some of these aspects.

A Case Study

In the study of effects of urbanization on rainfall mentioned earlier the following model was developed to investigate increased monthly rainfall at LaPorte, Indiana (Rao and Rao, 1974).

$$X_i = \alpha_1 \eta_{1i} + \frac{\alpha_2}{(1 - B^{12})} \eta_{2i} + \frac{\alpha_3}{(1 - B^{12})} \eta_{3i} + \frac{(1 - \theta_1 B)(1 - \theta_2 B^{12})}{(1 - B^{12})} a_i$$

$$\hat{\alpha}_1 = 0.865 ; \quad \hat{\alpha}_2 = -0.011 ; \quad \hat{\alpha}_3 = -0.002 \qquad (35)$$
$$(0.221) \qquad\qquad (0.010) \qquad\qquad (0.010)$$

$$\hat{\theta}_1 = -0.056 ; \quad \hat{\theta}_2 = 0.962$$
$$(0.034) \qquad\qquad (0.005)$$

The binary variables η_{1i}, η_{2i}, and η_{3i} were defined to be zero in the first period (1915–1928). η_{1i} was set equal to unity throughout the second period (1929–1968). η_{2i} and η_{3i} were set equal to unity during the warm (April–September) and cold (October–March) seasons of the year. α_2 and α_3 would therefore indicate, respectively, the increase in monthly rainfall in the second period and in the warm and cold seasons of the second period. The parameter estimates and their standard errors are listed following equation 35. After verifying that the a_i sequence was white, several investigations were carried out using this model, and these are briefly discussed below.

Inferences from the Parameter Estimates. All the parameter estimates in the model (eq. 35) except α_3 are larger than their standard errors. Only the estimate of α_3 is smaller than its standard error, which indicates that there is no increase in monthly precipitation in the cold seasons of the second period, compared with that of the first period. The precipitation increase in the warm seasons is also marginal, as the standard error of the parameter estimate α_2 is of the same magnitude as the parameter itself. The overall increase in monthly precipitation, as given by the model (α_1) is about 0.865 inch, which compares favorably with the observed increase of 0.69 inch. As there is interaction between the parameters, care must be exercised in interpreting the results. In the present case, for example, although the data indicate that the warm period precipitation increased by 0.64 inch, it is not reflected in the parameter α_2 directly but instead has influenced the parameter α_1.

Significance of the Dynamic Model. The residual variances from the models such as that shown in equation 35 with and without the dynamic models may be used to test the significance of the dynamic models. Let $\hat{\rho}_1$ be the variance of the residuals from the model without the dynamic term and $\hat{\rho}_2$ the variance of the residuals from the same model with the dynamic term. The statistic g_2 in equation 36,

$$g_2 = \frac{\hat{\rho}_1 - \hat{\rho}_2}{\hat{\rho}_2} (N - NP) \qquad (36)$$

is then F-distributed with (NP, N) degrees of freedom, where N is the number of samples used in estimating $\hat{\rho}_1$ and $\hat{\rho}_2$ and NP is the number of parameters in the model. If g_2 is less than $F_\alpha (NP, N)$, where α is the significance level, we consider the additional parameters as insignificant in reducing the residual variance. We can also perform this test with different forms of the dynamic model to resolve ambiguities related to certain parameters such as α_2 discussed above.

For the LaPorte data, the statistic g_2 obtained by two models, one with and the other without the dynamic model, was 19.235 and the $F_{0.05} (NP, N)$ value was 2.60, leading to the conclusion that the dynamic model is indeed significant.

Tests Based upon Multiple Step-Ahead Forecasts. In this test a model is fitted to the first period data such as that in equation 37,

$$X_j = \frac{(1 - \theta_1 B)(1 - \theta_2 B^{12})}{1 - B^{12}} a_j \cdot \begin{bmatrix} \theta_1 &=& -0.057 \\ & & (0.053) \\ \theta_2 &=& 0.930 \\ & & (0.014) \\ \sigma_a^2 &=& 2.653 \end{bmatrix} \tag{37}$$

and the residuals a_j are checked to be white. This model is then used to forecast the X_j values in the second period one step ahead, which are designated X_j.

$$\hat{X}_j = \hat{X}_{j-12} - \theta_1 e_{j-1} - \theta_2 e_{j-12} - \theta_1 \theta_2 e_{j-13} \tag{38}$$

The one-step-ahead forecast errors e_j in equation 38 are then computed by

$$e_j = X_j - \hat{X}_j \tag{39}$$

If this error sequence e_j is white, the process has not changed appreciably in the second period. In order to test whether one-step ahead forecast errors are white, we can compute the statistic g_3, which is given by $g_3 = \sum_{i=m}^{N} e_i^2/\sigma_e^2$, where σ_e^2 is the variance of the e_j sequence. If the error sequence e_j is white, g_3 is approximately χ^2 distributed with $(N - m)$ degrees of freedom (Box and Tiao, 1975). The decision rule is then as follows.

$$\text{If } g_3 \begin{cases} \leqslant \chi_\alpha^2 (N - m) \rightarrow \text{the process has not changed in the second period} \\ \\ > \chi_\alpha^2 (N - m) \rightarrow \text{the process has changed in the second period} \end{cases}$$

For the LaPorte data, the statistic g_3 was 379, whereas the corresponding $\chi_{0.05}^2$ ($\cdot$) value was 79, indicating a considerable change in the process. The observed and forecast monthly rainfall values and forecast errors are shown in Figure 16.

Weight Functions. Box and Tiao (1975) derived weight functions which represent the behavior of parameters in models such as equation 35 and hence of the process. If these weight functions decay rapidly, the change in the process does not persist too long (or the process has small "memory") and vice versa. The weight functions corresponding to the model in equation 35 for the LaPorte data are shown in Figure 17. The persistence in increased rainfall at LaPorte is quite long.

Discussion

By using some of the methods discussed in the present section, changes occurring in seasonal data, such as monthly or weekly data, may be characterized and their significance tested. The method is quite robust even if the time at which the process starts changing cannot be precisely established, as may be the case with many natural series (Rao and Rao, 1974). The method of analysis has been applied to a variety of data (Box

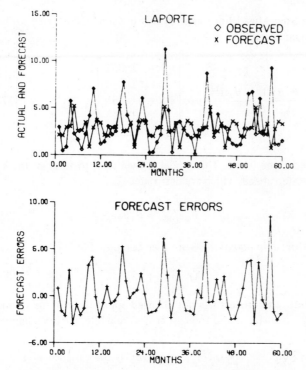

FIGURE 16 *Observed and one-month-ahead forecast monthly precipitation and forecast errors. The model (eq. 37) is based on 1898–1928 data. Forecast period is 1929–33.*

and Tiao, 1975; Hipel et al., 1975, 1977) and has considerable promise in threshold analysis. It is important to emphasize that the method can be used to analyze a series of changes, although the case of a process having only a single change has been considered in this chapter.

CONCLUSIONS

Concepts of thresholds in hydrologic time series and their quantification via stochastic models have been discussed in the preceding sections. Some of the related aspects and problems are considered herein.

In investigations of hydrologic time series and thresholds, periods with "longer" and "shorter" fluctuations must be distinguished. The longer periods correspond to glacial fluctuations. The shorter periods, sometimes referred to as "historical" fluctuations, are those with durations of decades and centuries (National Academy of Sciences, 1975) and correspond with the useful life of water resources structures. Consequently, analysis and modeling of shorter period fluctuations has received most of the attention. However, if the longer- and shorter-term fluctuations are in phase, or if the short-term fluctuations cross a threshold to another level, the hydrologic time series would change considerably

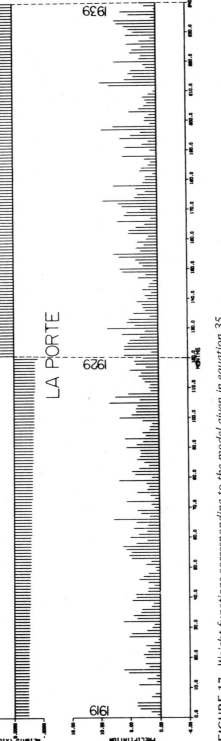

FIGURE 17 Weight functions corresponding to the model given in equation 35.

and would affect designs of water resource systems. Consequently, changes in hydrologic variables and thresholds associated with them deserve further investigation. Questions related to modeling hydrologic time series and understanding the physical reasons behind changes in these series are important aspects of these investigations.

Nonstationarity of the shorter-term behavior of hydrologic data, or the shorter-term fluctuations which may be due to natural or man-made causes, is being increasingly recognized as important, mainly as a result of investigations related to the Hurst phenomenon. O'Connell (1971) demonstrated that ARIMA (1, 0, 1) models reproduced Hurst's h. However, to adequately reproduce h, rather high values of the autoregressive coefficients, which may approach unity, are necessary. The behavior or ARIMA (1, 0, 1) models in which the autoregressive coefficients approach unity is such that "the attraction which the mean exerts on the series becomes less and less" (Box and Jenkins, 1970). The later work of Klemeš (1976) and Potter (1976) also substantiate the idea that nonstationarity of hydrologic series may be a significant cause of the Hurst phenomenon.

If hydrologic time series are considered to be nonstationary, questions related to the frequency and magnitude of changes in levels of these series must be analyzed and characterized. The constant-coefficient models such as those discussed earlier, along with Bayesian analyses, are useful for these purposes, although they have limitations. An alternative is to use slowly varying system models. The coefficients of these models may be recursively estimated (Kashyap and Rao, 1976). These models are more sensitive to changes in levels than the constant-coefficient models, although the properties of these models are yet to be clearly understood.

Whether constant- or variable-coefficient models are used to characterize the process under consideration, models that involve only the variable of interest, such as those considered in this chapter, do not yield any insight into the physical reasons for the changes observed. In order to develop prescriptive models of changing processes such as precipitation, it is important to know not only the causal variables effecting changes in these processes, but also to establish the relevant causal relationships. For example, little attention has been given in hydrologic analysis so far to the relationships between variation in atmospheric circulation and changing precipitation patterns (Kraus, 1955a, 1955b; Lamb, 1969). The important aspect in this particular problem is to establish the existence of causal connections between variables governing atmospheric circulation and precipitation. This problem is analogous in some respects to the investigation of causal structures of economic variables. These causality studies are needed to better understand and model the relationships between hydrologic and atmospheric and oceanographic variables.

In stochastic hydrology, most of the attention has been centered on linear systems analysis and modeling. However, the atmospheric processes that produce fluctuations in hydrologic variables of interest, such as precipitation and temperature, are nonlinear. It is well known that even simple nonlinear systems can produce trajectories which include pseudo-periodic and pseudo-random behavior. Lorenz (1968, 1969, 1970) has called attention to the "almost intransitive behavior" of nonlinear systems in which a process may remain in two (or more) quasi-stable states with transition periods in between them. When a system is intransitive, it may appear to be varying for shorter periods of time but may have stable statistics for longer time periods. The number of available observations of a series is thus important in defining the behavior of the system. The comment "what might appear to be a climatic transition on one time scale may become the natural noise level of a climatic state defined over a longer interval" (National Academy of Sciences, 1975) is also valid in the analysis of hydrologic thresholds. Investigations of alternative representations, including deterministic nonlinear models of hydrologic time series, are necessary to resolve some of these issues.

In conclusion, although considerable advances have been made in the analysis of hydrological thresholds, many important issues remain unresolved.

ACKNOWLEDGMENTS

The encouragement and administrative support of Dr. K. J. Mitchell, Research Director, Mr. D. A. Ferrier, Chief Scientist, Water and Soil Division, Ministry of Works and Development, Government of New Zealand, are appreciated. It is a pleasure to acknowledge the help, comments, and discussions with Drs. A. I. McKerchar, M. J. McSaveney, and R. P. Ibbitt, Water and Soil Division, Ministry of Works and Development, Government of New Zealand, during the preparation of the paper. Any remaining errors are my own.

The paper was written while the author was a Senior Research Fellow of the National Research Advisory Council of New Zealand. The work on LaPorte Anomaly was part of a research project "Urban Industrial Effects of Rainfall and Development of a Simulation Model for the Urban Rainfall-Runoff Process," OWRT-B-025-IND, supported by the office of Water Research and Technology, U.S. Department of the Interior. Support by these organizations is gratefully acknowledged.

REFERENCES CITED

Box, G. E. P., and Jenkins, G. M., 1970, Time series analysis, forecasting and control: San Francisco, Holden-Day.

——, and Tiao, G. C., 1965, A change in level of a nonstationary time series: Biometrika, v. 52, p. 181–192.

——, and Tiao, G. C., 1975, Intervention analysis with applications to economic and environmental problems: J. Amer. Stat. Assoc., v. 70, p. 70–79.

Changnon, S. A., Jr., 1968, The LaPorte weather anomaly—fact or fiction?: Bull. Amer. Meteorol. Soc., v. 49, p. 4–11.

——, 1969, Recent studies of urban effects on precipitation in the United States: Bull. Amer. Meteorol. Soc., v. 50, p. 411–421.

Glass, G. V., 1972, Estimating the effects of intervention into a nonstationary time series: Amer. Ed. Res. J., v. 9, p. 463–477.

Hipel, K. W., Lennox, W. O., Unny, T. E., and McLeod, A. I., 1975, Intervention analysis in water resources: Water Resources, v. 11, p. 855–861.

——, McLeod, A. I., and McBean, E. A., 1977, Stochastic modeling of the effects of reservoir operation: J. Hydrol., v. 32, p. 97–113.

Hurst, H. E., Black, R. F., and Simaika, Y. M., 1946, The Nile basin: the future conservation of the Nile: Cairo, Min. Public Works, Paper 41.

——, Black, R. P., and Simaika, Y. M., 1965, Long term storage: London, Constable.

Kashyap, R. L., and Rao, A. R., 1976, Dynamic stochastic models from empirical data: New York, Academic Press.

Klemeś, V., 1976, The Hurst phenomenon: a puzzle?: Water Resources Res., v. 10, n. 4, p. 675–688.

Kraus, E. B., 1955a, Secular changes of tropical rainfall regimes: Quart. J. Roy. Meteorol. Soc., v. 81, p. 198–210.

——, 1955b, Secular changes in the east-coast rainfall regimes: Quart. J. Roy. Meteorol. Soc., v. 81, p. 430–439.

Lamb, H. H., 1969, Climatic fluctuations: Flohn, H., ed., World in survey of climatology, vol. 2.: General climatology: New York, Elsevier, p. 173–249.

Lorenz, E. N., 1968, Climatic determinism: *in* Causes of climatic change: Meteorol. Monogr., v. 8, p. 1–3.

——, 1969, The predictability of a flow which possesses many scales of motion: Tellus, v. 21, p. 289–307.

——, 1970, Climatic changes as a mathematical problem: J. Appl. Meteorol., v. 12, p. 543–546.

National Academy of Sciences, 1975, Understanding climatic change: prepared by U.S. Committee for the Global Atmospheric Research Program, National Research Council, National Academy of Sciences, Washington, D.C., p. 1–239.

O'Connell, P. E., 1971, A simple stochastic modelling of Hurst's law: Mathematical models in hydrology: Proc. Warsaw Symp., IAHS-AISH Publ. 100, v. 1, p. 169–187.

Philip, J. R., 1975, Some remarks on science and catchment prediction: *in* Chapman, T. G., and Dunin, F. X., eds., Prediction in catchment hydrology: Canberra, Aust. Acad. Sci.

Potter, K. W., 1976, Evidence of nonstationarity as a physical explanation of the Hurst phenomenon: Water Resources Res., v. 12, n. 5, p. 1047–1052.

Rao, A. R., and Rao, R. G. S., 1974, Analysis of the effect of urbanization on rainfall characteristics—I: Water Resources Research Center, W. Lafayette, Ind., Purdue Univ. Tech. Rept. 50, p. 1–212.

Rao, R. G. S., Rao, A. R., and Kashyap, R. L., 1975, Aggregation of hydrologic data: School of Civil Engineering, W. Lafayette, Ind., Purdue Univ. Tech. Rept. CE-HYD-76-6, p. 1–66.

Wallis, J. R., and Matalas, N. C., 1971, In hydrology h is a household word: Mathematical models in hydrology: Proc. Warsaw Symp., IAHS-AISH Publ. 100, v. 1, p. 196–203.

Woodward, R. H., and Goldsmith, P. L., 1964, Cumulative sum techniques: *in* Mathematical and statistical techniques for industry: Edinburgh, Oliver Boyd (Imperial Chemical Industries Ltd., Monogr. 3).

Yevjevich, V., and Jeng, R. I., 1969, Properties of non-homogeneous hydrologic series: Fort Collins, Colo., Colorado State Univ. Hydrol. Paper 32.

10

THERMODYNAMIC APPROACH TO GEOMORPHIC THRESHOLDS

Iaakov Karcz

INTRODUCTION

The formulation of general theories of landform evolution two generations ago was followed by intense search for a more rigorous physical understanding of the earth sculpturing processes. Although the ensuing adoption and adaptation of mathematical, physical, and engineering concepts led to spectacular progress in modern geomorphology, many fundamental problems have remained unanswered. In some cases, field data appear to be incongruous, or even contradict the generally accepted principles of geomorphic development; in many others, the problem appears to exceed the bounds of the available analytical framework. Discussions of such situations often emphasize that the natural system is too complex to be amenable to analytic treatment beyond a somewhat perfunctory search for empiric relationships between the variables involved (Pattee, 1973). But even such search proves more fruitful when guided by some clear conceptual framework. It is fortunate, therefore, that in the past decade new ideas, concepts, and techniques of study of complex natural systems have been developed in various scientific disciplines (for references: physics and chemistry: Glansdorff and Prigogine, 1971; Haken, 1974; Nicolis and Prigogine, 1977; Prigogine, 1978; biology and ecology: Lewontin, 1969; Pattee, 1973; Prigogine, 1969; 1978; May, 1973; topology and structural stability: Thom, 1975; Zeeman, 1977). Although mathematically intricate, many of these concepts hold promise of direct applicability in geomorphology, similar to that of general systems theory, thermodynamics, and statistical mechanics in the past decade (e.g., Chorley, 1962; 1967; Leopold and Langbein, 1962; Scheidegger, 1964, 1970; Howard, 1965; Scheidegger and Langbein, 1966a, 1966b).

This chapter suggests the possible applicability and implications of some recently formulated thermodynamics principles—in extension of geomorphic steady-state doctrine—to account for threshold effects, transient dynamic equilibria, and episodic erosion.

BACKGROUND

Much of the geomorphological research on fluvial systems has focused on mechanics of interaction between the flow and the groundmass, and has followed the concepts and techniques developed in hydraulic engineering. Successful as it has been, beyond a certain point the classical engineering approach becomes insufficient. The limitations of experimental modeling, and the profound difficulties in formulating and solving equations governing the behavior of natural flow and its impact on complex boundaries, have dictated sweeping simplifying assumptions in analysis. For example, most treatments still assume unrealistic stable, near-equilibrium conditions of steady and uniform two-dimensional flow. In engineering research, such assumptions are justified as long as the obtained results are consistent with the hydraulic experience and satisfy practical needs. However, to date, engineers have worked in a range of conditions much narrower than that represented in the geomorphological and sedimentological record and dealt with time spans negligible in comparison. Furthermore, the engineering treatment often has emphasized the derivation of overall balance expressions rather than considering the more complex problem of kinematics of streambed deformation, which involves unsteady and nonuniform three-dimensional flow conditions.

The complexity of natural fluvial systems and the obvious limitations of a deterministic approach spurred a search for fundamental generalization in treatment of landform evolution. One trend of thought emanating from the works of Gilbert, Strahler, and Horton (for references: e.g., Hack, 1960; Howard, 1965; Carson and Kirkby, 1972; Melhorn and Flemal, 1976; Schumm, 1977) emphasized the consideration of energy distribution and expenditure within the geomorphic realm. This approach, combining intuitive reasoning with concepts from the theory of general systems (with its variational and thermodynamic overtones), led to the formulation of steady-state theory, which has dominated the study of fluvial geomorphology in the past three decades (e.g., Hack, 1960; Chorley, 1962; Howard, 1965; Scheidegger and Langbein, 1966b).

Geomorphological and hydraulic data have demonstrated a tendency toward equilibrium between the flow and the groundmass, reflected in exponential equations relating flow and channel parameters (for references: e.g., Leopold et al., 1964; Blench, 1965; Williams, 1977). Since the fluvial system is open, the attained "quasi-equilibria" (Langbein and Leopold, 1964) imply equality, or constant ratios of input and output achieved through an internal adjustment of the system (i.e., a steady or stationary state; e.g., Chorley, 1962; 1967; Emery, 1969; Glansdorff and Prigogine, 1971). The precise rules of adjustment, however, are not clear, since the system is complex and there are more unknowns than available governing equations. Strictly deterministic approach therefore fails, and it was suggested that any imposed set of flow conditions could be accommodated in a number of alternative ways. Leopold and Langbein (1962) and later also Scheidegger and Langbein (1966b) explored the use of the variational approach and, by intuitive reasoning and an analogy from statistical mechanics and thermodynamics, proposed several general principles. The steady-state river system was assumed to tend toward the most probable energy distribution, attained through a compromise between two opposing tendencies—one toward a uniformly distributed rate of energy expenditure, and the other toward minimum work, formulated in terms of Prigogine's rule of minimum entropy production. The most probable condition will be characterized by a minimum total variance of the system components. While accommodating a change in stream power, the channel may be expected therefore to tend toward an equal distribution in variables, subject, of course, to any constraints present. This minimization of variance in response to a change, theoretically derived by Smith (1974) and Knighton (1977), was related by Leopold and Langbein (1962) and many others to the Le Châtelier-Brown principle for steady states. Such a minimization implies the presence of a self-stabilizing, "negative" feedback, and is commonly equated with the assumption that geomorphic steady state is stable.

The steady-state doctrine is now generally accepted and has been employed in analysis of a wide range of geomorphic problems. Schumm and Lichty (1965) clearly indicated that, when viewed within the appropriate framework, it augments rather than replaces the Davisian cyclic model. A good measure of maturity of this concept is the growing recognition of need for its reexamination and theoretical expansion (e.g., Schumm 1973, 1976, 1977; Bull, 1975, 1976, 1978). It has been argued that the very frequent changes in geomorphic variables do not allow the system to reach a steady state, and induce a complex, often uneven response of the landforms (e.g., Schumm, 1973; Bull, 1975). Under certain conditions, relatively minor fluctuations in external variables may trigger major changes and transformations, suggesting the existence of thresholds, as well as far-from-equilibrium conditions. Similar thresholds, separating different modes, rates, or sense of operation, are identifiable also when input is constant, thus suggesting an inherent tendency of the system to develop in a series of discrete jumps, from one temporary dynamic equilibrium to another (Schumm, 1973, 1977). Obviously, such effects as thresholds, rapid landform transformation in response to minor fluctuations, episodic erosion (e.g., Schumm, 1976, 1977), self-enhancing feedbacks, and non-steady-state landforms (Bull, 1976, 1978) are not accounted for by the classical geomorphic steady-state theory. Their occurrence, however, rather than invalidating the existing approach, merely indicates the need for reassessment and expansion of the theoretical background.

Indeed, a closer inspection will show that the incongruence disappears once the simplifying assumptions inherent in the present formulation of geomorphic steady-state theory are relaxed. The work of Leopold and Langbein (1962) relied on the then-available thermodynamic concepts, largely restricted to the near-equilibrium, linear regime. Under such conditions, the minimum entropy production and Le Châtelier rule—the cornerstones of the steady-state doctrine—are valid. However, more recent studies (e.g., Glandsorff and Prigogine, 1971; Nicolis and Prigogine, 1977) show that in the nonlinear, far-from-equilibrium regime, the variational approach and the assumption of stability of the steady state do not necessarily hold. Thus, under conditions far removed from equilibrium, the question no longer is only whether the system tends to a steady state, but also whether the steady state is stable and persists, or whether it is unstable and evolves to a new temporary steady state, possibly with a greater degree of coherence.

THERMODYNAMICS OF GEOMORPHIC STEADY–STATE CONCEPT, REVIEW, AND EXTENSION

In a striking reductionist approach to natural systems, which preceded much similar work in biology, Leopold and Langbein (1962) applied thermodynamic concepts in derivation of the governing tendencies in fluvial geomorphology.

The thermodynamic background of the proposed framework (e.g., Prigogine, 1967; Katchalsky and Curran, 1965) will be briefly reviewed, assuming that the temperature-relief analogy is valid, as forcibly argued by Scheidegger (1964).

Since in a system at temperature T, exchanging dQ of heat with its environment, the change of entropy dS is

$$dS \geq \frac{dQ}{T} \qquad (1)$$

there must be an internal production of entropy within the system, and the change in entropy consists of two terms, one due to exchange with surroundings, and the other due to production by irreversible processes within the system

$$dS = d_eS + d_IS \tag{2}$$

d_eS may be negative or positive (and zero for a closed system), whereas d_IS is always positive definite, with a limit at equilibrium $d_IS = 0$, when no energy is dissipated. Obviously, for reversible processes $d_IS = 0$, while for irreversible processes $d_IS > 0$.

In a steady state, entropy does not change, $dS = 0$, and thus

$$d_IS = -d_eS \tag{3}$$

so that the internal production of entropy must be balanced by outflow of entropy.

Natural systems with no restraints, if left to evolve, will reach a condition of equilibrium in which all flows vanish and entropy is at maximum (i.e., a state of "death"). If, however, some forces are fixed, the flows due to the unrestrained forces vanish, whereas those due to the fixed forces assume constant values (i.e., a steady state is attained). At equilibrium, entropy is at maximum, and order is destroyed, whereas at steady state, the structure may be preserved once negative entropy flow into the system is maintained.

Commonly, evaluation of processes is through analysis of entropy production P, which may be expressed as a sum of contributions by the various irreversible processes, each term a product of a flow J_i and a force X_i:

$$P = \frac{d_1S}{dt} = \sum_i J_iX_i \tag{4}$$

Employing the thermodynamics-relief analogies (e.g., Leopold and Langbein, 1962; Scheidegger and Langbein, 1966b), that is, T temperature corresponding to elevation h, Q amount of heat to amount of mass, and J heat flux to mass flux), equation 4 may be rewritten for the geomorphic case in the more familiar form

$$P = \text{grad } \frac{T}{T^2} J = \frac{\text{grad } h}{h^2} J \qquad P = \int_{z_1}^{z_2} \frac{dh}{dz} \frac{1}{h^2} J\, dz \tag{4a}$$

where z is the distance.

Equation 4 indicates that the flow-force dependence is critically important for any study of entropy production. Onsager (e.g., Katchalsky and Curran, 1965) proposed that for near-equilibrium conditions, any flow is linearly dependent on its conjugate force,

$$J_i = L_iX_i \tag{5}$$

(as well as on all other forces operating within the system, i. e., $J_i = \sum_j L_{ij}X_j$). The L_i represent phenomenological coefficients (which may depend on system parameters but not on forces). For a system of two flows J_1 and J_2 driven by forces X_1 and X_2,

$$P = J_1X_1 + J_2X_2 \qquad (6)$$

where

$$J_1 = L_{11}X_1 + L_{12}X_2 \qquad (7)$$

and

$$J_2 = L_{21}X_1 + L_{22}X_2 \qquad (8)$$

so that four phenomenological coefficients would be required to specify the entropy production. However, according to Onsager's reciprocity rule (derived from statistical mechanics),

$$L_{12} = L_{21} \qquad (9)$$

This indicates coupling between the irreversible processes occurring within the system, and reduces the number of coefficients needed.

Assuming the previously considered system of equations 6–9, we insert 7 and 8 into 6 and obtain

$$P = L_{11}X_1^2 + (L_{12} + L_{21})X_1X_2 + L_{22}X_2^2 \qquad (10)$$

Keeping X_1 constant but X_2 unrestrained, differentiation of P with respect to X_2 yields

$$\frac{\partial P}{\partial X_2} = (L_{12} + L_{21})X_1 + 2L_{22}X_2 \qquad (11)$$

According to reciprocity rule 9,

$$\frac{\partial P}{\partial X_2} = 2L_{21}X_1 + 2L_{22}X_2 \qquad (12)$$

Substituting according to 8,

$$\frac{\partial P}{\partial X_2} = 2J_2 \qquad (13)$$

Since in a steady state, when X_2 is unrestrained its conjugate flow must vanish,

$$J_2 = L_{21}X_1 + L_{22}X_2 = 0 \tag{14}$$

$$\frac{\partial P}{\partial X_2} = 0$$

The latter indicates extreme P in steady state, and since P is positive definite, the extreme must be a minimum, so that we may conclude:

$$\frac{dP}{dt} = 0 \quad \text{at steady state} \tag{15}$$

$$\frac{dP}{dt} < 0 \quad \text{away from steady state}$$

or in geomorphic terms (following equation 4a), at steady state:

$$\frac{dP}{dt} = \frac{d}{dt}\int \frac{dh/dz}{h^2} J \, dz = 0 \tag{15a}$$

and at a point

$$\frac{dP}{dt} = \frac{d}{dt}\frac{dh/dz}{h^2} J = 0 \tag{15b}$$

This statement represents the minimum entropy production principle, one of the cornerstones of the Leopold and Langbein (1962) theory of geomorphic steady state. The conditions and arguments employed in its derivation lead also to the conclusion that steady state is stable (i.e., that it will respond to a disturbance so as to minimize its effects).

Assuming a fluctuation δX_2 in the unrestrained force X_2, and unperturbed flow J_2^o, the perturbed flow J_2 may be written as

$$J_2 = J_2^o + L_{22}\delta X_2 \tag{16}$$

Since in a steady state J_2^o will vanish (14),

$$J_2 = L_{22}\delta X_2 \tag{17}$$

Since $L_{22} > 0$,

$$L_{22}(\delta X_2)^2 > 0 \tag{18}$$

and

$$J_2\delta X_2 > 0 \tag{19}$$

Thus, the flow due to the perturbation must have the same sign as the perturbation, and tends to reduce it. This is the basis of the self-stabilizing effects referred to in geomorphical literature as "negative feedback," or the Le Châtelier moderation.

The Prigogine minimum entropy production and Le Châtelier moderation theorems for steady state clearly indicate that the geomorphic steady-state concept is inherently restricted to near equilibrium, where the regime is linear and Onsager rules are valid.

The concept of self-stabilizing steady state was generally accepted and applied ignoring those limitations, although it was recognized that steady-state conditions may exist only for relatively short time spans (e.g., Schumm and Lichty, 1965). At the same time, numerous phenomena related to flow of fluids, and the processes of sediment erosion, transport, and deposition clearly indicated that steady state and stability are not always synonymous. In fact, studies of origin of various sedimentary structures widely employ the concept of instability: the observed forming pattern is viewed as a manifestation of instability of the preexisting state. When the imposed force drives the system to a certain finite distance from equilibrium—a distance usually expressed in terms of a threshold value of some dimensionless number (e.g., the Reynolds, Froude, Rayleigh, Richardson numbers, etc.)—a fluctuation perturbing the system is not dampened but instead drives the system to a new configuration.

Details of development of the various hydrodynamic instabilities, as well as many aspects of appearance, transformation, and preservation of alluvial bedforms, suggest a self-organizing process in which the system is driven through an orderly series of configurations, each transformation a result of instability triggered when the corresponding threshold is exceeded. Such evolution, known in physics and biology as "order through fluctuations" or "range of dissipative structure," must therefore be added to concepts of "destruction of structure" at equilibrium and "perpetuation of structure" in the near-equilibrium, linear steady-state regime, mentioned in the geomorphic literature.

Search for an extension of the rules governing entropy production and stability has proceeded through consideration of time changes in entropy production. Following equation 4 it is possible to write the time change in terms of respective contributions due to the generalized flows and forces:

$$\frac{dP}{dt} = J_i \frac{dX_i}{dt} + X_i \frac{dJ_i}{dt} \quad \text{or} \quad dP = d_x P + d_J P \qquad (20)$$

Glansdorff and Prigogine (1971) show that under time-independent boundary conditions

$$d_x P = 0 \quad \text{for steady state}$$

$$d_x P < 0 \quad \text{away from steady state} \qquad (21)$$

(i.e., that the change in forces will always proceed so as to diminish the production of entropy). This "universal thermodynamic evolution criterion" holds in nonlinear as well as in linear regime, with one major difference.

In linear regime, symmetry between flows and forces (eq. 9) requires that the two terms on the right-hand side of equation 20 be equal.

$$d_x P = d_j P = \tfrac{1}{2} dP \qquad (22)$$

so that the universal criterion (equation 21) becomes the minimum entropy production criterion (eq. 15).

In a nonlinear regime, however, the symmetry relationships do not hold and thus $d_x P$ does not prescribe the sign nor magnitude of $d_j P$ (i.e., also the sign and magnitude of dP are unconstrained). Furthermore, $d_x P$ is not a total differential of a state function, except in the near-equilibrium linear regime. Thus, no true variational principle applies in the far-from-equilibrium conditions, and it becomes apparent that stability of the steady state is not ensured (Prigogine and Nicolis, 1971).

In terms of the geomorphic theory (Leopold and Langbein, 1962; Scheidegger and Langbein, 1966a) outlined in equations 4a, 15a, and 15b, we may then write

$$dP = d_x P + d_j P = Jd \frac{dh/dz}{h^2} + \frac{dh/dz}{h^2} dJ \qquad (20a)$$

and

$$d_x P = Jd \frac{dh/dz}{h^2} \qquad (21a)$$

Stability Criteria

The preceding arguments show that study of systems far removed from equilibrium must consider their stability and, if possible, predict the threshold beyond which an instability-driven transformation may be expected. Such studies occupy an increasingly important place in various fields of science and technology (for references: e.g., Minorsky, 1962; Hahn, 1967; Riggs, 1970; Glansdorff and Prigogine, 1971; May 1973; Orians, 1975) but so far are rarely employed in geomorphology. The discussion here is limited to an outline of the thermodynamic stability criteria formulated by Prigogine and his colleagues and is preceded by several brief general comments.

The stability of a system usually is defined in terms of its response to fluctuations. A stable system, if perturbed, tends to return to the unperturbed state and the fluctuation is dampened, whereas in an unstable system the fluctuation is amplified. When the perturbation is neither dampened nor amplified, the system is referred to as neutrally stable (Fig. 1). Stability concepts are conveniently portrayed by considering the system in n-dimensional state space, each dimension representing one of the state variables. Although the state-space techniques are mathematically useful for any number of variables, as a graphically descriptive tool they are limited by our ability to portray a curve in space and are particularly useful for systems with two state variables x and y, which may be represented by

$$\frac{dx}{dt} = f(x, y) \qquad \frac{dy}{dt} = g(x, y) \qquad (23)$$

This simple system is used therefore here to introduce the basic concepts of stability. The theory is equally applicable to n-dimensional state space such as required in treat-

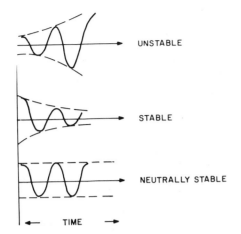

UNSTABLE

STABLE

NEUTRALLY STABLE

TIME

FIGURE 1 *Stability classification of perturbed mechanical systems, in coordinates of perturbation amplitude and time.*

ment of ecological or geomorphological systems (e.g., Lewontin, 1969; May, 1973). In this case, the instantaneous state of the system is represented as a point in the "phase plane" with coordinates x, y. A point for which equations 23 are zero is referred to as an equilibrium point or a singular point. Such points correspond to a steady-state solution for the system and thus stability of the steady state may be expressed in terms of stability of such a point. The time change of a system is represented by a trajectory traced in the phase plane by the continuous shift of the system from one position to another. Analytically, the trajectories are obtained by elimination of time from the solution of differential equations such as 23, that is,

$$\frac{dy}{dx} = \frac{g(x, y)}{f(x, y)} \qquad f(x, y) \neq 0 \tag{24}$$

If $f = 0, g \neq 0$, the roles of x, y may be interchanged. Equation 23 may admit periodic solutions yielding closing trajectories (orbits).

Any natural system is subject to small random fluctuations. It will be regarded stable if all trajectories in the neighborhood of a trajectory prescribed by the analytic solution to 24 remain in that neighborhood throughout. When, in addition, the distance between the prescribed and neighboring trajectories decreases with time, the system is defined as asymptotically stable. Points on different trajectories initially close to each other remain close in the former case and become closer in the latter (Fig. 2).

In the case of closing trajectories, the system may be either orbitally stable, that is, any representative point R on a neighboring trajectory remains close to the prescribed trajectory ϕ (but not to any specific points on ϕ), or may be unstable, when R moves away from ϕ (Fig. 2). Here stability is considered in terms of trajectories, so that two initially close points do not have to remain close to each other. When R approaches ϕ with time, the system possesses an asymptotic orbital stability and the asymptotically stable orbit is referred to as Poincaré's limit cycle (Fig. 3).

A possible criterion of stability would be the existence for all system points of a function describing a surface in the state space sloping toward the equilibrium point.

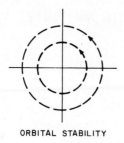

ORBITAL STABILITY

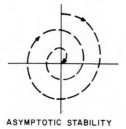

ASYMPTOTIC STABILITY

FIGURE 2 *Phase-plane portraits of orbitally stable, and asymptotically stable systems. The system evolution is described by trajectories in coordinates of variables x and y.*

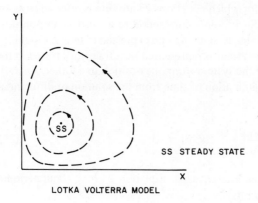

SS STEADY STATE

LOTKA VOLTERRA MODEL

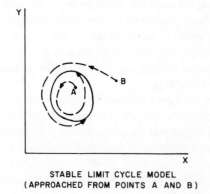

STABLE LIMIT CYCLE MODEL
(APPROACHED FROM POINTS A AND B)

FIGURE 3 *Phase-plane portraits of the Lotka-Volterra oscillating model and the stable limit cycle model.*

218

Thus, when a positive definite function x^2, representing the square of a distance in state space between two neighboring solutions, can be shown to have a time derivative $dx^2/dt \leqslant 0$, asymptotic stability is ensured. Such quadratic functions are referred to as Lyapunov functions (e.g., Minorsky, 1962; Prigogine, 1978).

Prigogine and his collaborators have shown that some thermodynamic functions represent Lyapunov functions and may be employed as stability criteria. For a linear steady state, entropy production P must be positive definite by virtue of the Second Law, and its time derivative must be a negative definite by the virtue of minimum entropy production principle, so that asymptotic stability is ensured. Since the $d_x P$ in the universal evolution criterion 20 is not a state function, another approach was used for the nonlinear regime: deviation in entropy production due to a perturbation in the acting forces (an external disturbance or an internal fluctuation), referred to as the excess entropy production $\delta_x P$. In a steady state with zero initial flux $\overset{\circ}{J_i}$, the perturbed flux $J_i = \delta J_i$, so that

$$\delta_x P = \Sigma J_i \delta X_i = \Sigma \delta J \delta X_i \qquad (25)$$

Since according to the evolution criterion 21,

$$d_x P \leqslant 0$$

if $\delta_x P$ of 25 is negative, the fluctuation will not be dampened. Therefore, stability requires that

$$\delta_x P = \Sigma \delta J_i \delta X_i \geqslant 0 \qquad (25a)$$

This may be shown also if entropy variation in this perturbed system is expressed by expanding around equilibrium and retaining terms up to the second order,

$$S = S_e + \delta S + \tfrac{1}{2} \delta^2 S \ldots \qquad (26)$$

then since at equilibrium, entropy is at maximum, the first-order term vanishes and the second-order term is a definite negative. The time derivative will be equal to excess entropy production $\delta_x P$ so that from equation 25,

$$\frac{d}{dt} \tfrac{1}{2} \delta^2 S = \delta_x P$$

$$\frac{d}{dt} \tfrac{1}{2} \delta^2 S = \sum_i \delta J_i \delta X_i > 0 \qquad (27)$$

That is, $\delta^2 S$ can be regarded as a Lyapunov function and a criterion of stability and demonstrates that instability will set in when the excess entropy production due to a fluctuation is negative. Furthermore, $\delta^2 S$ is related to the probability of occurrence of fluctuations, as demonstrated by Einstein's equation according to which fluctuation probability of proportional to $\exp (\delta^2 S/2k)$ (where k is the Boltzmann constant).

Following the geomorphic analogies of equations 4a, 15a, 15b, 20a, and 21a, we may assume that

$$\delta_x P = \delta J d \ \frac{dh/dz}{h^2} \qquad\qquad (25b)$$

which as a stability criterion

$$\delta_x P \geqslant 0$$

would yield

$$\frac{d(dh/dz)}{dh/dz} < \frac{2dh}{h} \qquad\qquad (27a)$$

Reorganization and Instability

The outline presented above shows that a system with imposed constraints is first driven away from thermodynamic equilibrium toward a steady state. As long as the imposed constraints are small, the system responds linearly and Onsager relationships, the Einstein fluctuation equation, and so on, are applicable. When constraints are stronger, the system may evolve smoothly into nonequilibrium steady states, governed by the minimum entropy production theorem. This continuation of equilibrium-like behavior is referred to as the "thermodynamic branch" (Glandsorff and Prigogine, 1971). With an increased distance from equilibrium and emergence of nonlinear relationships, the steady states along this branch are no longer necessarily stable. That critical distance from equilibrium at which instability may occur therefore represents a thermodynamic threshold, beyond which a perturbation introduced into the system does not decay to the steady state of the thermodynamic branch. Instead, the perturbation may evolve into a time-dependent regime, or may be shifted to a new steady state which does not belong to the thermodynamic branch. The limit of stability of thermodynamic branch may be regarded as a bifurcation point, expressed in terms of some critical value of a set of parameters E (reflected in the values of f and g in eq. 23) on which the system-governing equations depend. Below that value, only one positive bounded solution exists, so that in the phase plane, the changes in E would not be accompanied by any qualitative change in topological aspects of the trajectories. Above that point, the solution may no longer be unique and trajectories may change. This change is referred to as a loss of structural stability, or a *catastrophe*. The new regimes beyond the bifurcation of solutions of the thermodynamic branch may undergo still further bifurcations. Thom (1976) presented a comprehensive treatment of bifurcation based on analysis of structural stability of ordinary different equations for potential systems with a finite number of degrees of freedom. This elegant treatment, known as the "catastrophe theory," was applied in many studies in different disciplines (for references: e.g., Zeeman, 1977; Woodcock and Davis, 1978). Recent reviews indicate, however, that in many cases, the application involved fundamental theoretical misconceptions (e.g., Thom, 1977). It would appear that on the whole, the practical usefulness of the catastrophe theory is still in doubt, since so far it failed to predict any important experimental results (Thom, 1977). In our particular case, the geomorphic functions depend both on space and time and thus are governed by sets of partial differential equations, to which catastrophe theory is inapplicable. Furthermore, this theory would fail to account for the presence of systems with a periodic behavior, such as those discussed below (Nicolis and Auchmuty, 1974).

Instability of solutions of the thermodynamic branch heralds the onset of self-organization into steady-state dissipative structures, whose appearance and maintenance require a certain minimum level of dissipation within the system (e.g., Glansdorff and Prigogine, 1971). The statistical homogeneity of the steady state is thus broken and replaced by spatial structures often coupled with periodic behavior. The actual mechanism, analogous to phase transitions and governed by the bifurcation and branching of the solutions of nonlinear partial differential equations, is not fully understood, and thus discussions in literature are limited to specific cases, and are of a descriptive rather than a predictive nature (e.g., Haken, 1974; Nicolis and Prigogine, 1977).

Experimental and theoretical studies of system behavior close to and beyond the limit of thermodynamics branch (i.e., at, and beyond, the threshold) have identified three orderly regimes for semistable and unstable steady states, which may be applicable in geomorphological analysis. Two of these regimes show sustained fluctuations, whereas the third emphasizes instability transformations. In all, nonlinear relationships are evident, usually in association with positive feedbacks within the system, when the straight-forward probabilistic approach, such as employed by Leopold and Langbein (1962), becomes inapplicable (e.g., Glansdorff and Prigogine, 1971; Prigogine, 1978).

1. Close to the threshold, some systems display sustained, rather irregular ("incoherent") fluctuations, that is, a motion dominated by "noise-type" effects rather than a well-defined periodic motion of a specific amplitude and period. Such a system, with two independent variables, is governed by equations of the form

$$\frac{dx}{dt} = (A - By)x \qquad \frac{dy}{dt} = (Cx - D)y \qquad (28)$$

(where A, B, C, and D are positive constants), known as the Lotka–Volterra equations, originally developed for ecological prey-predator problems (e.g., May, 1973). Equilibrium points for the system in the phase plane would be at $(0, 0)$ and $(D/C, A/B)$. A phase plane portrait of the system (Fig. 3) based on elimination of dt (cf. eqs. 23 and 24) is

$$\frac{dy}{dx} = \frac{y(x - 1)}{x(y - 1)} \qquad (29)$$

where integration gives

$$K = x + y - \ln x - \ln y \qquad (30)$$

with K = a constant depending on the initial condition, which yields an infinite number of closed trajectories around the steady state, each corresponding to a certain value of K (Fig. 3).

It appears, therefore, that the system, although stable in the sense of orbital stability, lacks asymptotic stability (e.g., fluctuations are not dampened). Furthermore, a small fluctuation can easily shift the system from one orbit to another, and there exists no average orbit in the neighborhood of which the system is maintained. The trajectory geometry suggests also that small fluctuations when the system is close to the origin would lead to very considerable deviations in the later stage of system evolution.

2. The classic studies of Poincaré (e.g., Minorsky, 1962) demonstrated that equations 23 for systems with two independent variables admit solutions represented in the phase plane by closed trajectories at finite distances from another closed trajectory referred to as a limit cycle. If, with time, the limit cycle is approached by the neighboring trajectories, it possesses asymptotic orbital stability; that is, irrespective of initial conditions, the system evolves and approaches the same periodic solution with uniquely determined characteristics, such as amplitude and period. The phase plane diagram (Fig. 3) shows this asymptotic approach to the limit cycle irrespective of the initial conditions.

Stability analysis of steady states of various mechanical, ecological, and chemical systems (for references: e.g., Minorsky, 1962; May, 1973; Haken, 1974; Nicolis and Prigogine, 1977) demonstrated the frequent occurrence of this type of a coherent macro-

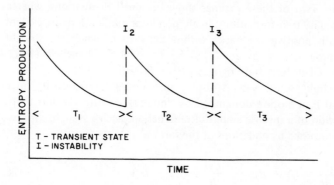

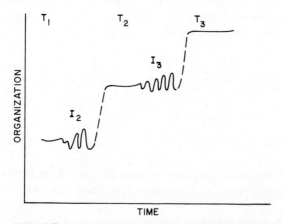

FIGURE 4 *Schematic diagram of evolution through successive instabilities, with intervening transient steady states and thresholds.*

scopic organization beyond the linear near-equilibrium regime of the thermodynamic branch. This type of ordered regime is believed to arise in response to fluctuations which affect the entire system evenly (e.g., Prigogine and Nicolis, 1971).

3. A third regime arises in response to spatially variable fluctuations (e.g., the system is no longer uniformly affected). Beyond the limit of the stability of the thermodynamic branch, the system may attain a new steady state, which differs from the previous one by the spontaneous breaking of the preexisting spatial uniformity and the emergence of an orderly distribution of the system-state variables (e.g., concentrations of x and y). The wave length of the spatial periodicity depends on the wavelength of the fastest-growing fluctuation beyond the onset of instability. Such a spatial configuration is referred to by Prigogine and his colleagues as a "giant fluctuation stabilized by the flow of energy and matter from the outside world," and possesses a lower entropy than the one corresponding to the thermodynamic branch.

Such spatial configurations, very commonly, are accompanied also by some degree of temporal organization, and the overall phenomenon displays the behavior of coupled nonlinear oscillators. The localization of effects provides a potential mechanism through which local thresholds may be overcome, as well as a mechanism for propagation of information. Both the spatial and temporal orderly regimes, as well as their coupled hybrids, may undergo further bifurcations, resulting in a complex series of successive, instability-triggered orderly configurations. The series would therefore consist of transient steady states, each developing instability to fluctuations and undergoing a transformation to the next, possibly transient, configuration (Fig. 4). Further complications may arise from the presence of more than one steady-state solution (i.e., the occurrence of multiple stable steady states).

SUMMARY

The main thrust of the concepts and results summarized above was to demonstrate the theoretical limitations of the current steady-state approach in geomorphology, and to consider the behavior of a system beyond those limits. The near-equilibrium, linear domain ("thermodynamic branch") of stable steady states and self-stabilizing feedback regime justifies the application of concepts and techniques of geomorphic dynamic equilibrium. Beyond that domain, orderly configurations arise and the system may evolve through a series of successive, instability-triggered transitions, leading to a more complex organized state. Such transitions are associated with increased entropy production and are characteristic of a nonlinear regime far removed from equilibrium. The occurrence of successive transitions implies that each involves a mechanism that promotes further instability by increasing nonlinearity and distance from equilibrium. Continuous dissipation allows the buildup and maintenance of ordered states, which, in turn, indicate that the ideas of chance, randomness, and probability stemming from the equilibrium thermodynamics and statistical mechanics are no longer applicable.

It appears, therefore, that evolution of systems beyond the stability threshold is dominated by self-organizing tendencies which head the system to an increased complexity. Such processes are strongly enhanced by autocatalytic effects within the system, and in some cases, depend also on the boundary conditions and size of the system. The mechanism of transition involves the occurrence of fluctuations, which may be related to entropy in the neighborhood of the thermodynamic branch by the Einstein fluctuation probability equation.

The instability-driven dissipative structures provide a convenient and attractive hypothetical framework for such phenomena as episodic erosion, complex response, and appearance of thresholds, and demonstrate, among others, definite inherent orderly changes in operation of the landform generating systems. Considerable difficulties in quantification of the landform parameters and processes need to be overcome, however, before the framework is shown to be an acceptable model rather than a caricature of the field evidence.

ACKNOWLEDGMENTS

It is a pleasure to acknowledge the gracious help of Dr. S. R. Dickman in the various stages of preparation of this manuscript. Critical comments and suggestions were also offered by Dr. M. Morisawa, and Dr. D. Coates. None carries the blame for the outcome.

REFERENCES CITED

Blench, T., 1965, Mobile bed fluviology: Univ. Alberta, Edmonton, Canada.

Bull, W., 1975, Allometric change of landforms: Geol. Soc. Amer. Bull., v. 86, p. 1489–1498.

——, 1976, Landforms that do not tend towards steady state: in Melhorn, W., and Flemal, R. C., Theories of landform development: Proc. 6th Geomorphol. Symp., Binghamton, N.Y., p. 111–128.

——, 1978, Threshold of critical power in streams: Geol. Soc. Amer. Bull., (in press).

Carson, M. A., and Kirkby, M., 1972, Hillslope form and processes: New York, Cambridge Univ. Press, 475 p.

Chorley, R. J., 1962, Geomorphology and general systems theory: U.S. Geol. Survey Prof. Paper 500-A, p. 1–10.

——, 1967, Models in geomorphology: in Chorley, R. J., and Haggett, P., eds., Models in geography: London, Methuen, p. 21–38.

Emery, F. E., ed., 1969, Systems thinking: Hammondsworth, England, Penguin, 398 p.

Glandsdorff, P., and Prigogine, I., 1971, Thermodynamic theory of structure, stability, and fluctuations: New York, Wiley–Interscience, 306 p.

Hack, J. T., 1960, Interpretation of erosional topography in humid temperate region: Amer. J. Sci. v. 258A, p. 80–97.

Hahn, W., 1967, Stability of motion: Berlin, Springer, 442 p.

Haken, H., ed., 1974, Cooperative effects: Amsterdam, North-Holland, 301 p.

Howard, A. D., 1965, Geomorphological systems, equilibrium, and dynamics: Amer. J. Sci. v. 263, p. 302–312.

Katchalsky, A., and Curran, P. F., 1965, Non-equilibrium thermodynamics in biophysics: Cambridge, Mass., Harvard Univ. Press, 248 p.

Knighton, A. D., 1977, Alternative derivation of the minimum variance hypothesis: Geol. Soc. Amer. Bull, v. 88, p. 364–366.

Langbein, W. B., and Leopold, L. B., 1964, Quasi-equilibrium states in channel morphology: Amer. J. Sci., v. 262, p. 782–794.

Leopold, L. B., and Langbein, W. B., 1962, The concept of entropy in landscape evolution: U.S. Geol. Survey Prof. Paper 500A, p. 1–20.

——, Wolman, M. G., and Miller, J. P., 1964, Fluvial processes in geomorphology: San Francisco, W. H. Freeman, 522 p.

Lewontin, R. C., 1969, The meaning of stability: *in* Diversity and stability in ecological systems: Brookhaven Symposium in Biology, 22, p. 13–24.

May, R., 1973, Stability and complexity in model ecosystems: Princeton, N. J., Princeton Univ. Press, 235 p.

Melhorn, W. N., and Flemal, R. C., eds., 1976, Theories of landform development: Proc. 6th Geomorphol. Symp., Publications in Geomorphology. SUNY, Binghamton, N.Y., 306 p.

Minorsky, N., 1962, Non-linear oscillations: Van Nostrand, Princeton, 714 p.

Nicolis, G., and Auchmuty, 1974, Dissipative structures, catastrophes, and pattern formation—a bifurcation analysis: Proc. Nat. Acad. Sci. USA, v. 71, p. 2748–2751.

——, and Prigogine, I., 1977, Self-organization in non-equilibrium systems: New York, Wiley–Interscience, 491 p.

Orians, G. H., 1975, Diversity, stability, and maturity in natural ecosystems: *in* Van Dobben, W., and Lowe-McConnell, R. H., eds., Unifying concepts in ecology: The Hague, Junk, p. 139–151.

Pattee, H. H., 1973, Hierarchy theory, the challenge of complex systems: New York, Braziller, 156 p.

Prigogine, I., 1967, Introduction to thermodynamics of irreversible processes: New York, Wiley, 320 p.

——, 1969, Structure, dissipation, and life: *in* Marois, M., ed., Theoretical physics and biology: Amsterdam, North-Holland, p. 23–52.

——, 1978, Time, structure, and fluctuations: Science, 201, p. 777–785.

——, and Nicolis, G., 1971, Fluctuations and the mechanism of instabilities: *in* Proc. 3rd Int. Conf. from Theoretical Physics to Biology: Basel, Karger, p. 89–109.

Riggs, D. S., 1970, Control theory and physiological feedback mechanisms: Baltimore, Md., Williams & Wilkins, 599 p.

Schneidegger, A., 1964, Some implications of statistical mechanics in geomorphology: Int. Assoc. Sci. Hydrol. Bull., v. 9, p. 12–16.

——, and Langbein, W., 1966a, Probability concepts in geomorphology: U.S. Geol. Survey Prof. Paper 500C, p. 1-14.

, and Langbein, W. B., 1966b, Steady state in stochastic theory of longitudinal river profile development: Int. Assoc. Sci. Hydrol. Bull., v. 11, n. 3, p. 43–49.

——, 1970, Theoretical geomorphology: 2nd ed., Berlin, Springer.

Schumm, S. S., 1973, Geomorphic thresholds and response of drainage systems: *in* Morisawa, M., ed., Fluvial geomorphology: Proc. 4th Geomorphol. Symp., Binghamton, N.Y., p. 299–310.

——, 1976, Episodic erosion, a modification of the geomorphic cycle: *in* Melhorn, W., and Flemal, R., eds., Theories of landform development: Proc. 6th Geomorphol. Symp. Binghamton, N.Y., p. 69–87.

——, 1977, The fluvial system: New York, Wiley–Interscience, 338 p.

——, and Lichty, R. W., 1965, Time, space, and causality in geomorphology: Amer. J. Sci., v. 263, p. 110–119.

Smith, T. R., 1974, A derivation of the hydraulic geometry of steady state channels from conservation principles and sediment transport laws: J. Geol., v. 82, p. 98–104.

Thom, R., 1976, Structural stability and morphogenesis, an outline of a general theory of models: Reading, Mass., W. A. Benjamin.

——, 1977, What is catastrophe theory about?: Lecture notes, International Workshop on Synergetics, Schloss Elmau, Bavaria, West Germany, 5 p.

Williams, G., 1977, Hydraulic geometry of river cross sections, theory of minimum variance: U.S. Geol. Survey Prof. Paper 1029, 47 p.

Woodcock, A., and Davis, M., 1978, Catastrophe theory: New York, E. P. Dutton, 152 p.

Zeeman, E. C., 1977, Catastrophe theory, selected papers 1872–1977: Reading, Mass., Addison-Wesley, 675 p.

11

THRESHOLDS IN RIVER REGIMES

Alan D. Howard

INTRODUCTION

The bed of natural stream channels can be composed of bedrock, fine-grained alluvium, or a coarse-grained gravel pavement. This classification appears to have a "natural" basis, in that most natural channels, or segments thereof, can be assigned unequivocally to one of the three classes. Downstream transitions between any two of the three channel types are abrupt, as are temporal changes resulting from changed hydraulic regime.

Thus natural discontinuities, or *thresholds*, occur between these channel types, and, as will be shown, each type has a distinctive pattern of response to controlling factors. Although thresholds are generally defined as an abrupt *temporal* change in morphology or dynamics of landforms, similar abrupt *spatial* transitions are common (stream banks, excarpment brinks, riffles and pools, stream junctions, etc.), and the transitions between channel types discussed here can be of either type. In evolving landscapes spatial thresholds may be analogous to temporal changes at a point ("substitution of space for time"), such as in the case of an advancing gully network or a retreating escarpment; this is often true for the channel bed transitions discussed here.

Temporal thresholds between channel types are rarely mentioned in geomorphic literature, largely because the generally slow time scales of fluvial response limit the number of directly observable cases. However, a few scattered cases have been mentioned. Howard (in preparation, a) cites examples where badland washes alternate between bedrock and fine-bed alluvial channels as a result of seasonal changes in relative discharge and sediment supply. Bed degradation below dams often converts fine-bed alluvial channels to coarse-bed channels by the armoring process, or in some cases completely scours the alluvium to bedrock (Gessler, 1971; Hammad, 1972; Simons et al., 1975).

The spatial thresholds are better known, although not usually remarked upon as such. Many otherwise alluvial channels have steep rapids or falls where especially resistant rocks crop out. Howard (in preparation, a) discusses abrupt downstream transitions in badland channel networks between bedrock and fine-bed alluvial channels. Steep, shale-floored rills and washes on badland slopes commonly abut fine-bed alluvial pediments (Smith, 1958; Schumm, 1956; Howard, 1970). Yatsu (1955) describes striking downstream transitions from gravel- to sand-bed rivers in Japan, and Shaw and Kellerhals (1977) cite similar transitions in Alberta. The alternating coarse-bed riffles and fine-bed pools common to many rivers may also be examples.

The median grain sizes at 158 stream cross sections throughout the United States sampled by Williams (1978a) are indirect evidence for a threshold between fine- and coarse-bed alluvial beds. The frequency distribution of the medians is bimodal; 65% of the medians lie in the 3ϕ range between 0.125 and 1 mm, 20% in the 3ϕ range from 16 to 128 mm, but only 8% occur in the intermediate range from 2 to 16 mm (Williams, 1978a, Table 11). Other indirect evidence for thresholds between the two types of alluvial channels will be presented in this chapter.

The regime of streams will be considered herein primarily in terms of their long-term adjustment to discharge pattern and to the size, grading, and quantity of sediment supplied by slope erosion. These factors will be termed the *hydraulic regime*, and they will be considered to be the independent (or external) variables, whereas the hydraulic geometry of the channel (in particular, the channel gradient) constitutes the dependent, or internal, variables. This contrasts with the short-term engineering view wherein gradient, bed material size, channel dimensions, and discharge are the independent variables and the bed material load is the dependent variable, and with very long geologic time scales, where tectonism, global climate, and the nature of the bedrock are the independent variables and the entire landscape becomes a dependent set of variables (Schumm and Lichty, 1965.).

Schumm (1974) and Schumm and Beathard (1976) distinguish between *extrinsic* thresholds, in which abrupt changes in morphology or dynamics occur as a result of gradual (progressive) change in the independent variables, and *intrinsic* thresholds, in which the threshold occurs without any necessary change in the external variables. In general, the thresholds discussed here are extrinsic in that gradual downstream or temporal change in hydraulic regime triggers transformations between the channel types. However, erosion of bedrock channels without change in hydraulic regime can reduce the gradient to the point that it is transformed to an alluvial bed, an intrinsic threshold.

Alluvial Channels

Alluvial channels are characterized by one of two types of beds: (1) coarse gravel or cobbles that move only near bankfull discharge, and (2) sand-bed channels in which transport occurs at all but perhaps the lowest flows. The former channels are distinguished morphologically as *coarse-bed* and the latter as *fine-bed*. Hydraulically, these correspond to "threshold" or "stable" channels on the one hand, and to "live-bed" or "regime" channels on the other. The recognition of this dichotomy is not new. For example, Maddock (1969) presents separate regime formulas for gravel- and sand-bed channels, and other authors have recognized that coarse-bed channels are near the threshold of motion for peak discharges (Kellerhals, 1967; Li et al., 1976). Similarly, many studies have been made of sediment transport relationships applicable to natural sand-bed streams. However, it does not seem to be widely recognized that transitional forms between these channel types are rare, and that a hydraulic threshold exists. In fact, these two types of alluvial channels have distinct hydraulic geometries (particularly the *down-stream* relationships). Henderson (1966, p. 472) recognized that fine- and coarse-bed channels have different exponents, relating channel gradient to dominant discharge and that these exponents:

> may be used to distinguish between those river channels which have shaped themselves according to live-bed criteria and those others (probably in coarse material) which have shaped themselves according to conditions obtaining at the threshold of motion. But it must be conceded that so far no complete theory has been presented explaining the mechanics of the process.

A major theme in this paper is an attempt toward an explanation of this threshold behavior as well as an examination of the factors determining which of the two channel types will occur in a particular stream.

Bedrock Channels

Bedrock channels are defined here as channel segments whose beds lack a coherent cover of active alluvium, even at low flow. "Bedrock" is employed somewhat loosely, for steep rills on soil slopes and steep channels dissecting cohesive fluvial deposits fall within this definition. The qualification of coherency of the alluvial cover is included to allow for scattered pockets of alluvium within scour holes in otherwise bedrock-floored channels (Einstein, 1964). Short bedrock channel segments are common along many rivers at the outcrop of resistant rock, but large channel systems that are exclusively bedrock are rare. Such channels seem to be common in headwater sections where relief ratios are high, often forming the fingertip tributaries of otherwise alluvial channel systems.

Bedrock channels remain free from alluvial deposition because of their steep gradient; they are always steeper than alluvial channels with the same hydraulic regime imposed from upstream. Whereas the gradient of alluvial streams is determined by its hydraulic regime, the gradient of bedrock channels can be considered to be an independent variable. For example, Howard (in preparation, a) shows that bedrock channels are steeper than alluvial channels in an area of badland topography, and whereas the gradient of the alluvial channels shows a strong correlation with drainage area (due to consistent downstream changes in the controlling hydraulic regime), there is no such correlation for bedrock channels. Rather, the gradient of these channels may have been determined by a number of factors, including the original land slope (for newly dissected topography), the resistance of the bedrock to erosion, and the erosional history of the area.

Bedrock channels have no alluvial deposition at low flows, even though they may transport considerable bed material load and, in the case of ephemeral washes, may have periods of no flow. The reason for the lack of alluvial bed is apparent for bedrock channel sections downstream from alluvial stream sections; the bedrock channel, by virtue of its steeper gradient, has a greater transporting capacity than the upstream alluvial channel for any stage, so that the bedload will be preferentially deposited on the alluvial channel during waning flow. For headwater bedrock channels the upstream sediment supply diminishes more rapidly than the transporting capacity during waning stages because that supply occurs primarily during the runoff stage, whereas transport in the channel continues during base flow recession.

In the following sections the behavior of alluvial and bedrock streams is more closely examined in order to demonstrate the existence of the thresholds and to explore the implications of these thresholds to fluvial sedimentology and to river response to changes in hydraulic regime. This introductory section closes with a discussion of two important concepts, the dominant discharge and the selection of independent and dependent variables of hydraulic geometry.

Dominant Discharge and Status of Variables of Hydraulic Geometry

The role of discharge in sediment transport and bed erosion will be represented by a *dominant* discharge, that is, a steady discharge (not necessarily of continuous duration) that performs the same action as the natural sequence of discharges. For bed material transport this is a constant discharge that would transport the same bedload and suspended load as the long-term average of the actual discharges. This discharge can be determined by summing and weighting the natural discharges according to their transporting capacity, generally using an appropriate transport formula (Shen, 1971b). In some channels the armoring by the coarse fraction at low flows (Emmett, 1976; Bagnold, 1977) and "hiding factors" for small grains resting between larger ones (Einstein, 1950) would have to be taken into account in calculating the dominant discharge for fine-bed material.

The dominant discharge is a function of grain size, with higher values for coarser fractions. Thus, there is actually a suite of dominant discharges for sediment transport that may in turn be considerably different than the dominant or formative discharges determining channel width or meandering characteristics of the stream. However, the mean annual flood or the bankfull discharge is often assumed to approximate the dominant discharge for all these aspects of fluvial behavior.

For reasons that will become apparent, the dominant discharge for bedrock channels is considered to be a constant discharge producing a rate of erosion equivalent to the naturally varying discharges. For some channels where weathering must precede erosion, the concept of a dominant discharge may be inapplicable.

In accord with the time perspective adopted here, the channel gradient for alluvial channels and the rate of bed erosion for bedrock channels are considered to be the primary dependent variables. The hydraulic regime also affects the channel cross-sectional characteristics (in particular, the channel width) and meandering behavior, and these can be considered to be either dependent or independent variables (Maddock, 1969; Schumm, 1969, 1971b). However, for the following reasons these will be considered here to be independent variables which can affect the adjustment of channel gradients or erosion rates:

1.   The channel width and degree of meandering are determined by different hydraulic properties than is the channel gradient. In particular, the wash load and the finer suspended load play an integral role in determining channel width and sinuosity (Schumm, 1960, 1971b), whereas the fine sediment has only a secondary effect upon channel gradient. In addition, bank vegetation is a variable factor affecting channel width.

2.   Some alluvial channels are constricted by bedrock or talus banks, as the Colorado River in the Grand Canyon. In many other channels the development of free meanders is constrained by the valley walls. These constraints may not directly influence the gradient, but they may do so indirectly through their effects upon the width and sinuosity.

3.   Regime formulas for gradient of alluvial channels are presented in their most general form in terms of unit discharges of water and sediment.

4.   Channel width can respond rapidly to change in hydraulic regime, in that single large floods can dramatically widen a channel (Schumm and Lichty, 1963; Burkham, 1972), whereas gradient generally responds more slowly. Thus, alluvial channel gradients are adjusted to long-term averages of channel width.

5.   Changes of channel width and sinuosity in response to changes in hydraulic regime often have counteracting effects upon stream gradient. For example, channel widening by floods, which should cause steepening of alluvial channel gradients, is often accompanied by decrease in sinuosity, which has an opposite effect. This balance

of effects is cited by Schumm (1969) as the reason for the lack of aggradation or entrenchment of the Murrumbidgee River in Australia following a change in hydraulic regime which drastically changed channel patterns.

REGIME AND HYDRAULIC GEOMETRY OF ALLUVIAL CHANNELS

Natural stream channels are primarily alluvial, that is, they have at low flow stages a bed composed of unconsolidated sediment transported into place by the stream. Except under highly unsteady flow or if bed material is completely scoured during high flows, an alluvial channel carries a capacity load of the size ranges dominating the bed. This assumption is implicit in the use of engineering formulas to calculate the bedload or total sediment load from flow characteristics. Their low-flow alluvial covers occur because transport capacity decreases during falling states; alluvium transported at peak or dominant discharges must be redeposited at low flows. The thickness of the active layer, h, is approximately related to the high-discharge flow depth, D, by (assuming negligible transport at low flows)

$$h = \frac{CD}{S_s(1 - p)} \tag{1}$$

where C is the concentration of the bed-material load, S_s is the specific gravity of the sediment, and p is the bed porosity. The active bed thickness is small, ranging from only a few grains to as much as 2 cm for the Colorado River in the Grand Canyon (maximum flow depths of 10 m and sand-sized total load concentrations of 0.4%). Therefore, the much greater seasonal depth of scour and fill reported in sand-bed streams (up to 4.5 m in the Grand Canyon; Leopold and Maddock, 1953) must be due to other factors, such as (1) migration of bedforms (Foley, 1976), (2) a rapid-weir control mechanism (Silverston and Laursen, 1976), or (3) in small streams at least, to seasonal or flood-by-flood differences in rate of sediment supply from slope erosion.

Alluvial streams with stable hydraulic regimes have gradients that are just sufficient (or graded) to carry the supply of bed material in the size range dominating the bed (after Mackin, 1948). The evidence for this is largely indirect. Under steady flows and a constant sediment supply, flumes, which are presumably models of natural streams, adjust their gradients so that input and output of bed material load are equal. In natural streams the situation is complicated by variable discharges and fluctuating sediment supply, but the same type of adjustment presumably occurs on a statistical basis (Howard, in preparation, a). Alluvial channel gradients respond to changes in hydraulic regime in a manner that is predictable by sediment transport relationships. The gradient of alluvial channels is predictable from bed-material transport formulas using suitably averaged values of discharge and sediment load. Finally, the gradients of alluvial channels decrease downstream in a regular manner as would be predicted by downstream changes in discharge, sediment load, and grain size. The degree of uniqueness of alluvial channel grade is a subject of continuing discussion, largely beyond the scope of this chapter. The nature of alluvial channel regime is given closer attention in Howard (in preparation, a), although much of the data presented in this paper confirms the utility of the concepts of regime and grade.

The occurrence of abrupt spatial (downstream) and temporal transitions between fine- and coarse-bed channels, and the lack of beds with intermediate grain sizes is a consequence of the nonlinear rate laws for sediment transport, and, in particular, of the existence of a critical threshold stress for initiation of motion, which is the greater the coarser the grain size (although additional factors, such as differential abrasion and lack of supply of intermediate grain sizes may also in part account for the paucity of alluvial beds of intermediate grain size). Although the gradient of alluvial channels is roughly balanced at the minimum value sufficient to transport the sediment load supplied from upstream, in some channels the sparse load of coarse cobbles requires the steepest gradient, and in others the abundant fine bedload is the critical factor. The reasons for this dichotomy and the factors influencing which channel type will occur in a given situation will be illustrated by using sediment transport equations to predict the gradient necessary to transport each size fraction of the load supplied from upstream.

Calculation of Required Gradients

The procedure used is a variation of the normal engineering use of transport formulas; rather than calculating the quantity of bed material in transport as a function of velocity (or shear) and grain size of the bed material, the formulas are used to calculate the gradient needed (the *required gradient*) as a function of the size and quantity of bed material supplied from upstream with the dominant discharge. Since the sediment transport functions generally relate the quantity of sediment in transport to either the bed shear or the flow velocity and depth, a flow resistance relationship is needed in addition to relate these parameters to discharge and gradient. For most formulas the Manning equation was assumed:

$$V = \frac{R^{2/3}S^{1/2}}{n} \quad \text{(mks units)} \tag{2}$$

where V is the average velocity, R the hydraulic radius, and S the gradient. For fine grain sizes the resistance coefficient, n, was given an arbitrary fixed value, for the coarser grain sizes the grain resistance was assumed to dominate and the Manning–Strickler relationship was used:

$$n = 0.04d^{1/6} \quad \text{(mks units)} \tag{3}$$

The most critical assumption made in the present calculations is that the spectrum of grain sizes supplied from upstream can be divided into size fractions, each with its characteristic long-term rate of supply, and that the required gradient can be calculated independently for each size fraction. In essence, the assumption is made that each size fraction is transported independently of each other. This requires justification, for there is obviously some interaction between all sizes in transport [e.g., Einstein's (1950) "hiding factor" for small bed-material grain sizes and Maddock's (1973) discussion of the influence of the wash load on bed-material transport rates]. However, in natural streams the grains dominating the bed are a small fraction of the total sediment through-flow. For example, the channel bed of the Colorado River at the Grand Canyon and Lee's Ferry gaging stations is a uniform sand, whereas the river transports a very well-graded mixture to Lake Mead (Fig. 1). The size fractions represented on the bed are presumably

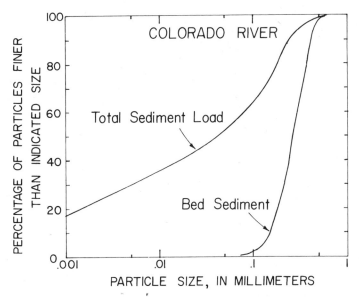

FIGURE 1   *Cumulative histograms of grain size for bed material and total sediment load of the Colorado River in the Grand Canyon. Bed sediment is average of several determinations at Lees Ferry and Grand Canyon gaging station. Total load curve from sedimentation in Lake Mead. (After Smith et al., 1960.)*

those that at that section of the channel are the most difficult to transport, that is, which require the steepest gradient. The proportion of the total sediment load which is finer than the bed material is transported in suspension at below-capacity concentrations (i.e., the wash load). Material coarser than the dominant bed material is carried in such small quantities that it does not accumulate, but remains dispersed on the bed, or is areally sorted (e.g., into riffles or cobble bars). Thus, there appears to be a small range of grain sizes selected from the total range in transport which dominates the bed and which is carried by the stream in capacity quantities. Correspondingly, it seems reasonable to assume that this same grain size range determines the required gradient, and that the critical range of grain sizes (i.e., the resultant dominant bed material grain sizes) from the total range of sizes supplied from upstream can be determined by finding the grain size range that gives the greatest required gradient given the quantity of supplied sediment for that size range.

The assumption of independence between transport of different size ranges certainly is not valid for very high sediment concentrations (i.e., for mudflows). The threshold between fine- and coarse-bed alluvial channels probably ceases to exist for mudflows, for its deposits are essentially unsorted.

The quantity of sediment supplied from upstream as a function of grain size has only rarely been determined for natural channels, the distribution being known most often where all the load came to rest in a reservoir or natural delta, as in the case of the Colorado River (Fig. 1). For illustration purposes in the following calculations the quantity of supplied sediment was assumed to follow a log-normal frequency distribution with respect to grain size, so that the distribution can be completely characterized by its (logarithmic) mean and variance (Krumbein and Graybill, 1965). The log-normal distribution (or a more complex combination of such distributions) of grain sizes of sediments is commonly assumed to result from the weathering and erosional processes that supply sediment to stream channels as well as by the subsequent processes of abrasion and sorting that act during transport (Spencer, 1963; Visher, 1969; Middleton, 1976). The sediment transported by the Colorado River is clearly not log-normally distributed (Fig. 1), but the basic consequences of the model will hold for a wide range of distributions of

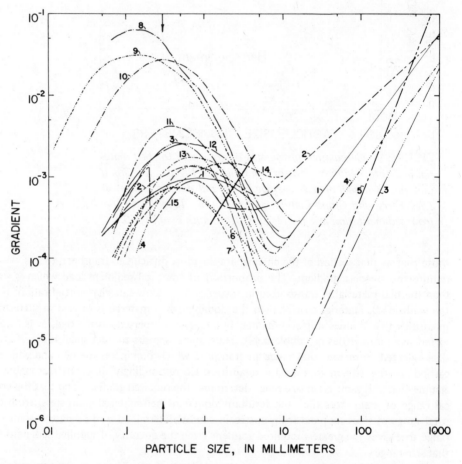

FIGURE 2   Curves of required gradient versus grain size for several transport theories for nominal values of hydraulic variables. Median grain size is 0.3 mm (arrows). Curves for various transport formulas identified as follows (see Tables 1 and 2 for references): 1. Einstein-Brown with Shields criterion for threshold of motion; 2. Maddock; 3. Yang; 4. Ackers and White; 5. Baker and Ritter (threshold-of-motion criterion); 6. Shen; 7. Engelund-Hansen; 8. Bagnold bedload; 9. Meyer Peter; 10. Yalin; 11. Bagnold total load; 12. Kalinske; 13. Laursen; 14. Bogardi; 15. Peterson. Heavy solid line gives threshold of suspension.

supplied sediment, so that the particular distribution assumed is not critical. The sediment load is broken into equal logarithmic size ranges, each with a characteristic grain size and quantity of supply, and the gradient required to transport that fraction has been calculated by the use of various bed- and total-load equations for the assumed hydraulic regime.

Only a subset of the numerous transport formulas were used in these calculations (generally the most well known or most recent ones), and a few could not be used because they cannot be easily solved for the required gradient. Some of the formulas are restricted in their range of applicability to fine grain sizes and high transport rates (e.g., the Shen and Bogardi relationships); that is, they are not valid for low transport rates near the threshold of motion. Others, however, like the Einstein, Maddock, and Yang relationships, include the threshold of motion criterion, and a few relationships have been proposed which are concerned solely with initiation of motion (e.g., Baker and Ritter, 1975). Some of the selected formulas were "bedload" and some were "total load" (the latter including the proportion of the bed material transported in suspension as opposed to saltation and rolling).

The variable input parameters to the gradient calculations are the log-mean grain size, d, the variance of the grain size distribution, θ, the water discharge, Q, the concentration of the sediment (sum of all grain sizes), C, the channel width, W, and the roughness, n. In all calculations the sediment specific gravity was assumed to be 2.65, and the water viscosity was held constant. A wide channel was also assumed, so that the hydraulic radius, R, equals the flow depth, D. Variations in the six basic parameters were considered independently with reference to a nominal set of parameter values ($d = 0.3$ mm; $Q = 1000$ m^3/s; $W = 100$ m; $\theta = 0.87$; $C = 0.01$; $n = 0.02$). A graph of the required gradient versus grain size is shown in Figure 2 for the various transport formulas using the nominal parameter values. Figures 3 and 4 show, for one of the transport formulas (the Einstein-Brown relationship), the effect of variation of individual parameters with the rest fixed at their nominal values.

Factors Determining Bed Material Size

The curves of required gradient versus grain size for the various transport formulas under nominal conditions show the same basic pattern (Fig. 2): a local maximum lying within the range 0.1 to 2 mm, a local minimum near 5 to 10 mm, and a continuous increase for grain sizes beyond 10 mm, where the threshold of motion is the determining factor. The channel gradient at equilibrium, or in-regime, will presumably be the highest required gradient. But since the required gradient increases without limit for increased grain size in the threshold domain, the question arises why natural channels have finite gradients and, in fact, often have gradients determined by the maximum in the fine-grained range. The limiting factor must be one of supply. The required gradient for threshold conditions in natural channels will be determined by the supply of coarse grain sizes. Above some *critical* grain size the supply of sediment will be insufficient to form a coherent bed. If the required gradient for this size is greater than the fine-grained maximum, the bed will be dominated by this coarse grain size (because the flow will carry a below-capacity load of the size range corresponding to the fine-grained maximum, there will be little tendency toward deposition of the fine grains during waning flows). On the other hand, if the critical grain size is such that the gradient at initiation of motion is below the required gradient of the fine-grained maximum, the low-flow bed will be dominated by grain sizes near the fine-grained maximum (termed for convenience the *capacity* grain size), diluting the representation of the coarse grain sizes. The availability of coarse grain sizes (and hence the critical grain size) depends upon bedrock type, weathering and slope transport processes, upstream sorting and abrasion, and flow

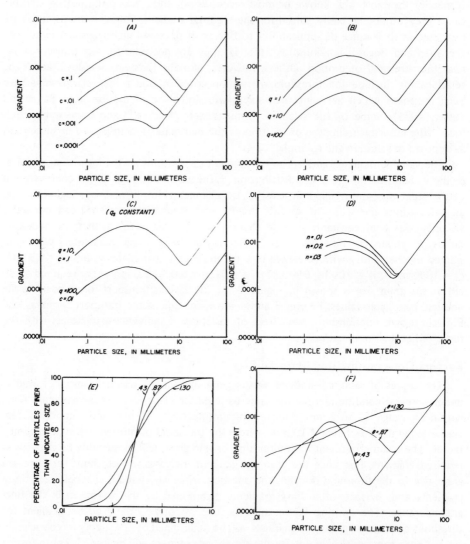

FIGURE 3 *Curves of required gradient versus grain size, showing effect of variation of single variables of the hydraulic regime with other variables constant at nominal values: (A) variation in concentration; (B) variation in specific discharge; (C) variation in specific discharge and concentration with specific sediment discharge fixed; (D) variation in roughness; (E) grain size distributions for different values of variance, θ; (F) required gradient curves for different values of variance.*

frequency characteristics (because the dominant discharge for the coarse grains will be higher than for the finer particles).

The effect of variations of the input parameters and of the derivative parameters $q = Q/W$, $q_s = C \cdot Q/W$, and $Q_s = C \cdot Q$ was investigated for the Einstein-Brown transport function (Figs. 3 and 4). Except as noted below, the behavior for other equations was

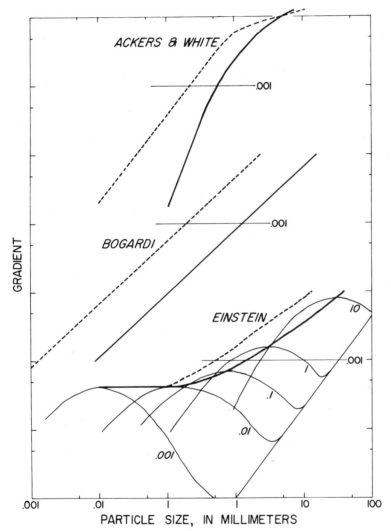

FIGURE 4 *Variation of maximum required gradient for fine-bed alluvial channels with average particle size for three transport relationships. Dashed lines show relationships between average particle size in transport and channel gradient, whereas solid lines show relationship between particle size of bed material and channel gradient. Representative curves of required gradient shown for several average grain sizes for the Einstein-Brown relationship.*

TABLE 1 Comparison of Downstream Gradient Changes of Fine-Bed Streams with Predictions Using Sediment Transport Formulas[a]

A. Predicted relationships for $S \propto q_s{}^{\alpha} d^{\beta} q^{\gamma}$

	α	γ	Value of β for indicated size of bed sediment			
			0.1 mm	0.3 mm	1 mm	3 mm
Bedload formulas						
1. Bagnold and Yalin	0.95	−0.86	0	0	0	0
2. Meyer-Peter	0.81	−0.86	0	0	0	0
3. Einstein-Brown	0.48	−0.86	0	0.31	0.54	0.64
Total-load formulas						
4. Bagnold	0.84	−1.09	1.48	0.77	0.15	0.16
5. Shen	0.54	−0.95	0.70	0.28	0.26	0.14
6. Laursen	0.60	−0.90	4.54	1.95	0.78	0.45
7. Yang	0.71	−0.99	1.83	1.02	0.41	0.11
8. Maddock	0.47	−0.81	0.58	0.53	0.54	0.49
9. Peterson	0.40	−0.79	—	—	—	—
10. Bogardi	0.35	−0.86	0.91	0.91	0.91	0.91
11. Kalinske	0.57	−0.86	0.57	0.57	0.57	0.57
12. Engelund and Hansen	0.87	−1.48	0.43	0.43	0.43	0.43
13. Ackers and White	0.91	−1.20	2.49	2.31	1.26	0.81

B. Empirical data for fine-bed streams

Geographic Area	Median grain size[b] (m)	$W \propto A^r$ r^c	$Q \propto A^e$ e^c	$Q_s \propto A^p$ p^c	$S = KL^a d^{\beta}$ (m)			
					K	a	β	R^{2d}
Virginia Badlands[e]	0.003	0.25	0.9	1	0.22	−0.25	—	0.91
Utah Badlands[f]	0.003	0.24	0.9	1	2.15	−0.32	0.15	0.94
Great Plains[g]	0.004	0.13	0.71	0.85	3.31	−0.34	0.16	0.61
Ephemeral, New Mexico[h]	0.02	0.41	0.8	1	0.43	−0.25	—	0.85
Ephemeral, New Mexico[i]	0.008	0.4	0.8	0.9	0.19	−0.19	—	0.41

238

C. Comparison of predicted versus empirical values for dependence of gradient on equivalent length

$$S \propto L^a$$

Area	Predicted values of a for indicated transport formula[a]													Av.[j]	Empirical value
	1	2	3	4	5	6	7	8	9	10	11	12	13		
Virginia	0.20	0.06	-0.27	-0.11	-0.29	-0.18	-0.15	-0.23	-0.29	-0.40	-0.18	-0.42	-0.13	-0.24	-0.25
Utah	0.20	0.06	-0.27	-0.11	-0.29	-0.18	-0.15	-0.23	-0.29	-0.40	-0.18	-0.42	-0.13	-0.24	-0.32
Great Plains	0.21	0.10	-0.18	-0.03	-0.19	-0.11	-0.07	-0.15	-0.20	-0.29	-0.10	-0.27	-0.05	-0.15	-0.34
New Mexico	0.38	0.24	-0.09	0.12	-0.09	0.01	0.06	-0.06	-0.12	-0.22	?	-0.11	0.12	-0.03	-0.25
New Mexico	0.21	0.10	-0.18	-0.03	-0.19	-0.11	-0.07	-0.15	-0.20	-0.29	-0.10	-0.27	-0.05	-0.15	-0.19

[a]References for sediment transport formulas: 1. Bagnold bedload (Yalin, 1977, p. 117–122); Yalin bedload (Yalin, 1977, p. 143–148); 2. Meyer-Peter bedload (Shen, 1971a, p. 11–22); 3. Einstein-Brown bedload (Henderson, 1966, p. 440–441); 4. Bagnold total load (Yalin, 1977, p. 143–148); 5. Shen total load (Shen, 1971b); 6. Laursen total load (Graf, 1971, p. 213–214); 7. Yang total load (Yang and Stall, 1976); 8. Maddock total load (Maddock, 1976); 9. Peterson (Parker and Anderson, 1977); 10. Bogardi (1974, p. 324); 11. Kalinske (Henderson, 1966, p. 141–142); 12. Engelund and Hansen (ASCE, 1975, p. 208–209); 13. Ackers and White (1973).

[b]Average of individual median values.

[c]Two-decimal values of r, e, and p are estimated from empirical data, while one-digit figures are assumed.

[d]Coefficient of determination.

[e]Unpublished data from Howard (in preparation, a).

[f]Data from Howard (1970).

[g]Data from Schumm (1960, Table 1).

[h]Data from Leopold and Miller (1956, Appendices A and B).

[i]Data from Renard and Laursen (1975).

[j]Omits formulas number 1 and 2 from averages.

239

similar. However, because of the irregular behavior of the Peterson relationships with grain size, the following generalities do not pertain to these relationships:

1. Although the Einstein-Brown relationship is supposedly a bedload formula, it gives results more similar to the various total load relationships than to the other bedload formulas (Meyer-Peter, Bagnold, and Yalin), both in terms of predicted required gradients and corresponding capacity grain size (Fig. 2) and in the patterns of response to parameter variation (Table 1). The higher required gradients and finer capacity grain sizes predicted by the other bedload formulas apparently occur because they do not take into account suspension of bed material at high flow rates and high sediment concentrations, conditions that occur for the finer grain sizes in the nominal parameter values [the threshold of suspension is indicated on Figure 2 using the criterion of equality of shear velocity and fall velocity (Middleton, 1976)]. The total load equations, apparently including the Einstein-Brown relationship, are more suitable to regime analysis than the bedload equations because appreciable bed material moves in suspension in natural fine-bed streams. This will be apparent in later discussion comparing the downstream hydraulic geometry of fine-grained streams with the predictions based upon the various transport equations (Table 1).

2. Increase in sediment concentration (equivalent also to increase in Q_s since Q and W are fixed) causes a corresponding increase in the required gradient at the capacity grain size, whereas, of course, the initiation of motion gradients are unaffected (Fig. 3A). The capacity grain size is little affected by change in concentration.

3. Increase in the unit discharge, q, with other input parameters constant (equivalent to an increase in Q with W fixed or a decrease in W with Q fixed) reduces the required gradient at the capacity grain size less than the threshold gradient. The capacity grain size remains nearly constant (Fig. 3B).

4. Keeping q_s constant while q increases or C decreases (Fig. 3C) affects both the required gradient at the capacity grain size and the threshold gradients by the same amount, while leaving the capacity grain size unaffected. However, other combinations of threshold criteria and bedload or total load formulas will give slight differences in the rate of change of threshold and fine-grain maximum gradients to variations in q (Tables 1 and 2).

5. Change in roughness, n, through the range of 0.01 to 0.03 (Fig. 3D) reduces the required gradient at the capacity grain size by a factor of 2 or less (although the Yang, and Ackers and White formulas indicate a similar *increase* through this same range). The Maddock formula has no variable resistance parameter. Threshold values of coarse grain sizes are unaffected because of the assumption of the Manning-Strickler relationship.

6. Increase in the variance of grain sizes in transport (with the total concentration constant) broadens the curve of the fine-grained maximum while simultaneously reducing the depth of the valley at intermediate grain sizes (Fig. 3F). Simultaneously, the size of the capacity bed material is increased, whereas the required gradient is relatively unaffected. However, the various formulas differ somewhat in the detail of these variations.

7. The behavior of the predicted values of the required gradient with change of average grain size shows the greatest discrepancy in the various formulas (Fig. 4). In general, however, several conclusions can be made: (1) the capacity grain size is coarser than the average grain size (dashed lines in Fig. 4 show the relationship between the *average* grain size and the required gradient of the fine-grained maximum, whereas the solid lines show

the dependency of the required gradient upon the capacity grain size, the latter presumably being the dominant size of the bed material); (2) the threshold values of the required gradient are unaffected; (3) the width and depth of the "valley" between the fine-grained maximum and the threshold curve decreases as the grain size increases; and (4) the capacity grain size increases as the average grain size increases. It is with respect to this last observation that the various formulas show the greatest differences. Four basic patterns occur: (a) the required gradient is essentially unaffected by the average grain size (the Meyer-Peter, Yalin, and Bagnold bedload formulas, not shown on Fig. 4); (b) the required gradient increases more rapidly for coarser average grain size (the Einstein-Brown relationship); (c) the rate of change of gradient is independent of the average grain size (the Bogardi relationship shown in Fig. 4, and also the Kalinske formula); and (d) the required gradient increases more rapidly for fine than for coarse grain sizes (the Ackers and White relationship on Fig. 4 as well as all remaining total load formulas tested).

These conclusions, based upon the sediment transport formulas listed in Tables 1 and 2, may be partially inaccurate if Bagnold (1977) is correct in suggesting that bedload transport relationships should include a relative depth term, d/R, and that initiation of motion should be specified in terms of a critical power, $(\rho RSV)_c$, rather than a critical shear stress.

Downstream Hydraulic Geometry

The reality of the existence of the division of natural alluvial channels into fine-bed and coarse-bed channels is confirmed by regional empirical data on downstream hydraulic geometry. The bedload transport equations discussed above can be translated into regime equations for channel gradient. For the live-bed regime of fine-bed channels, these equations can be summarized as follows:

$$S = K_1 \frac{q_s^\alpha f(d)}{q^\gamma} \quad \text{or} \quad S = K_1 \frac{Q_s^\alpha f(d) W^{\gamma - \alpha}}{Q^\gamma} \tag{4}$$

For a limited range of bed material size $f(d)$ can be replaced by $K_2 d^\beta$. For threshold of motion conditions and coarse grain sizes, the channel regime can be represented by

$$S = K_3 \frac{d^{\beta'}}{q^{\gamma'}} \quad \text{or} \quad S = K_3 \frac{d^{\beta'} W^{\gamma'}}{Q^{\gamma'}} \tag{5}$$

In a climatically and physiographically homogeneous area, statistical downstream hydraulic relationships can be used to express the interrelationships among Q, W, Q_s, and the drainage area, A:

$$Q = K_4 A^e \tag{6}$$

$$Q_s = K_5 A^p \tag{7}$$

$$W = K_6 Q^b \tag{8}$$

$$W = K_7 A^r \quad \text{(implying that } r = e \cdot b) \tag{9}$$

TABLE 2 Comparison of Downstream Gradient Changes of Coarse-Bed Streams with Initiation of Motion Criteria

A. Predicted relationships for initiation of motion formulas[a,b]

Formula	$S = K_1 d^{\beta''} R^{\epsilon}$			$S = K_2 d^{\beta} q^{\gamma'}$		
	K_1	β''	ϵ	K_2	β	γ'
1. Shields	0.078	1.00	−1.00	0.42	1.29	−0.86
2. Yang	0.048	1.33	−1.33	0.46	2.00	−1.33
3. Maddock	0.019	0.50	−0.75	0.22	0.90	−0.60
4. Baker and Ritter	0.18	1.85	−1.00	1.15	2.50	−0.86

B. Regional empirical data for coarse-bed streams using bankfull depth; numbered columns give average ratio of predicted gradient to observed gradient, using formulas listed above

Geographic area	K_1	β''	ϵ	$S_p/S_o{}^k$				Median grain size[c]	R^{2d}
				1	2	3	4		
Virginia and Maryland[e]	0.022	0.48	−0.64	0.82	0.21	0.61	0.15	0.050	0.17
Pennsylvania[f]	0.026	0.56	−1.33	0.71	0.19	0.49	0.13	0.050	0.60
Idaho[g]	0.015	0.15	−1.49	0.19	0.046	0.16	0.023	0.043	0.57
New Mexico[h]	0.104	0.96	−1.19	0.57	0.23	0.24	0.20	0.116	0.47
Various U.S. streams[i]	0.0083	0.30	−0.78	1.03	0.22	0.88	0.17	0.048	0.30

C. Regional empirical data for coarse-bed streams using estimated discharge, with comparison to predictions of initiation of motion formulas and equivalent length regressions

Geographic area	K_2	β'	γ'	R^2	S_p/S_o				$S = K_3 L^c d^\beta$			
					1	2	3	4	K_3	c	β	R^2
Virginia and Maryland[m]	0.039	0.42	−1.03	0.81	0.83	0.083	1.67	0.059	4255	−0.79	0.50	0.85
Pennsylvania[m]	0.067	0.47	−1.08	0.72	0.62	0.064	1.24	0.056	3376	−0.77	0.45	0.75
Idaho[j,l]	0.031	0.10	−0.89	0.37	0.21	0.021	0.44	0.009	116	−0.53	−0.01	0.38
New Mexico[m]	0.11	0.63	−0.48	0.39	1.13	0.36	1.20	0.22	41	−0.37	0.61	0.48
Various U.S. streams[l]	0.017	0.48	−0.36	0.20	1.53	0.15	3.17	0.10	0.051	0.08	0.63	0.12

[a]Grain size in meters, depth in meters, and q in m²/sec.
[b]References for initiation of motion formulas: 1. Shields (Henderson, 1966); 2. Yang (Yang and Stall, 1976); 3. Maddock (1976); 4. Baker and Ritter (1975).
[c]Average of individual median values.
[d]Coefficient of determination.
[e]Data from Hack (1957, Table 8).
[f]Data from Brush (1961, Appendix A).
[g]Data from Emmett (1975, Tables 9, 12, 13).
[h]Data from Miller (1958).
[i]Data from Williams (1978b).
[j]Values for K_2, β', and γ' become more similar to other areas listed if d_{90} is used instead of d_{50}. These become 0.29, 0.52, and −1.00, respectively.
[k]Ratios averaged logarithmically.
[l]Uses bankfull discharge.
[m]Uses the mean annual flood.

Defining the ratio of $A/W = L$, the "equivalent length" (Howard, 1970), the regime equations can be expressed in terms of either q or L using the foregoing relationships. For live-bed conditions:

$$S = K_8 L^a f(d) \quad \text{or} \quad S = K_9 q^b f(d) \tag{10}$$

where

$$a = \alpha \left(\frac{p - r}{1 - r} \right) - \gamma \left(\frac{e - r}{1 - r} \right) \tag{11}$$

$$b = \alpha \left(\frac{p - e \cdot b}{e - e \cdot b} \right) - \gamma \tag{12}$$

For threshold of motion conditions:

$$S = K_{10} L^c d^{\beta''} \quad \text{or} \quad S = K_{11} q^{-\gamma'} d^{\beta'} \tag{13}$$

where

$$c = -\gamma' \left(\frac{e - r}{1 - r} \right) \tag{14}$$

The regime of channels at the threshold of motion can also be related to the grain size and hydraulic radius (depth):

$$S = K_{11} d^{\beta'} R^\epsilon \tag{15}$$

These relationships are compared in Tables 1 and 2 with empirical hydraulic geometry relationships derived from various sets of areal hydraulic geometry data. Data for coarse-bed alluvial channel networks have been used to estimate by linear regression (using logarithmic transforms of the data) the exponents of equations 13 and 15. These have been compared with the predictions of four formulas for the initiation of motion in Table 2. A similar approach was used for fine-bed channels (median bed material size less than 10 mm), empirically estimating exponents for equation 10 and comparing them with 13 bed material transport formulas (Table 1). Both the basic data and the summary comparisons in Tables 1 and 2 indicate that natural alluvial stream systems fall into two distinct ranges of size of bed material. In addition, the gradients *within* each of the two types of channel systems are similarly related to downstream changes in equivalent length (or unit discharge or channel depth) and grain size, despite widely different locations and climatic regimes, whereas there is a considerable difference for both predicted and empirical gradient relationships *between* the two types of channels. Finally, at least some of the sediment transport formulas give reasonable predictions of the regimes of the two types of streams.

The fine-bed stream systems are quite similar in the equivalent length regression (Table 1), despite widely different environments. By contrast, the predicted values of the equivalent length and grain size exponents (Table 1A, C) are quite varied, both for different transport formulas and because of the different dependencies of W, Q, and Q_s on A in the various areas. The observed dependency of gradient upon grain size is smaller than most of the transport formulas would suggest, but the range of variation of grain size in the empirical data is quite small. The predicted values of the equivalent length exponent are very sensitive to changes in the values of r, e, and p in equation 11. Since the hydraulic geometry relationships of equations 6–9 are estimated with small amounts of data showing a high degree of scatter (or in some cases are merely given assumed values), the high degree of variability in the regional predictions for the equivalent length exponent should be expected. However, the similarity among the regional empirical gradient relationships and the high coefficients of determination (Table 1B) can leave little doubt that the fine-bed streams are regime channels.

Coarse-bed stream systems are also similar to each other in their downstream gradient relationships (Table 2B, C) both in the exponents and in the constants, K_j. The coefficients of determination for the regional data regressions are generally high except that data from streams selected from throughout the United States show less consistent relationships than do the homogeneous areas, as would be expected. The eastern United States data have the greatest degree of regularity. As was the case with fine-bed streams, the fitted hydraulic geometry exponents fall close to at least some of the values predicted by threshold of motion formulas (Table 2A). In addition, since the sediment load does not enter into the formulas, the threshold relationships also predict the magnitude of the multiplicative constants, K_j. The closeness of the predicted threshold gradients, S_p, to the observed values, S_o, can be compared by calculating the average value of S_p/S_o (Tables 2B, C). If the channels are at the threshold of motion at high discharges (specifically, the bankfull discharge or the mean annual flood), these ratios should average unity. They are close to unity for the Shields and Maddock criteria, but are less than unity for the Yang, and the Baker and Ritter formulas. Overall, the values tend to be somewhat less than unity, suggesting that material of median bed size should move before bankfull discharge. However, imbrication and packing affect movement of coarse-bed material, so that the most representative bed material size for threshold conditions is coarser than the median (and is often assumed to be d_{84}), which would bring S_p/S_o values closer to unity. The Idaho data for streams fed by meltwater and draining high mountains may be intermediate in behavior between coarse- and fine-bed streams; these data have the lowest values of S_p/S_o, suggesting that at high flows these streams may be under live-bed rather than threshold conditions.

The upstream (drainage basin) conditions that determine whether a particular alluvial stream will be in "live-bed" or "threshold" regime are illustrated in Figures 2-4. In particular, live-bed streams occur where sediment concentrations are high but the range of grain sizes supplied is narrow. Other factors encouraging live-bed channels are large specific discharge, narrow channels, and a fine average grain size of supplied alluvium. Such conditions are met in most badlands and in the semiarid Great Plains, where erosion of poorly consolidated sedimentary rocks supplies high sediment yields but little coarse detritus. On the other hand, coarse-bed threshold channels occur where sediment loads are relatively low, but where coarse grain sizes are well represented. Thus high mountain areas with resistant bedrock, steep slopes, and a predominant role of physical weathering are likely to form threshold alluvial channels.

DYNAMICS OF BEDROCK CHANNELS

Although the *gradient* of bedrock channels is not directly determined by the hydraulic regime, their *rate of erosion* does depend upon the flow of water and sediment over the bed. Thus bedrock channels have been described by Einstein (1964) as "erosional" and by Gilbert (1877, 1914) as "corrasional." However, strictly descriptive characterization as "bedrock" channels is preferred here, for alluvial channels may also slowly erode their beds (Howard, in preparation, a).

Erosion of bedrock channels requires both weathering and detachment, processes that often go hand in hand. The type of process that is responsible for bed erosion varies with rock type and hydraulic regime. Resistant granites, for example, may be primarily eroded by sediment abrasion. In this case the rate of erosion would vary with the flow velocity, as well as with grain size, density, hardness, and quantity of the sediment supplied from upstream. Soluble rocks such as limestone may be eroded primarily by chemical attack (particularly in subterranean channels) so that the rate of erosion would be dependent upon flow velocity as well as the chemical composition of the water. A similar situation may occur with some resistant rocks which must be weathered before grain detatchment can occur. In some cases, such as erosion of rills and gullies on soil or in shale badlands, weathering may be of secondary importance to detatchment, so that erosion rates are related either to the abrasion capacity of the bedload or simply to the exceedence of a critical shear stress on the channel bottom.

Under appropriate circumstances the erosion of bedrock channels can be quantitatively modeled. For example, Howard (in preparation, a) finds that erosion rates in badland bedrock channels are functionally related to the channel gradient and to the drainage area in a manner that is consistent with the assumption that erosion rates are proportional to the bed shear stress during high flows. In particular, the average erosion rate, E, increases in proportion to drainage area, A, and gradient, S, as follows:

$$E = KA^\phi S^\sigma \tag{16}$$

where ϕ and σ are constants and the factor K includes the effects of inherent bed erodibility and of the magnitude and frequency characteristics of the flow. If erosion rates were directly proportional to the bed shear stress, the parameters ϕ and σ would have the values 0.38 and 0.81, respectively, assuming also that bed resistance and channel shape (but not size) remain constant downstream (Howard, 1970; in preparation, a). Empirical measurements in badland topography estimate $\phi = 0.44$ and $\sigma = 0.68$ (Howard, in preparation, a), which is in reasonable agreement with the model.

Bedrock streams have no simple downstream hydraulic geometry, because bed and bank erodibility vary, and the gradient is a semiindependent variable. However, if the erosion rate is nearly uniform throughout a drainage network of bedrock channels on homogeneous rock (such as would occur during long-continued erosion in a high-relief area) and erosion rates follow the rate law above, then a consistent downstream hydraulic geometry results (Howard, 1970). In particular,

$$S \propto A^{-\phi/\sigma} \tag{17}$$

Factors Determining Occurrence of Bedrock Versus Alluvial Channels

The circumstances promoting bedrock instead of one of the two types of alluvial channels are complex, for the gradients of bedrock channels depends upon the past erosional history of the stream system. However, the interactions between alluvial and bedrock channels can be illustrated by numerically simulating possible erosional histories of a stream channel using equation 16, which is a partial differential equation, since the erosion rate, E, is $\partial y/\partial t$ and the gradient, S, is $\partial x/\partial t$, where y is the vertical and x the horizontal directions. The numerical simulations follow the erosion of a representative stream profile, constructed with the assumption that drainage area increases as the square of the distance downstream from the divide. The profiles shown in Figure 5 were evaluated downstream from the point at which the drainage area was unity to the downstream

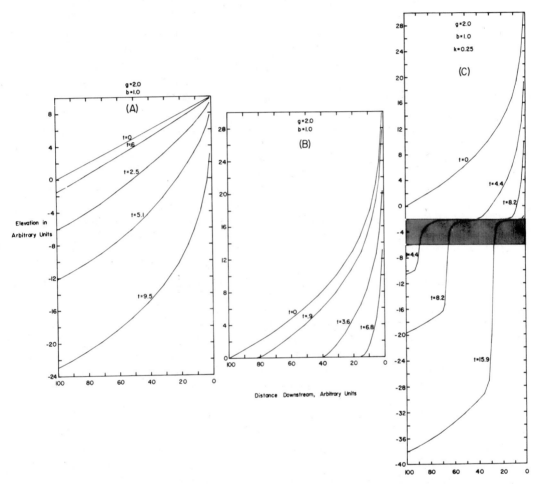

FIGURE 5 *Simulated erosion of bedrock channels under different assumed boundary conditions and bed erodibility. Bed erodibility, k, assumed to be unity except for the dark band in (C), which is four times less erodible. (A) Erosion with constant rate of downcutting of downstream terminus of stream (at position 100, time shown in arbitrary units); (B) erosion to a constant-elevation base level; and (C) erosion of a resistant layer. The value of b is the exponent in the equation $E = k\tau^b$, where τ is the bed shear, E the erosion rate, and k the erodibility. The value of g is the exponent in the downstream hydraulic geometry equation $R = CA_x^g$, where R is the hydraulic radius, A_x the cross-sectional area, and C a constant.*

end at x equal to 100 and a drainage area of 10,000. The boundary conditions are the initial stream profile and the vertical position of its downstream terminus through time. The erodibility (K in eq. 16) can be specified as a function of position (vertically and areally) and time. The numerical methods are outlined in Howard (1970). Several representative simulations were conducted:

Case 1.   Uniform rate of lowering of the downstream end of the channel (Fig. 5A). An initially straight profile was assumed and the erodibility was uniform. Near the end of the simulation, a concave profile develops in which the rate of erosion is constant through time and equal throughout the drainage basin, that is, in a steady-state or dynamic equilibrium (Hack, 1960). In steady-state erosion headwater, bedrock channels might change downstream into alluvial streams, and, under sufficiently low overall erosion rates, the channel system might be entirely alluvial. To illustrate these relationships, equations 4–9 and 16 can be used to express the relative gradients required for a bedrock channel, S_b, for the given erosion rate and the gradient required for an alluvial channel, S_a. For this comparison the following additional assumptions are made:

$$K_5 = E \quad \text{(from eqs. 6 and 16)} \tag{18}$$

$$p = 1 \quad \text{(eq. 6)} \tag{19}$$

$$d = K_{12}A^u \tag{20}$$

Here E is the steady-state erosion rate. The ratio of gradients, S_b/S_a, for a live-bed alluvial channel is

$$\frac{S_b}{S_a} = K_{13} \frac{E^{1/\sigma - \alpha}}{A^{\alpha(1-r) - \gamma(e-r) + \beta u + \phi/\sigma}} \tag{21}$$

Downstream grain size changes in fine-bed alluvial channels are generally slight, so βu is essentially zero. For reasonable values of the remaining parameters in the exponents in equation 21, both E and A are raised to positive powers. This suggests that bedrock streams will be favored for high erosion rates ($S_b/S_a > 1$), but that, because of the drainage area term, a downstream transition to a live-bed alluvial channel might occur if $S_b/S_a < 1$ (this assumes the ability of alluvial streams to erode the underlying bedrock up to a finite rate limit while keeping pace with overall basin erosion; such behavior is documented by Howard (in preparation, a) for badland channels in Virginia). Downstream transitions from bedrock rills and gullies to fine-bed alluvial channels are common in badlands (Smith, 1958; Schumm, 1956; howard, 1970; Howard, in preparation, a).

For threshold alluvial channels the equivalent gradient ratio is

$$\frac{S_b}{S_a} = K_{14} \frac{E^{1/\sigma}}{A^{\beta'u + \phi/\sigma - \gamma'(e-r)}} \tag{22}$$

In this case the exponent u is likely to have an appreciable negative value, as a result of downstream sorting and abrasion, so that the area exponent in equation 22 is likely to have a very small positive value, or, more likely, a negative value. Under these circumstances the stream system is likely to be either entirely coarse-bed alluvial or bedrock, depending upon the rate of overall erosion. A complicating factor is that the maximum grain size of bed material for threshold channels may increase with erosion rates, both due to increased rate of supply of all size ranges and due to production of coarser detritus on steeper slopes. This would increase the range of erosion rates over which threshold channels would be expected.

Conditions of essentially steady-state erosion are most closely approximated in areas with high relief or in the headwater areas of large drainage networks where base level influence is minimal (or, more precisely, is felt only over time scales longer than those considered in this paper).

Case 2. Base level constant (Fig. 5B). In this simulation the elevation of the downstream end of the profile was held constant. The initial profile was constructed to be steady-state for an arbitrary constant rate of base level lowering. Erosion to base level occurs in a finite time. More realistically, the lower parts of the stream network would become alluvial as the gradients dropped to the minimum necessary to transport the supplied sediment. The alluvial sections would gradually extend headward at the expense of the bedrock channels. If Schumm's (1963) speculation that orogeny is characterized by short periods of uplift followed by long periods of stability is correct, then during the periods of stability, alluvial channels should gradually extend headward at the expense of bedrock channels in a manner similar to this case.

Case 3. Effect of a resistant layer (Fig. 5C). Like case 1, base level was considered to be lowering at a uniform rate, initially with a steady-state profile in bedrock of uniform erodibility. However, the erodibility (K in eq. 16) was decreased in a horizontal layer, corresponding to a resistant layer. The rate of erosion in a bedrock stream encountering a resistant layer varies from place to place and through time. Through, and immediately below, the resistant layer stream gradients are higher than the steady-state case, whereas just above the resistant layer the gradient and rate of erosion are small. If the gradient becomes sufficiently small, a segment of alluvial stream may form, graded to the outcrop of the resistant layer. Howard (1970; in preparation, a) has outlined several cases in badland topography where alluvial stream segments have formed upstream from a control point in a resistant rock layer.

In summary, the simulations indicate that alluvial channel segments will develop if the base level remains constant or lowers very slowly. Likewise, alluvial channel will form upstream from resistant horizons. Although not illustrated in the simulations, alluvial channels will also occur where the original gradient of the stream channels on a constructional land surface is lower than the value required fo transport of sediment supplied from upstream.

CONCLUSIONS, APPLICATIONS, AND LIMITATIONS

Theoretical and empirical evidence has been presented for the occurrence of three distinct types of channel beds (bedrock, fine-bed alluvial, and coarse-bed alluvial) whose areal and temporal interactions are characterized by abrupt thresholds. In this concluding section some practical consequences of these thresholds to river behavior are detailed. In addition, a few more speculative implications are discussed, and the limitations in the application of the threshold concept are examined.

River Response

The response of rivers to a natural or man-induced change in the hydraulic regime will depend upon the preexisting nature of the bed. Under certain circumstances a small or gradual change in hydraulic regime can trigger a change to a different channel type. A few of many possible examples are discussed below; these examples concern primarily temporal thresholds, although some spatial (downstream) examples are considered.

Although the downstream hydraulic geometry of both alluvial and bedrock channels has been considered, one additional aspect remains for discussion. In going downstream in a drainage basin the unit discharge, q, increases while the sediment concentration remains nearly constant or decreases slightly. In addition, sorting and abrasion decrease coarser grain sizes more rapidly than finer, although Hack (1957) and Brush (1961) show that bed particle size of coarse-bed streams does not necessarily decrease downstream, but may actually increase, depending upon the geologic and physiographic situation. These factors tend to reduce required threshold gradients more rapidly than the live-bed gradients, so that downstream transitions from coarse- to fine-bed streams should be common. This may be illustrated by comparing the required gradient of a coarse-bed stream, S_c, to that of a fine-bed stream, S_f, using equations 4, 5, 21, and 22:

$$\frac{S_c}{S_f} = K_{15} \frac{d^{\beta' - \beta}}{q_s{}^{\alpha} q^{\gamma' - \gamma}} = K_{16} E^{-\alpha} A^{-\alpha(1 - r) - (\gamma' - \dot{\gamma})(e - r) + u\beta'} \tag{23}$$

In the second part of the equation, downstream changes in grain size of the fine material has been discounted. Since $\gamma' - \gamma$ is approximately zero (Tables 1 and 2), substitution of reasonable values of the other parameters suggests that coarse-bed channels may yield downstream to fine-bed channels, as observed by Yatsu (1955) and Shaw and Kellerhals (1977).

Equation 23 suggests that fine-bed channels would be favored by high erosion rates, so that changes in vegetation cover or runoff characteristics leading to higher sediment yields (e.g., by deforestation or construction activities) might convert coarse-bed channels to fine-bed channels by aggradation. However, in other cases where under natural conditions $S_c \gg S_f$, appreciable increases in sediment yield might not raise S_f to the point that aggradation is required. If increase in sediment yield occurs because of increasing relief within a drainage basin, then the indicated effect of erosion rates in equation 23 may be offset by increases in the critical grain size for coarse bed streams, owing both to the increased quantity of coarse load delivered and a probable increase in the average size of coarse detritus due to steeper slope gradients.

Changes in the hydraulic regime can also lead to conversion of an alluvial channel to a bedrock channel, or vice versa. Seasonal alternations between live-bed and bedrock channels in Virginia badlands resulting from higher winter sediment concentrations have been documented by Howard (in preparation, a).

An increase in channel width increases required threshold gradients more than fine-bed gradients. Thus, a local or systematic increase in channel width might trigger conversion to a coarse-bed channel. For example, the Colorado River in the Grand Canyon is a sand-bed channel where the width is constricted by talus or resistent bedrock, but cobble bars are common where less resistant rocks allow channel widening (Howard, in preparation, b). In streams where $S_f \approx S_c$, coarse and fine material may be areally sorted into riffles and pools, respectively. This may in part be due to width effects; Keller and Melhorn (1973) indicate that riffles are associated with divergent flow and pools with convergent flow.

Although much has been written on the response of rivers to changes in hydraulic regime (Schumm, 1969, 1971a; Santos-Cayudo and Simons, 1972; Simons and Senturk, 1976), the effects of channel type upon the pattern and rate of adjustment have been largely ignored. In particular, bedrock, fine-bed, and coarse-bed channels have considerably different characteristic time constants (or "relaxation times"; see Howard, 1965; Chorley and Kennedy, 1971; Allen, 1974) for response to changes in hydraulic regime. Generally, the slowest to respond are the bedrock channels, which in many cases (excepting badlands and soil gullying) can be considered to be noneroding over engineering time scales. Threshold channels may be very slow to respond to changes in regime because of the slow average rate of motion of the coarse bed pavement and the prominant roles of sorting and in situ weathering and abrasion in downstream changes in grain size (Schumm and Stevens, 1973). In fact, it is possible that some of the coarse cobble pavements of high mountain streams originated during periods of accentuated physical weathering and mass movement during the late Pleistocene or to later erosion of this coarse debris (Brush, 1961; Miller, 1958). Live-bed channels respond much more rapidly to change in regime, with response times measured in months, years, or decades, depending upon the size of the drainage network (time scales for aggradation and entrenchment are approximately related to the square of the length of the alluvial channel; Gessler, 1971).

The difference in characteristic time scales between threshold and live-bed channels can result in short-term adjustments to changed hydraulic regime which are different than the ultimate response (i.e., the response can be divided into primary and secondary responses; Howard, 1965). For example, an increase in discharge, a decrease in sediment load, or a man-made increase in gradient due to meander cutoff should require a decrease in gradient of fine-bed alluvial channels by degradation. However, if the channel transports a small percentage of coarse detritus, the scouring of the underlying alluvium may produce a gravel pavement (threshold channel), owing to the slower transport rates of the coarse fraction. Such transitions are frequent below dams, where in addition to vastly reduced sediment supply, the competency of the flow is reduced by clipping of flood peaks. In such circumstances the amount of scouring required to develop a pavement can be calculated from the grain size distribution of the alluvium (Gessler, 1971; Pemberton, 1976; Simons and Senturk, 1976).

The inverse situation may occur if a gradient increase is required through aggradation. If a threshold alluvial channel receives a greater influx of sediment without change in grain size distribution (resulting from, say, deforestation or upstream urbanization), the long-term equilibrium response might be the development of a steeper coarse-bed channel due to increase in the critical grain size. However, if the increased sediment load also increases the required gradient for the fine-bed material above the existing channel gradient, the immediate response may be a temporary conversion of the channel to a fine-bed. An even more dramatic transition will occur if a rise in local base level (such as by construction of a dam) requires upstream alluviation of a threshold channel (assuming the hydraulic regime is unchanged). Because of the slow delivery of the coarse fraction forming the bed, the initial alluviation will occur by deposition of the fine sediment load at a gradient smaller than the ultimate threshold equilibrium. For practical purposes the long-term equilibrium may not be established during the lifetime of the dam.

Deficiencies of Intermediate Bed Particle Sizes

Natural alluvial streams normally carry less than a capacity load of the grain sizes finer than those represented on the bed (i.e., the wash load), although the grain size dividing the suspended and wash load varies with the discharge (Partheniades, 1977). It appears to be universally accepted that the grain sizes represented on the bed are carried at capacity (e.g., Partheniades, 1977; Shen, 1971a; Simons and Senturk, 1976). However, in fine-bed streams grain sizes coarser than the median bed material (the capacity grain size) are also carried in less-than-capacity amounts, although they are represented on the bed because they move by saltation or rolling. This is because the required gradient for transport of this coarse material is less than the actual gradient, which is determined by the capacity grain size (Fig. 2).

This phenomenon may help to explain the commonly observed deficiency of grain sizes in the range of about 1–10 mm in fluvial sediments (Church and Gilbert, 1975; Slatt and Hoskins, 1968; Emmett, 1976; Williams, 1978a; Yatsu, 1955). This grain size range corresponds to the valley in the required gradient curves in Figure 2. Material in this size range, being carried below capacity as bedload, will only be deposited at very low flows; thus, fluvial sediments should be relatively depleted in this size range. By contrast, the still coarser grain sizes near the critical value move much more slowly on the average, since they are moved only during high discharges, so that they are well represented on the bed. Assuming spherical particles, the percentage of the bed area, P, covered by particles of grain size d supplied in quantity q_{sd} and traveling with an average velocity V_p (including periods of no motion) is given by

$$P = 100 \times \frac{6}{4} \frac{q_{sd}}{dV_p} \qquad (24)$$

where q_{sd} is measured in units of volume of solids per unit time and channel width. Thus, the strong representation of fine bed material near the capacity grain size is due to high q_{sd} and small d, despite relatively high V_p. On the other hand, the grain sizes near the critical grain size have nearly zero V_p, giving a high P despite large d and small q_{sd} The deficient grain size range has high V_p compared to the supply rate.

The usual alternative explanations for the deficiency of intermediate bed material grain sizes has been either that the original weathering processes produce little detritus in this size range or that abrasion and breaking of grains is rapid for these coarser grains but is less so for sands, which have smaller momentum relative to the fluid drag. Another possible explanation for the deficient grain sizes are the techniques used for size measurement of bed material: seiving for finer grain sizes and transect measurements for pebbles and cobbles. The intermediate grain sizes may possibly be underestimated in abundance, since they fall between techniques. Distinguishing between these possible explanations should be possible by examining the size range of all material deposited by rivers in estuaries, lakes, and oceans. If the bimodal distribution of fluvial sediments is due to the sorting process described above, the missing intermediate grain sizes should be found in the still-water environments in their original proportions, whereas if it is due to source deficiencies or differential abrasion, the paucity should also occur in the downstream sediment sink. In addition, if differential sorting is the mechanism producing the bimodal distribution, the grain size range of the deficient sizes will vary from environment to environment in conjunction with differences in median grain size and variance of the detritus supplied by slope erosion (see Figs. 3 and 4). This may help to account for the apparent lack of bimodal grain size distributions in aggregate averages of grain size distributions from diverse fluvial environments (Shea, 1974).

Miscellaneous Aspects of Alluvial Channel Thresholds

Few "live-bed" alluvial streams (with gradients determined by the capacity grain size) have beds coarser than a few millimeters. The main reason for this is the large quantity of bed material that must be supplied to maintain live-bed conditions for coarse bed sizes. In the Einstein-Brown transport relationship two channels will have equivalent bed dynamics if their values of the parameter Φ are equal:

$$\Phi = \frac{q_s}{wd} \cong \frac{q_s}{Kd^{3/2}} \quad \text{(for coarse-bed material)}$$

Thus, live-bed (large Φ) conditions for coarse-bed sizes requires larger volume of suppled load than for fine-bed sizes, a situation that seems to occur only in proglacial streams (e.g., Church and Gilbert, 1975; Fahnestock, 1963) or in streams draining alpine areas characterized by meltwater flood peaks and abundant coarse debris produced by physical weathering (Emmett, 1976). Such areas are also characterized by a high variance in the distribution of grain sizes in the bed material load. These streams are likely to have a correspondingly large range in the size of bed material, because the required gradient varies very slowly with grain size for large variance (Fig. 3F).

The sharp transitions between bed types may in part be due to the difference in ease of transport of bed material depending upon the type of underlying surface. In eolian transport fine saltating particles rebound much higher when moving over a fixed surface or over coarse grains (Bagnold, 1941; Ellwood et al., 1975). As a result, Bagnold noted that the capacity of the wind to transport saltating grains diminishes where the sand passes onto a loose sand surface (the saltation height is included in theoretical bedload transport relationships of Bagnold, 1973; 1977); as a result, downwind transitions from fixed or coarse-grained beds to fine sand are abrupt. A similar mechanism may help to account for abrupt downstream transitions between bedrock and sand, bedrock and gravel, or gravel and sand beds.

Coarse-bed alluvial channels may occur in two intergrading types. Along downstream portions of channels in areas of homogeneous high relief most of the gravel forming the bed has been transported from headwater areas and has been subject to systematic sorting and abrasion. However, in places (particularly in headwater tributaries), the bed may be steepened by very local additions of coarse debris from adjacent slopes which is not moved downstream except after considerable in situ weathering and abrasion. The smallest tributaries in areas dominated by physical weathering (e.g., in mountains of the Mojave desert) are often floored by coarse detritus only a few degrees less steep than its angle of repose. The distinction between local and basin-wide source for coarse-bed material is very marked along the Colorado River in the Grand Canyon, where riffles composed of rounded gravel in transport along the river contrast sharply with steep rapids in angular boulders added locally at the entrance of small, steep tributaries (Howard, in preparation, b).

Uncertainties and Limitations

Although the present analysis seems to *explain* why thresholds occur between channel types, it offers little aid in making *a priori predictions* of which type of bed would occur in a particular segment of a particular stream except in cases where the stream is clearly not operating close to any of the thresholds between channel types. Prediction of alluvial channel gradients and bed type requires knowledge of the quantity and size distribution of the sediment supplied from slope erosion; since the details of sediment supply cannot be easily predicted from morphological characteristics of the drainage basin, intrastream measurements are required. The capacity grain size and required gradient for fine-bed streams should be calculable to reasonable accuracy from long-term, systematic measurements of bed and suspended loads. However, the critical grain size for coarse-bed streams (and the corresponding required gradient) depends upon the relatively small rate of supply of coarse grain sizes and upon the frequency of flood peaks. Sampling of coarse bedload is notoriously difficult. The problem is further complicated by the variable porosity and fabric of coarse beds, as well as by the action of sorting and abrasion. Finally, because of the long time constant of coarse-bed streams, it is possible that the channel may be disequilibrium with the present hydraulic regime. The occurrence of bedrock channels depends upon past erosional history and resistance of the bedrock to erosion, factors difficult to quantity. However, the measurement of the channel gradient will tell whether it is steep enough that an alluvial bed does not form, except for gradients near the threshold to an alluvial channel.

Maddock (1969) feels that resistance due to bedforms is an independent method by which streams may adjust to changes in hydraulic regime. The resistance relationship used in this paper (the Manning equation) is certainly an oversimplification of the actual dependence of resistance on the hydraulic regime, particularly for fine-bed alluvial streams. The effect of the development of bedforms upon the capacity grain size and the required gradient needs to be further investigated. Actual curves of required gradient may more resemble the complex curve of the Peterson relationship than the simpler peaks in the fine-grain range indicated by the other formulas (Fig. 2), and it is possible that additional thresholds due to change in bed resistance may occur within fine-bed alluval channels. The situation is further complicated by the dependence of bedforms upon river stage and the disequilibrium that may occur between bedforms and the hydraulic regime due to the finite relaxation time of bedforms (Allen, 1974).

The main variable of hydraulic geometry considered in this paper has been the gradient. However, other variables of hydraulic geometry probably also exhibit different behavior, depending upon the type of channel bed. In particular, this should be true of channel width and sinuosity, which up to this point in the paper have been considered to be independent variables. Bedrock channels are commonly narrow and straight, because bedrock also forms the channel banks. Fine-bed channels are quite variable in width and sinuosity, depending upon the quantity of suspended and wash load and the amount of vegetation (Schumm, 1960). The extent of variation of width, sinuosity, and other hydraulic geometry variables (except gradient) with bed type is uncertain, yet it is clear that hydraulic geometry relationships would be more consistent if the data were stratified according to channel type. For example, Li et al. (1976) have derived essentially complete equations of hydraulic geometry for networks of nonbraided coarse-bed channels with banks of the same composition as the bed which appear to fit the Pennsylvania streams measured by Brush (1961).

In conclusion, the occurrence of three distinct types of channel beds (bedrock, as well as fine- and coarse-bed alluvial) is the result of thresholds in sediment transport mechanics. Because temporal or spatial variations in the controlling hydraulic regime often trigger change from one channel type to another, each with a distinct hydraulic geometry and pattern of river response, the threshold behavior should be incorporated into future modeling of fluvial mechanics. Further study of threshold behavior is desirable, including field studies, flume experimentation, and theory.

ACKNOWLEDGMENTS

The author greatly appreciates the comments of Robert Dolan, who read an earlier draft of the paper.

REFERENCES CITED

Ackers, P., and White, W. R., 1973, Sediment transport: new approach and anaylsis: J. Hydraulics Div., Amer. Soc. Civil Engineers, v. 99, p. 2041–2060.

Allen, J. R. L., 1974, Reaction, relaxation, and lag in natural sedimentary systems: general principles, examples, and lessons: Earth Sci. Rev., v. 10, p. 263–342.

American Society of Civil Engineers, 1975, Sedimentation engineering: V. A. Vanoni, ed., New York, 745 p.

Baker, V. R., and Ritter, D. F., 1975, Competence of rivers to transport coarse bedload material: Geol. Soc. Amer. Bull., v. 86, p. 975–978.

Bagnold, R. A., 1941, The physics of blown sand and desert dunes: London, Methuen, 264 p.

——, 1973, The nature of saltation and bed-load transport in water: Proc. Roy. Soc., ser. A, v. 332, p. 473–504.

——, 1977, Bed load transport by natural rivers: Water Resources Res., v. 13, p. 303–312.

Bogardi, J., 1974, Sediment transport in alluvial streams: Budapest, Akadémiai Kiadó, 826 p.

Brush, L. M., Jr., 1961, Drainage basins, channels, and flow characteristics of selected streams in central Pennsylvania: U.S. Geol. Survey Prof. Paper 282-F, 145–181.

Burkham, D. E. 1972, Channel changes of the Gila River in Safford Valley, Arizona, 1846–1970: U.S. Geol. Survey Prof. Paper 655-G, 23 p.

Chorley, R. J., and Kennedy, B. A., 1971, Physical geography: a systems approach: London, Prentice-Hall, 370 p.

Church, M., and Gilbert, R., 1975, Postglacial fluvial and lucustrine environments: *in* Glaciofluvial and glaciolacustrine sedimentation: Soc. Econ. Paleontol. Mineral., Spec. Publ. 23, p. 22–100.

Einstein, H. A., 1950, The bed-load function for sediment transportation in open channel flows: U.S. Dept. Agric., Soil Cons. Serv., Tech. Bull. 1026, 78 p.

——, 1964, Sedimentation: Part II. River Sedimentation: *in* Chow, V. T., ed., Handbook of applied hydrology: New York, McGraw-Hill, p. 17-35–17-67.

Ellwood, J. M., Evans, P. D., and Wilson, I. G., 1975, Small scale eolian bedforms: J. Sed. Petrol., v. 45, p. 554–561.

Emmett, W. W., 1975, The channels and waters of the Upper Salmon River area, Idaho: U.S. Geol. Survey Prof. Paper 870-A, 116 p.

———, 1976, Bedload transport in two large, gravel-bed rivers, Idaho and Washington: Proc. Third Fed. Inter-Agency Sedimentation Conf., p. 4-101 to 4-114.

Fahnestock, R. K., 1963, Morphology and hydrology of a glacial stream—White River, Mount Ranier, Washington: U.S. Geol. Survey Prof. Paper 422-A, 70 p.

Foley, M. G., 1976, Scour and fill in an ephemeral stream: Proc. Third Fed. Inter-Agency Sedimentation Conf., p. 5-1 to 5-12.

Gessler, J., 1971, Aggradation and degradation: in Shen, H. W., ed., River mechanics: Fort Collins, Colo., Water Res. Publ., p. 8-I to 8-24.

Gilbert, G. K., 1877, Report on the geology of the Henry Mountains: U.S. Geol. Survey, Rocky Mtn. Region Rept., 160 p.

———, 1914, The transportation of debris by running water: U.S. Geol. Survey Prof. Paper 86, 263 p.

Graf, W. H., 1971, Hydraulics of sediment transport: New York, McGraw-Hill, 513 p.

Hack, J. T., 1957, Studies of longitudinal stream profiles in Virginia and Maryland: U.S. Geol. Survey Prof. Paper 294-B, p. 45–97.

———, 1960, Interpretation of erosional topography in humid-temperate regions: Amer. J. Sci., v. 258-A, p. 80–97.

Hammad, H. Y., 1972, River bed degradation after closure of dams: J. Hydraulics Div., Amer. Soc. Civil Engineers, v. 98, p. 591–607.

Henderson, F. M., 1966, Open channel flow: New York, Macmillan, 522 p.

Howard, A. D., 1965, Geomorphological systems—equilibrium and dynamics: Amer. J. Sci., v. 263, p. 302–312.

———, 1970, A study of process and history in desert landforms near the Henry Mountains, Utah: unpubl. Ph.D. dissertation, Johns Hopkins University, Baltimore, Md., 198 p.

———, in preparation, a, Channel dynamics in badlands.

———, in preparation, b, Geomorphology and sedimentology of the Colorado River in the Grand Canyon.

Keller, E. A., and Mehorn, W. N., 1973, Bedforms and fluvial processes in alluvial stream channels: selected observations, in Morisawa, M., ed., Fluvial geomorphology: Publ. Geomorphol. SUNY Binghamton, N.Y., p. 253–283.

Kellerhals, R., 1967, Stable channels with gravel-paved beds: J. Waterways Harbors Div., Amer. Soc. Civil Engineers, v. 93, p. 63–84.

Krumbein, W. C., and Graybill, F. A., 1965, An introduction to statistical methods in geology: New York, McGraw-Hill, 475 p.

Leopold, L. B., and Maddock, T., Jr., 1953, The hydraulic geometry of stream channels and some physiographic implications: U.S. Geol. Survey Prof. Paper 252, 57 p.

———, and Miller, J. P., 1956, Ephemeral streams—hydraulic factors and their relation to the drainage net: U.S. Geol. Survey Prof. Paper 282-A, 37 p.

Li, R., Simons, D. B., and Stevens, M. A., 1976, Morphology of cobble streams in small watersheds: J. Hydraulics Div., Amer. Soc. Civil Engineers, v. 102, p. 1101–1117.

Makin, J. H., 1948, Concept of the graded river: Geol. Soc. Amer. Bull., v. 59, p. 463–512.

Maddock, T., Jr., 1969, The behavior of straight open channels with movable beds: U.S. Geol. Survey Prof. Paper 622-A, 70 p.

———, 1973, A role of sediment transport in alluvial channels: J. Hydraulics Div., Amer. Soc. Civil Engineers, v. 99, p. 1915–1931.

———, 1976, Equations for resistance to flow and sediment transport in alluvial channels: Water Resources Res., v. 12, p. 11–21.

Middleton, G. V., 1976, Hydraulic interpretation of sand size distributions: J. Geol., v. 84, p. 405–426.

Miller, J. P., 1958, High mountain streams—effects of geology on channel characteristics and bed material: New Mexico Bur. Mines Mineral Resources, Memoir 4, 53 p.

Parker, G., and Anderson, A. G., 1977, Basic principles of river hydraulics: J. Hydraulics Div., Amer. Soc. Civil Engineers, v. 103, p. 1077–1087.

Partheniades, E., 1977, Unified view of wash load and bed material load: J. Hydraulics Div., Amer. Soc. Civil Engineers, v. 103, p. 1037–1057.

Pemberton, E. L., 1976, Channel changes in the Colorado River below Glen Canyon Dam: Proc Third Fed. Inter-Agency Sedimentation Conf., p. 5-61 to 5-73.

Renard, K. G., and Laursen, E. M., 1975, Dynamic behavior model of ephemeral stream: J. Hydraulics Div., Amer. Soc. Civil Engineers, v. 101, p. 511–528.

Santos-Cayudo, J., and Simons, D. B., 1972, River response: in Shen, H. W., ed., Environmental impact on rivers, Fort Collins, Colo., Water Resources Publ., p. 1-1 to 1-25.

Schumm, S. A., 1956, The role of creep and rainwash on the retreat of badland slopes: Amer. J. Sci., v. 254, p. 693–706.

———, 1960, The shape of alluvial channels in relation to sediment type: U.S. Geol. Survey Prof. Paper 352-B, 30 p.

———, 1963, The disparity between present rates of denudation and orogeny: U.S. Geol. Survey Prof. Paper 454-H, 13 p.

———, 1969, River metamorphosis: J. Hydraulics Div., Amer. Soc. Civil Engineers, v. 95, p. 255–272.

———, 1971a, Fluvial geomorphology: channel adjustment and river metamorphosis: in Shen, H. W. River mechanics: Fort Collins, Colo., Water Resources Publ., p. 5-1 to 5-22.

———, 1971b, Fluvial geomorphology: the historical perspective: in Shen, H. W., ed., River mechanics: Fort Collins, Colo., Water Resources Publ., p. 4-1 to 4-30.

———, 1974, Geomorphic thresholds and complex response of drainage systems: in Morisawa, M., ed., Fluvial geomorphology: Publ. Geomorphol., SUNY, Binghamton, N.Y., p. 299–310.

———, and Beathard, R. M., 1976, Geomorphic thresholds: an approach to river management: in Rivers '76: New York, Amer. Soc. Civil Engineers, p. 707–724.

———, and Lichty, R. W., 1963, Channel widening and flood-plain construction along Cimarron River in southwestern Kansas: U.S. Geol. Survey Prof. Paper 352-D, p. 71–88.

———, 1965, Time, space, and causality in geomorphology: Amer. J. Sci., v. 263, p. 110–119.

———, and Stevens, M. A., 1973, Abrasion in place: a mechanism for rounding and size reduction of coarse sediments in rivers: Geology, v. 1, p. 37–40.

Shaw, J., and Kellerhals, R., 1977, Downstream grain size changes in Albertan rivers (Abst.): First Int. Symp. Fluvial Sedimentology, Calgary, Alberta, Canada.

Shea, J. H., 1974, Deficiencies of clastic particles of certain sizes: J. Sed. Petrol., v. 44, p. 985–1003.

Shen, H. W., 1971a, Wash load and bed load: *in*, Shen, H. W., ed., River mechanics: Fort Collins, Colo., Water Resources Publ., p. 11-1 to 11-30.

——, 1971b, Total sediment load: *in* Shen, H. W., ed., River mechanics: Fort Collins, Colo., Water Resources Publ., p. 13-1 to 11-30.

Silverston, E., and Laursen, E. M., 1976, Patterns of scour and fill in pool-rapid rivers: Proc. Third Fed. Inter-Agency Sedimentation Conf., p. 5-125 to 5-136.

Simons, D. B., and Senturk, F., 1976, Sediment transport technology: Fort Collins, Colo., Water Resources Publ., 807 p.

——, D. B., Richardson, E. V., and Mahmood, K., 1975, One-dimensional modeling of alluvial rivers: *in* Mahmood, K., and Yevjevich, V., eds., Unsteady flow in open channels: Fort Collins, Colo., Water Resources Publ., p. 813–877.

Slatt, R. M., and Hoskins, C. M., 1968, Water and sediment transport in the Norris Glacier outwash area, upper Taku Inlet, southeastern Alaska: J. Sed. Petrol., v. 38, p. 434–456.

Smith, K. G., 1958, Erosional processes and landforms in Badlands National Monument, South Dakota: Geol. Soc. Amer. Bull., v. 69, p. 975–1008.

Smith, W. O., Vetter, C. P., Cummings, G. B., et al., 1960, Comprehensive survey of sedimentation in Lake Mead, 1948–49: U.S. Geol. Survey Prof. Paper 295, 248 p.

Spencer, D. W., 1963, The interpretation of grain size distribution curves of clastic sediments: J. Sed. Petrol., v. 33. p. 180–190.

Visher, G. S., 1969, Grain size distributions and depositional processes: J. Sed. Petrol., v. 39, p. 1074–1106.

Williams, G. P., 1978a, Hydraulic geometry of river cross sections–theory of minimum variance: U.S. Geol. Survey Prof. Paper 1029.

——, 1978b, Bankfull discharge of rivers: Water Resources Res., v. 14, p. 1141–1154.

Yalin, M. S., 1977, Mechanics of sediment transport: 2nd ed., Oxford, Pergamon, 298 p.

Yang, C. T., and Stall, J. B., 1976, Applicability of unit stream power equation: J. Hydraulics Div., Amer. Soc. Civil Engineers, v. 102, p. 559–568.

Yatsu, E., 1955, On the longitudinal profile of the graded river: Trans. Amer. Geophys. Union, v. 36, p. 655–663.

12

GEOMORPHIC THRESHOLDS AS DEFINED BY RATIOS

William B. Bull

INTRODUCTION

During the last hundred years, particularly since about 1950, many insights about the operation of geomorphic open systems have been gained by those geomorphologists who have used conceptual frameworks centered about the idea that processes and landforms interact in a manner that tends toward a state of equilibrium between all the variables involved. This concept has been described by words such as "equality of action," "graded stream," "steady state," "quasi-equilibrium," and "dynamic equilibrium." Geomorphic equilibrium occurs when self-regulating feedback mechanisms cause an adjustment among the variables of part of a system, such that changes in landscape morphology do not occur with time. Once this adjustment occurs, the landscape under consideration has a time-independent configuration in that statistically similar landform assemblages are maintained as long as changes in the independent variables do not occur.

Some workers prefer to try to understand the mechanics of change (Bull, 1975) rather than the mechanics of equilibrium, because equilibrium conditions are unlikely to be maintained for long periods of time. Changes in uplift rates, climate, erodibility of surficial materials, and human impacts produce changes of landscapes instead of equilibrium. Long time lags of response and adjustment for hillslope subsystems to such changes result in long time spans for adjustment of stream subsystems. Most fluvial systems now are responding to several changes in independent variables, each with its own time lag needed to approach a new equilibrium condition. Other landforms, such as deposits and topographic inversions, do not even tend toward equilibrium configurations (Bull, 1976).

Adjustments with fluvial systems are further complicated by thresholds and feedback mechanisms that produce complex responses (Schumm, 1973) of fluvial systems to changes in either independent or dependent variables. A geomorphic threshold is a point or period of time that separates different modes of operation within part of a landscape system. The threshold conceptual framework includes within it the concept of equilibrium, because equilibrium occurs during time periods of no net change in landforms. Points in time that separate reversals in modes of operation are thresholds, but are not equilibrium situations unless an adjustment to a time-independent landform assemblage has occurred. An example would be where a stream reverses from an aggradational to a degradational mode of operation in response to a perturbation. Although thresholds generally can not be forecast a priori, the passage of a threshold is readily identifiable in the field.

Several advantages accrue from the use of the thresholds approach. Thresholds can be used in studies involving investigations that range from minutes to millions of years, and for spaces of equally great contrast. The threshold approach tends to focus attention on those variables and complex responses that are likely to cause the mode of system operation to change. Study of self-enhancing feedback mechanisms generally is encouraged by a threshold approach, whereas study of self-regulating feedback mechanisms generally is encouraged by an equilibrium approach. Thus, the threshold conceptual framework seems to be particularly well suited for studies involving the interaction of humans with their environment.

THE USE OF RATIOS TO DEFINE THRESHOLDS

A threshold may be regarded as a balance between opposing tendencies. Ratios are a convenient way of describing thresholds if the numerator and denominator describe the opposing tendencies. The numerator consists of those variables that if increased favor a change to a new mode of operation. The denominator consists of those variables that if increased resist the tendency for change, or favor a change in process that is different than that described by the numerator. The part of the system under consideration may be considered to be at a threshold or equilibrium condition when the ratio is equal to 1.0. The advantages of using such a format for defining thresholds are several. The components of the threshold are identified and compared to each other. The numerical index defines the relative condition that must be met in order to cross the threshold and change the mode of system operation. Ratios provide an allometric approach to defining thresholds, because the relative importances of two or more aspects of the system are used to define the threshold. The variables included in the ratio may be simple or complex, depending upon the degree of complication of the subject being studied, or on the desire of the geomorphologist to simplify the complexity of the real world.

A threshold that describes changes in dominant geomorphic processes is the hillslope runoff threshold.

$$\frac{\text{factors that promote runoff}}{\text{factors that promote infiltration}} = 1.0 \tag{1}$$

Consider the situation of rain falling on an outcrop. Two factors that if increased would tend to promote runoff include rainfall intensity and outcrop steepness. Factors that if increased would tend to cause infiltration include variables such as outcrop permeability, outcrop roughness, and vegetation density. The hillslope runoff threshold separates different dominant processes, but no change in landscape morphology generally is involved during the time of a single rainfall nor is the attainment of equilibrium likely.

A single component numerator and denominator may be defined to describe the threshold of movement of a boulder on a stream bed.

$$\frac{\text{fluid shear stresses tending to move boulder}}{\substack{\text{bed material shear strength tending to prevent} \\ \text{movement of boulder}}} = 1.0 \tag{2}$$

The threshold as described by equation 2 may be regarded as an attempt to simplify natural complexities. For this threshold both a change in dominant process and a change of stream-bed morphology are involved if the threshold is passed. Again equilibrium conditions are highly unlikely.

A more complex and versatile threshold involves the availability of stream power to do work. Bedload transport rate is a function of stream power, $\gamma\,QS$, where γ is a specific weight of water, Q is stream discharge, and S is slope (Bagnold, 1973, 1977). It is convenient to consider the total power supply per unit area of stream bed, ω, where

$$\omega = \frac{\gamma QS}{w} = \gamma dSu = \tau u \tag{3}$$

where w, d, and u are the mean flow width, depth, and velocity, respectively, and τ is the mean boundary shear stress.

The availability of stream power to transport bedload is important in determining whether a reach of the stream is aggrading, degrading, or at grade (equilibrium). Thus, the important threshold separating the modes of erosion and deposition in streams may be defined in terms of Bagnold's equation for stream power. Bull (1979) calls this the threshold of critical power in streams.

$$\frac{\text{stream power}}{\text{critical power}} = 1.0 \tag{4}$$

The available stream power was selected as the numerator because sediment transport is highly sensitive to changes in discharge and slope of stream flow. Critical power is the stream power needed to transport the sediment load supplied to a reach of a stream.

Increases in stream power tend to promote sediment transport, and increases in critical power (sediment load and size, and hydraulic roughness) tend to promote deposition. The term "critical power" is a shorthand expression (through the continuity equation, $Q = wdv$) for variables such as width, depth, and velocity that affect hydraulic roughness and channel morphology. All these variables interact to determine the capacity and competence of the stream to transport sediment.

Critical power includes hydraulic roughness, and, like the useful concept of hydraulic roughness, it cannot be measured directly in the field. Despite this apparent drawback, the ratio definition of the threshold is substantially more versatile than erosion-deposition thresholds stated merely in terms of available channel slope. The critical-power threshold involves changes in both dominant process and landscape morphology. Periods of equilibrium may coincide with this threshold, and occur during time periods when stream power is equal to critical power.

The threshold of critical power in streams occupies a key position in the complex interactions between the hillslope and stream subsystems, and it is affected by feedback mechanisms and complex responses operating in either subsystem. Evidence for active downcutting by a stream and lack of net alluviation show that the threshold is being exceeded. Reaches that are close to the critical-power threshold have alluvium in amounts that exceed that scoured by large discharges, and have floodplains that are narrower than the valley-floor widths. Parallel fill terraces suggest fluctuating hillslope conditions and periodic return to similar threshold conditions along a given reach of a stream. Self-enhancing feedback mechanisms may tend to keep a reach of a stream on either the erosional or depositional side of the threshold (Bull, 1979). Recognition of how far removed a reach of a stream is from the critical-power threshold should aid human-beings in their attempts to manage hillslope and stream subsystems.

The factors that determine whether a snout of a debris flow continues to move or comes to a halt have been described by Hooke (1967). He states (p. 453) that "because mud has a finite yield strength, debris flows stop when the shear stress on the bed no longer exceeds the yield strength of the mud, τ_o, or $\tau_o = \rho g d S / \tau_c$, where g is the gravitational acceleration and ρ, d, and S are the density, depth, and hydraulic gradient of the flow, respectively." Hooke's equation expressed as a ratio-type threshold is

$$\frac{\rho g d S}{\tau_c} = 1.0 \tag{5}$$

The yield strength of the mud becomes greater with increase in debris load, loss of water, or both. Temporary stagnation of the snout or margins of a debris flow may occur repeatedly during a given flow event as surges of mud are produced by the source area. When this threshold is passed, both the form and dominant process affecting the mass of mud and rock fragments change. Equilibrium conditions are highly unlikely.

Definition of thresholds by comparison of driving and resisting forces has been used for the exceedingly complex systems involved in landslides (Leopold et al., 1964, p. 340).

$$\frac{\text{factors promoting landslide movement}}{\text{factors resisting landslide movement}} = 1.0 \tag{6}$$

A host of variables may be included in the numerator and denominator, depending on the local situation. Some of the variables in the numerator might include precipitation, hillslope steepness, clay mineralogy, shearing and jointing, removal of material from the footslope, infiltration capacity of surficial materials, groundwater seepage stresses acting in the downslope direction, and increase in weight due to absorption of water. Some of the variables in the denominator include overall rock strength, processes of cementation, deposition by stream or hillslope processes on the footslope, drainage of water from sub-surface materials, and removal of part of the weight of soil and rock from the landslide source area. If the landslide threshold is passed, both the dominant process and the landscape morphology are changed. Equilibrium conditions are highly unlikely.

In conclusion, geomorphic thresholds may be defined in terms of ratios, the numerators and denominators of which describe opposing tendencies, and which may be simple or complex depending on the needs of the investigator or the complexity of the real world. The chief advantage of the threshold approach is to focus attention on feedback mechanisms and complex responses that are likely to cause the mode of system operation to change.

REFERENCES CITED

Bagnold, R. A., 1973, The nature of saltation and bed-load transport in water: Proc. Roy. Soc. London, ser. A, p. 473–504.

———, 1977, Bedload transport by natural rivers: Water Resources Res., v. 13, p. 303–312.

Bull, W. B., 1975, Allometric change of landforms: Geol. Soc. Amer. Bull., v. 86, p. 1489–1498.

———, 1976, Landforms that do not tend toward a steady state: *in* Flemal, R., and Melhorn, W., eds., Theories of landform development: Binghamton, N.Y., State Univ., New York, Publ. Geomorphol., p. 111–128.

———, 1979, The threshold of critical-power in streams: Geol. Soc. Amer. Bull., v. 90, p. 453–464.

Leopold, L. B., Wolman, M. G., and Miller, J. P., 1964, Fluvial processes in geomorphology: San Francisco, W. H. Freeman, 522 p.

Schumm, S. A., 1973, Geomorphic thresholds and complex response of drainage systems: *in* Morisawa, M., ed., Fluvial geomorphology: Binghamton, N.Y., State Univ., New York, Publ. Geomorphol., p. 229–310.

IV

THRESHOLDS IN OTHER GEOMORPHIC PROCESSES

13

FREQUENCY, MAGNITUDE, AND SPATIAL DISTRIBUTION OF MOUNTAIN ROCKFALLS AND ROCKSLIDES IN THE HIGHWOOD PASS AREA, ALBERTA, CANADA

James S. Gardner

INTRODUCTION

Rockfalls and rockslides are important and, in some cases, spectacular geomorphic events on the steep and exposed slopes of high mountain areas. Their occurrence is influenced by a complex of variables, some of which are intrinsic to the rock mass while others are extrinsic, involving environmental factors of meteorological, hydrological, and biological origin. Rockfalls and rockslides are discrete events signaling the fact that a critical value, condition, or threshold has been exceeded. In simple terms, this may be the result of a decrease in resistance to movement and/or an increase in force encouraging movement.

This chapter describes the frequency, magnitude, and spatial distribution of rockfalls and rockslides in the Highwood Pass area in the Canadian Rocky Mountains (Fig. 1) (115°6'W, 50°35'N). Together with information on past and present environmental conditions in the area, these data provide a basis for explaining rockfalls and rockslides and their geomorphic significance in the area.

The theoretical and general significance of the chapter rests mainly in the data on frequency and magnitude of rockfall and rockslide occurrence. Analysis of frequency and magnitude of geomorphic events has been an important theme in geomorphology, as exemplified by the landmark paper by Wolman and Miller (1960) and the more recent works of Gretener (1967) and Gage (1970). This theme relates to the more general concern of uniformitarianism and catastrophism (see, e.g., Albritton, 1967; Gould, 1965) and the explanation of existing landforms. Most empirical evidence indicates that geomorphic events or forces can be placed somewhere on a continuum between high frequency/low magnitude and low frequency/high magnitude. Floods provide a good example in which the relationship between frequency and magnitude is a negative exponential one. However, in the assessment of the geomorphic work or change accomplished by these events, relationships are not quite so clear. For example, Wolman and Eiler (1958), Dury (1973), and Gardner (1977a) report minor geomorphic effects following rare, high-magnitude floods. On the other hand, Schumm and Lichty (1963) report some major geomorphic effects as a result of high-magnitude flooding. Taking both magnitude and frequency over extended periods of time into consideration, Leopold, et al. (1964) conclude that floods of moderate frequency and magnitude account for a larger proportion of sediment transport than the rare, catastrophic floods. In the case of individual rockfalls and rockslides, high-magnitude events are of greater geomorphic significance than are low-magnitude events. But again, over extended periods it is difficult

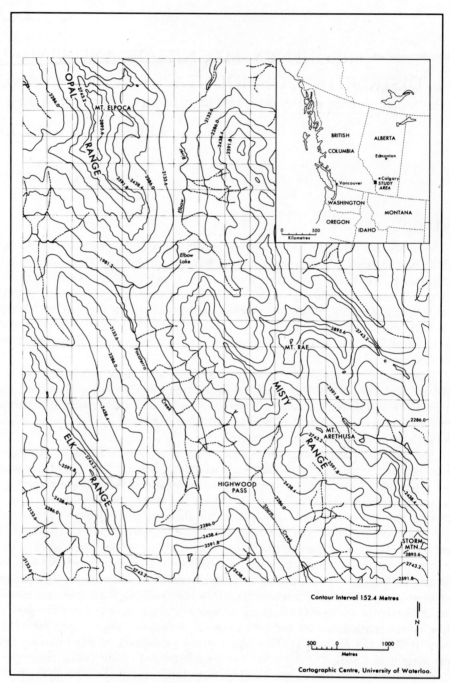

FIGURE 1 *Study area and its location.*

to say which have the greater geomorphic significance. This chapter addresses this issue.

The threshold concept is expecially useful in understanding individual events and rationalizing their geomorphic consequences. Why, for example, do some floods have a minor geomorphic impact, whereas others of like magnitude have catastrophic impacts? Why under conditions of given rainfall intensity do numerous debris flows occur, whereas at another time under similar conditions none occur? What factors initiate high-magnitude rockfalls and rockslides in seemingly unchanging environmental conditions? In summary, we find that a widespread event or force may not produce a similarly widespread geomorphic response, and events or forces of similar magnitude occurring at different times may not produce a similar response in the geomorphic system. The concept of geomorphic threshold, particularly as explained by Schumm (1973), helps to rationalize these issues. He explicitly recognized conditions extrinsic as well as intrinsic to a geomorphic system and noted that whereas extrinsic input may be constant, changes, responses, or adjustments may not be. Responses may occur abruptly at a threshold state controlled by factors intrinsic to the geomorphic system, whether it be a slope, stream channel, or beach. This concept seems particularly appropriate in the case of rockfalls and rockslides.

The practical significance of this chapter rests in the data describing the frequency and spatial distribution of rockfalls, and the explanatory inferences that can be made from the results. Rockfalls and rockslides can be damaging and disrupting to a variety of human activities and land uses in mountainous regions. Obvious examples include catastrophic events such as occurred at Frank, Alberta, in 1903 and at Hegben Lake, Montana, in 1959. In addition, small and frequent rockfalls along highways and railroads create a hazard for transportation. And information on the probability, frequency, location, and magnitude of occurrence plus likely causal factors has some value in land-use planning and regulation (Simpson, 1973).

DEFINITIONS AND RELATED WORKS

Several of the terms used in this chapter have been used in a variety of ways in the literature, with the result that their meaning is ambiquous. As used here, the distinction between the terms "rockfall" and "rockslide" is one of the characteristics of initial motion. Rockfalls are initiated along zones of weakness, such as joints, which do not reflect sedimentary or petrographic boundaries such as bedding planes. Secondary rockfalls occur when loose or already weathered material falls from its position. In rockslides, the initial movement is a sliding along a zone of weakness such as a bedding plane. This distinction has been made before in the literature (Bjerrum and Jorstad, 1968). Once in motion, the form of movement, which is rapid, may be highly complex. Rockfall usually implies some free-falling motion, often accompanied by bouncing and rolling of individual particles. Rockslides can be translated into free fall and bouncing and rolling, depending on the morphology of the surface over which the rock mass is moving.

An area of considerable debate in the literature relates to the type of motion involved in high velocity/high magnitude rockfalls and rockslides subsequent to the initial failure (Hsu, 1975). Some have proposed a flowing-type motion with compressed air and dust providing the medium (Heim, 1932; Kent, 1966), whereas others have proposed a sliding motion over a body of compressed air (Shreve, 1966, 1968). The significance of this from the geomorphic viewpoint lies in the interpretation of landforms produced. Heim

(1882), for example, concluded that the Elm rockfall and debris stream flowed because he observed geometric flowlines and ridges and furrows on the surface of the deposit. Howe (1909) attributed rock streams in the San Juan mountains to "landslides," whereas the same landforms are interpreted as rock glaciers today.

In the literature, different terms are applied to rockfalls and rockslides of greatly different magnitudes and apparently different characteristics. For example, the term "landslide" appears often and is widely used in popular literature. The term "bergsturz" is often used to describe rockfalls and rockslides of a very large magnitude that display the flowing and/or sliding motion discussed above. Many of the terms and early classifications are discussed in Ladd (1935) and more recently in Hsu (1975). Whalley (1974) has used terminology and a classification using magnitude as a primary definition criterion. The terms are: debris fall (<10 m^3), boulder fall and block fall (<100 m^3), cliff fall (>100 m^3), and bergsturz (10^6 m^3). A similar breakdown is used in this chapter in conjunction with the idea that initial motion can be a fall or slide. In all cases, the events discussed in this chapter involve rock particles and apparently do not include large amounts of fines, water, and/or snow and ice.

The geotechnical, geological, and geomorphological literature on rockfalls and rockslides had emphasized the rare, high-magnitude events. The reasons for this are obvious. Despite their relatively low frequency, high-magnitude events have caused loss of life and property damage, in addition to creating some spectacular landforms. Among the better known examples are the Elm rockfall (Heim, 1882); the Vaiont rockslide (Muller, 1964); the Sherman rockslide (Shreve, 1966); the Blackhawk rockslide (Shreve, 1968); the Madison Canyon rockslide (USGS, 1964); the 1970 Huascaran rockfalls and rockslides (Browning, 1973); the Hope, B.C., rockslide (Mathews and McTaggart, 1969); the Saidmarreh rockslide in Iran (Harrison and Falcon, 1938); and the Frank rockfall/rockslide (McConnell and Brock, 1904; Cruden and Krahn 1973). The Madison, Frank, and Hope events are examples of observed high-magnitude occurrences in mountainous area near the area discussed in this paper. Indeed, Turtle Mountain, the source of the Frank event, displays geologic conditions similar to those in the study area. Increasingly, there is evidence, primarily in the form of deposits, of ancient high-magnitude rockfalls and rockslides in the Canadian Rocky Mountains and adjacent mountain areas (see, e.g., Dishaw, 1967; Duffy, 1967; Cruden, 1976; Eisbacher, 1971, 1977; Gardner, 1977b; Locat and Cruden, 1977). To date, the most detailed analysis of the deposits and landforms created by such events has been that of Mudge (1965). Of course, many of these works have examined the causal factors in rockfall and rockslide occurrence and most stress the role of intrinsic geological conditions, whereas others (including Schumm and Chorley, 1964) have noted the importance of moisture and temperature conditions acting in concert with the geological circumstances.

Relatively little work has been done on the high frequency/low magnitude rockfalls and rockslides. Ladd (1935) noted their importance in causing disruptions and inconveniences for transportation. Indeed, most long-term data regarding these events come from railroad and highway records such as those used by Bjerrum and Jorstad (1968) and Rapp (1960a). Most other studies of high frequency/low magnitude rockfall and rockslide have been done in association with studies of debris slope development (Rapp, 1960b; Gray, 1972; Gardner 1968; Stock, 1968). Inventories of rockfalls and rockslides of this magnitude have been completed by Gardner (1970a) and Luckman (1976). Several of these works have discussed the causal factors in low-magnitude rockfalls and rockslides, stressing the role of extrinsic variables such as fluctuating temperatures, freeze and thaw, and the presence of water. Gardner (1971) found an association between rock surface temperature fluctuations and rockfall frequency on north-facing mountain walls.

THE STUDY AREA

While rockfalls and rockslides are not restricted to high mountain areas, they and their results are probably best displayed in such areas. High mountain areas are usually regarded as high-energy geomorphic environments in which there is a propensity for catastrophic events to accomplish much of the geomorphic work (Hewitt, 1972). In addition, erosion or denudation rates are relatively high in such environments (see, e.g., Corbel, 1959).

High mountain areas have a number of conditions favorable to rockfalls and rockslides. Most of the earth's alpine regions are geologically youthful mountain ranges which mark zones of crustal instability, making them prone to earthquakes which trigger catastrophic events. The geology of many mountain ranges is complex, the rocks having being disturbed by folding and faulting during orogenesis. Very steep and long slopes are prevalent. Glaciers are, or have been, widespread, and their erosional and hydrological effects can contribute to slope instability through oversteepening, groundwater fluctuations, and removal of physical support during valley glacier recession and thinning. Precipitation in mountain areas can be copious and intense, thereby providing the moisture which is often seen as an important causal factor in rockfalls and rockslides. Temperatures can be highly variable and freeze-thaw frequencies relatively high.

The study area displays many of these characteristics. The study from which this chapter is derived has attempted to document the mass transfer and frequency and magnitude of processes generally operating on the mountain slopes. Centered near Highwood Pass in the front ranges of the Canadian Rockies, the study area is about 80 km west-southwest of Calgary (Fig. 1). The pass is situated close to treeline at an altitude of 2185 m above sea level. The study area encompasses about 100 km^2 of alpine terrain ranging in elevation from 1970 m to 3200 m.

The topography of the area, and most of the front ranges, strongly reflects the geological history, as opposed to more recent geomorphic processes. A series of northwest to southeast trending thrust faults are reflected in the long parallel mountain ridges and valleys (Fig. 2). For the most part, the prominent ridges are composed of Paleozoic dolomites, limestones, and shales. Of particular note are the Rundle and Palliser Formations. These have been thrust over Mesozoic shales, siltstones, sandstones, and coal deposits, of which the Kootenay Formation is of particular interest as a rockfall source. Dips are generally toward the southwest and are steep. Nearly vertical attitudes are evident in the Opal Range in the northern part of the study area (Fig. 1). Scarp faces, some 800 m high, are generally northeast-facing, although where dips are near vertical, free faces are exposed on west- and southwest-facing slopes as well (Fig. 3). The scarp slopes are the principal sources of rockfall and the steep dip slopes are the principal sources of rockslides.

The glacial history of the study area is poorly understood, largely because of a lack of tills, well-preserved ash, and datable organic material. The topography contains evidence of classical alpine glaciation in the form of elliptical valleys, truncated spurs, horn peaks, aretes, and cirques. The low valleys in the area have probably been free of ice for

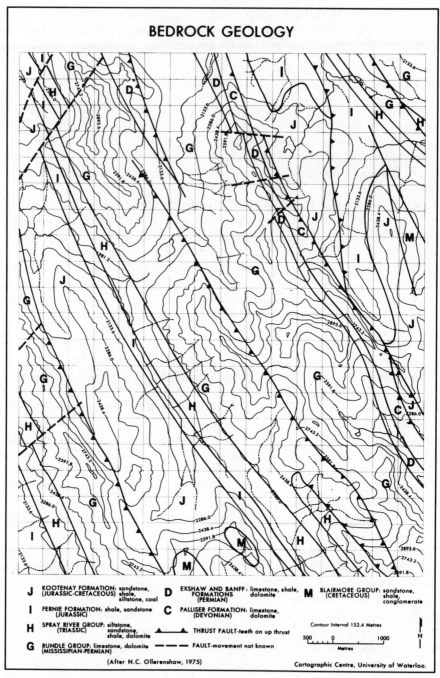

BEDROCK GEOLOGY

J KOOTENAY FORMATION: sandstone,
(JURASSIC-CRETACEOUS) shale,
siltstone, coal

I FERNIE FORMATION: shale, sandstone
(JURASSIC)

H SPRAY RIVER GROUP: siltstone,
(TRIASSIC) sandstone,
shale, dolomite

G RUNDLE GROUP: limestone, dolomite
(MISSISSIPIAN-PERMIAN)

D EXSHAW AND BANFF: limestone, shale,
FORMATIONS dolomite
(PERMIAN)

C PALLISER FORMATION: limestone,
(DEVONIAN) dolomite

▲——▲ THRUST FAULT-teeth on up thrust

— — — FAULT-movement not known

M BLAIRMORE GROUP: sandstone,
(CRETACEOUS) shale,
conglomerate

Contour Interval 152.4 Metres

500   0   1000
Metres

N

(After N.C. Ollerenshaw, 1975)

Cartographic Centre, University of Waterloo.

FIGURE 2 *Bedrock geology of the study area. Note in particular the northwest to southwest trending units and thrust faults.*

FIGURE 3 *Southwestern part of the study area, showing east-and northeast-facing cliffs and debris slopes along the Lewis trust fault. The lower hills in the left center of the photo are composed of the Kootenay Formation.*

8000 to 10,000 yr (L. Jackson, personal communication). There is little evidence for more recent ice advances during the Holocene, as has been postulated for the main ranges of the Canadian Rockies (Osborn, 1975; Harris and Howell, 1977), with the exception of Little Ice Age glaciation on some north- to northeast-facing slopes of the higher peaks such as Mt. Rae. Remnants of Little Ice Age glaciers remain on Mt. Rae today.

Climatological data for the study area are sparse. Some generalizations can be made on the basis of records from nearby locations and ecological similarities to better known sites elsewhere in Canadian Rockies. Mean annual temperature at Highwood Pass is probably slightly below freezing. Freezing temperatures occur there in every month of the year, while summer maximum temperatures (based on 3 yr of records) rarely exceed 20°C. Subzero ground temperatures were discovered 120–170 cm beneath alpine tundra vegetation at an altitude of 2360 m during August 1977. In addition, active sorted polygons of various sizes are found in moist locations at a variety of elevations. Most of the evidence points to rigorous periglacial temperature conditions with freeze-thaw cycles in spring and fall and permafrost at favorable sites above treeline.

Precipitation in the area comes as winter, spring, and fall snow and summer rain. Rain comes in two ways: frontal storms of long duration and low intensity; and convective storms of short duration and sometimes high precipitation intensity. The greatest intensity recorded during this study occurred in July 1975 when 26 mm fell in 8 hr. Snow has been recorded at Highwood Pass in every month of the year, and it accounts for more than half of the approximate mean annual precipitation of 70 cm. Snowcover depths at Highwood Pass are at a maximum in April and average about 150 cm (based on occasional Atmospheric Environment Service measurements). However, snowcover depths and durations vary greatly throughout the area depending on altitude, exposure to wind and sunlight, topography, and vegetation cover. For example, in February 1977, during a low-snowfall winter, photographs taken in the area show south-, southwest-, and west-facing slopes which are exposed to sunlight and the warm, chinook (foehn) winds completely devoid of snow. In general, a snowcover persists until mid-May, whereas in some years, the Highwood Pass road remains closed by snow until mid-June. In gullies and depressions above treeline and on north- to northeast-facing slopes, snowpatches persist through June, July, and part of August (Fig. 4). This discontinuous snowcover is an important factor in explaining the occurrence of low-magnitude rockfalls.

STUDY METHODS

There are serious difficulties in the acquisition of data for the frequency and distribution of rockfalls and rockslides. Part of the difficulty stems from the fact that they are discontinuous or discrete events. In addition, theory tells us that the greater the magnitude of rockfall and rockslide, the more likely they are to be discontinuous or infrequent. The problem of sampling discontinuous phenomena in time has been well discussed in Thornes and Brunsden (1977). Similar problems arise in gaining a representative sample of phenomena which are discontinuous or nonuniformly distributed in space. Without continuous observation, as is possible with temperature and stream discharge, the representative qualities of a sample are subject to question.

Sampling problems are partly of logistical origin. While one might design an optimal

FIGURE 4 *Discontinuous snow cover typical for the study area in late spring to early summer (mid-June to mid-July).*

sampling program, difficulties of access to the study area, weather conditions, and related factors mitigate against carrying out the program. This is particularly true for the high mountain area described in the paper.

Two methods were used to gather data for the frequency and magnitude of rockfalls and rockslides in this study. The first consisted of an inventory of occurrences observed in the field. Similar inventory methods have been used in the past by Stock (1968), Gardner (1970a), and Luckman (1976). The study area, being remote and of little consequence in past or present transportation and other land uses, does not have a backlog of rockfall and rockslide disruptions, as is the case for some parts of the Alps and Scandinavian mountains. The data presented in this chapter were derived from inventories in July, and August 1975; June, July, and August 1976; and July, August, and September 1977. With such a sample it is obvious that an adequate measure of seasonality in rockfall and rockslide occurrence cannot be made.

This inventory does contain several advances on previous work. Besides the number of observations, the inventory took into account both rockfall occurrence and nonoccurrence, and provided information on location of occurrences. The method involves continuous observation while in the field. A grid made up of cells with the equivalent of 500 m sides was superimposed on a 1:50,000 scale map of the study area (Fig. 1). The cells were given identification numbers. The time during which each cell was under observation in the field was recorded, providing a continuous record of observation time. If a rockfall or rockslide occurred, its cell location, time of occurrence, and size characteristics (i.e., number of rocks, size or rocks) were recorded. All observed rockfalls and rockslides were low-magnitude debris falls (<10 m^3). Most occurrences were first identified by the sound of falling rocks, after which the movement was usually seen. Specific attempts were made to observe at unusual times, such as during the night, but the bulk of the observations were made incidental to doing other fieldwork in the area. Thus, the sample is biased toward daylight hours. However, a conscious effort was made to observe in all weather conditions so as not to bias the sample toward warm, dry days.

The second method of data collection consisted of field mapping of existing landforms and deposits which were interpreted as being the products of rockfalls and rockslides. Talus slopes, for example, provide evidence of rockfall and rockslide activity, probably of low magnitude. Characteristics of high-magnitude rockfall and rockslide deposits have been described in the literature (Mudge, 1965). Criteria used for identifying these deposits include a heterogeneous mixture of coarse, angular material; the presence of some exceptionally large blocks on the surface; a hummocky and porous surface lacking in vegetation, soil, and standing water; a lobate or fan shape for the deposit; the presence of a scar on the mountain slope above; the greatest thickness of debris at or near the distal end of the deposit; a break in slope or depression between the deposit and valley-side slope; and occasionally the presence of large ridge, furrow, and flowline patterns on the surface of the deposit. At present, no absolute method of dating the mapped deposits in the study area is available. However, rough estimates of relative age have been made based on extent of weathering and soil development, vegetation, including lichen cover, stratigraphic position, and geomorphic situation of the deposit.

STUDY RESULTS

The study results are presented in three sections: a brief description of landforms

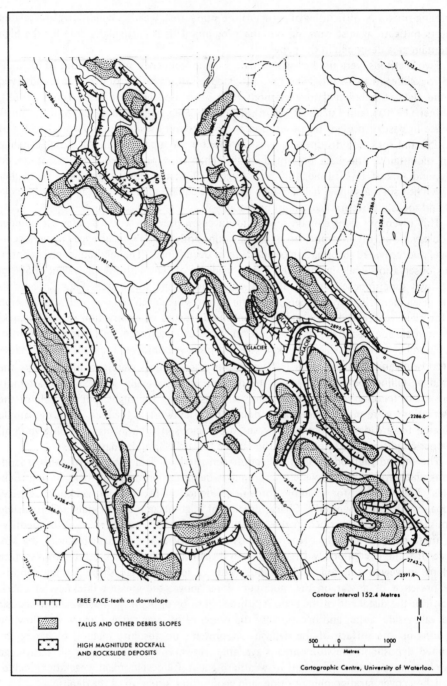

FIGURE 5 *Distribution of deposits and landforms related to rockfall and rockslide occurrence.*

276

and deposits associated with rockfalls and rockslides, a description of the temporal characteristics of rockfall and rockslide occurrence, and a description of the spatial distribution of rockfall and rockslide occurrence.

Landforms

Field and aerial photograph mapping revealed a number of landforms produced at least in part by rockfalls and rockslides. Figure 5 summarizes the distribution of these features in the study area. Free faces, usually in the form of scarp slopes along the major thrust faults, are common elements of the landscape and provide the source areas for rockfalls. Less frequent are free faces on dip slopes, although some fine examples are located on the west-facing slopes of the Opal and Misty Ranges (Fig. 1). Without exception, talus and other debris slopes are found where free faces are located (Figs. 4 and 5). Gravity sorting on many of these features indicates that rockfalls and rockslides are important contributors of material. The sorting and size of the material, in addition to field observations of the processes involved, suggest that low magnitude/high frequency rockfalls and/or rockslides are the primary contributors to the debris slopes associated with free faces. It should be recognized, however, that talus shift, debris flow, snow avalanches, and fluvial action may modify the landforms and deposits. Not all debris slopes in the study area are associated with free faces. Some, for example, are located at the foot of

FIGURE 6 *Deposit produced by blockfalls and/or low-magnitude cliff falls in the study area.*

FIGURE 7 *Large, high-magnitude rockfall deposit (No. 1 in Fig. 5). The rockslide "splashed" up the opposing slope and moved downvalley, leaving a train of debris in the valley bottom over a distance of 1.5 km.*

gullies as avalanche fans. Others appear as steep alluvial fans dependent on intense rainstorms and snowmelt which produce rapid runoff and torrents on the steep mountain slopes.

Single or small groups of blocks or boulders, usually located on valley bottoms or cirque floors and isolated from debris slopes, provide evidence for rockfalls and rockslides in the block fall and boulder fall magnitude (10–100 m^3) range (Whalley, 1974). Figure 6 illustrates such a feature. No events of this magnitude have been directly observed in the study area.

Deposits created by high-magnitude rockfalls and rockslides are among the most striking landforms in the study area. Eight deposits of this type have been mapped and are shown in Figure 5. They range in size from about 10^3 m^3 to greater than 10^6 m^3 of debris, with the largest being located below prominent scarp slopes in the western part of the study area (Fig. 5). Surface vegetation cover and degree of weathering of the material suggest that the features were deposited by single or several contemporaneous events. The deposits display many of the morphological characteristics previously summarized from the literature (Figs. 7 and 8). The surfaces of the larger deposits contain pockets of fine material and some soil development among large and angular blocks. The soil pockets support vegetation, both woody and herbaceous, and exposed rock surfaces support lichen growth. Lichen growth is limited to a few species by the cal-

FIGURE 8 *Large deposit (No. 2 in Fig. 5) which represents one or several high-magnitude rockfalls, as well as subsequent modification by rock glacier movement and/or glacier ice ablation.*

careous nature of the bedrock. Weathering of the large limestone blocks is evident on the surface of the deposits. Superimposed on the margins of some of the deposits are well-developed talus or debris slopes. With the exception of features 3 and 5 (Fig. 5), the deposits are located on valley or cirque floors isolated by a break in slope from the valley walls. Several deposits extend up the opposing valley slope a short distance (Fig. 7).

Frequency of Occurrence

The inventory method has provided data on rockfall and/or rockslide occurrence only in the low-magnitude range. Over three field seasons, 827 events were observed in 993 hr of observation giving an average frequency of 0.83 event per hour of observation. This is slightly greater than a previously published rockfall frequency of 0.7 from the Lake Louise area of the Rocky Mountains (Gardner, 1970a) and a value of 0.66 from the Suprise Valley area studied by Luckman (1976). However, of the 993 observation hours, rockfall and/or rockslide events were recorded in only 217 hr. This gives an average probability of observing one or more events in any hour of 0.22, somewhat less than Luckman's probability value of 0.30. The temporal frequency distribution of observed rockfall and/or rockslide occurrences is given in Figure 9. The shape of the curve indicates that

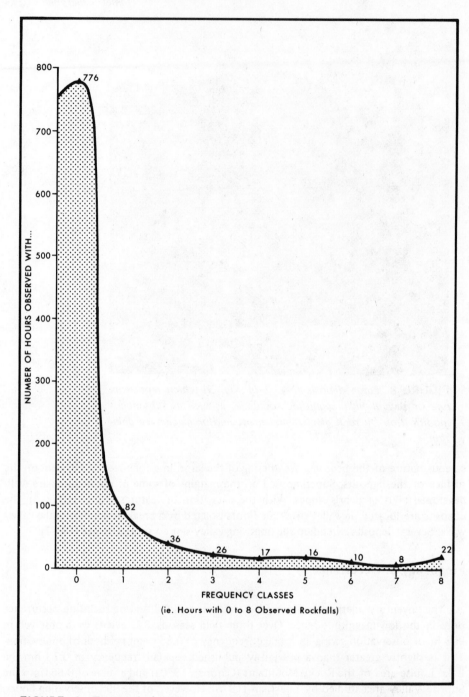

FIGURE 9 *Temporal frequency distribution of rockfall/rockslide occurrence (1975–77) showing the number of hours of observation in which 0, 1, 2, . . . 8+ were observed.*

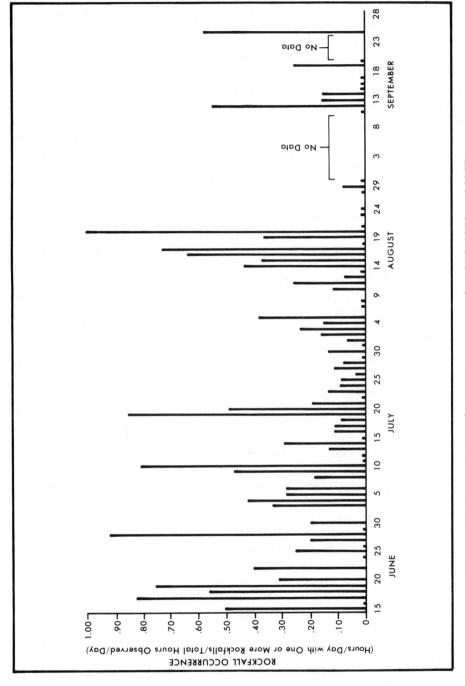

FIGURE 10 *Time series of aggregated rockfall/rockslide occurrences for 1975, 1976, and 1977.*

281

events tended to be concentrated in particular hours so that no events were recorded in 776 hr, 82 hr had one event, and 22 hr had eight or more events.

There are no apparent trends in rockfall/rockslide frequency during the study periods. Aggregated data for three field seasons are illustrated in Figure 10. The index of frequency used is the total number of hours per day in which one or more rockfalls were observed divided by the total number of hours of observation for the day in question. A similar summarization of relative frequency (total occurrences per day/total hours of observation per day) also shows no apparent trends over the study period. In most years, the period shown would include part of the spring to summer, and summer to fall transitions in the study area. Day-to-day variation in rockfall/rockslide occurrence may be greater than variation related to seasonality within the study period. However, the lack of data for winter, spring, and fall precludes a quantitative analysis of seasonability in rockfalls/rockslides.

The diurnal occurrence of rockfalls/rockslides does show some pattern, confirming previous findings. The data are summarized in Figure 11 in the form of an "average day," in recognition of the problems that exist in generalizing about the daily pattern of occurrences when day-to-day variation is so great. Two frequency indices are used: the relative frequency or the total number of occurrences observed in any given hour of the day divided by the total number of hours that hour was observed, and the probability or the number of hours with one or more rockfalls in any given hour of the day divided by the number of hours that hour had been observed. Prior to 0700 and after 2200 hr, the sample size or number of observation hours was small (<10 hr) and does not warrant discussion at this time. Both distributions show an increase in frequency and therefore probability, other things being equal, through the midday and early afternoon hours. Both early morning and evening hours on the average have relatively low rockfall frequencies.

Assessing and describing the frequency of occurrence of greater magnitude events without direct observations is a more difficult and speculative matter. That they have occurred in postglacial time (8000–10,000 yr) is evidenced by landforms in the study area and the periodic occurrence of high-magnitude events elsewhere in the Cordillera within historical time. Most studies of high-magnitude slope failure in deglaciated mountain areas concede that the time spectrum is postglacial. However, Whalley (1974) explicitly points out, and several other researchers imply, that the time spectrum should include the period of deglaciation given that on removal of valley-side support as the glaciers thin and recede, slope failures can occur (Duffy, 1967; Eisbacher, 1971). Several rockfall and rockslide deposits have been tentatively interpreted as being deposited initially on waning glacial ice. With these qualifications and considering the lack of absolute dating in the study area, a reasonable time spectrum for this study is a maximum of about 10,000 yr.

There is no evidence, nor can it be assumed, that the events which produced the deposits mapped in the study area occurred uniformly over the last 10,000 yr. In discussing the frequency of major rockslides and rockfalls in the Alps, Gretener (1967) suggests a figure of 1 in 1000 yr. Very little if anything has been published on the frequency of these major events in the Rockies, although a number of studies have identified ancient deposits produced by high-magnitude rockfalls and rockslides (e.g., Dishaw, 1967; Cruden, 1976). In the Canadian Rockies, one observed high-magnitude event,

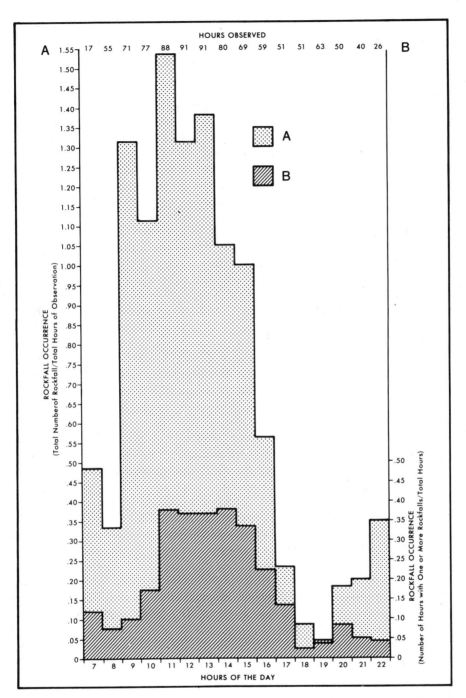

FIGURE 11 *Observed rockfall/rockslide occurrence on the "average day," using data from 1975, 1976, and 1977. A shows the frequency of rockfalls for each hour. B shows the probability of observing one or more rockfalls/rockslides for each hour.*

the Frank rockslide, has occurred in the last 80 years. Osborn and Jackson (1974) in describing the Mt. Indefatigable slide deposit, located about 15 km from the study area, suggest that it was produced sometime in the "last several hundred years." However, field examination of that deposit revealed that the surface supported mature Englemann spruce and lodgepole pine, the rocks were extensively weathered, and pockets of fines with a surface organic horizon were present. These characteristics suggest a much greater age for the deposit. The Indefatigable deposit is similar in these characteristics to the features in the study area. These characteristics, plus the development of overlapping debris or talus slopes and, in one case, the presence of a small neoglacial (perhaps Little Ice Age) protalus rampart or valley-side moraine on the free-face side of the deposits, suggest an early Holocene age for most of the features examined. Most evidence to date suggests that the frequency of high-magnitude slope failure was relatively great during and immediately after deglaciation. At that time, such events were probably important factors in post/glacial slope stabilization. Under present environmental conditions in the study area, a frequency of once in 1000 years is liberal, whereas, considering changes in environmental conditions during the last 10,000 yr, once in 10,000 yr is too conservative. Considering other factors, such as seismic activity and the occurrence of recent events such as the Frank slide, a very tentative estimate of once in 5000 yr for the 100-km^2 study area may not be unrealistic.

Spatial Distribution

The distribution of related landforms provides important clues about the spatial distribution of rockfalls and rockslides. High-magnitude slope failures can occur anywhere steep slopes are present, and the resulting deposits can block major valleys. Thus, one might expect that they could occur virtually anywhere in the study area. Requisite conditions for low-magnitude rockfalls and rockslides are more restrictive. Free faces are necessary. Thus, one expects their location to reflect the distribution of free faces and to be reflected by the distribution of talus slopes. Figure 5 shows the distribution of free faces, talus and other debris slopes, and major rockfall/rockslide deposits.

The inventory also provided data on the location of low-magnitude rockfall/rockslide occurrence. When mapped, these locations reflect the distribution of major free faces in the study area. Within the study area shown in Figure 1 are 278 grid cells or quadrants. Of these, 175 have been monitored for varying lengths of time. The total hours of observation for three field seasons was 993 and the total number of observed events was 827. It is possible to observe more than one cell in any given hour, and the observations were largely incidental to, or in the course of, other fieldwork. Therefore, the sample was not random, but there is no indication that it was spatially biased toward potentially favorable rockfall/rockslide locations. This is illustrated, visually at least, in Figure 12, which shows the distribution of cells for which there are data. These cover peaks, free faces, valley sides, and valley bottoms. The data show no relationships between either number of rockfalls recorded, relative frequency, or probability of observing one or more rockfalls per hour and the total number of hours of observation per grid cell.

Figure 12 shows the distribution by grid cell of rockfall frequencies in the study area. The frequencies, which represent the number of rockfalls per hour of observation, range from 0 to 7. Highest frequencies are associated with the major free faces, which

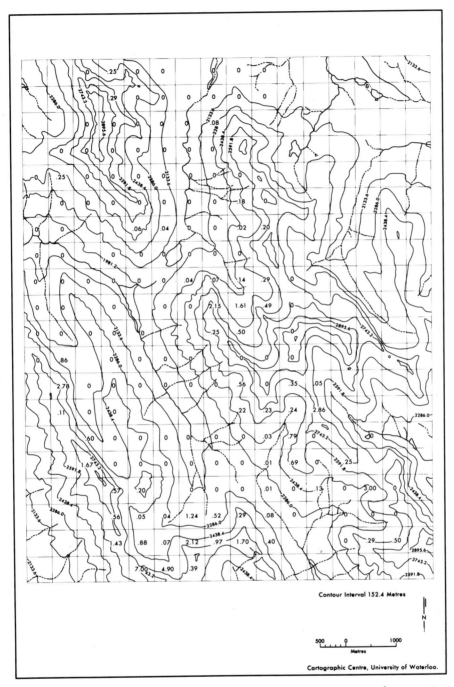

FIGURE 12 *Spatial distribution of rockfall/rockslides frequency (i.e., total observed rockfalls/number of hours of observation for each grid cell).*

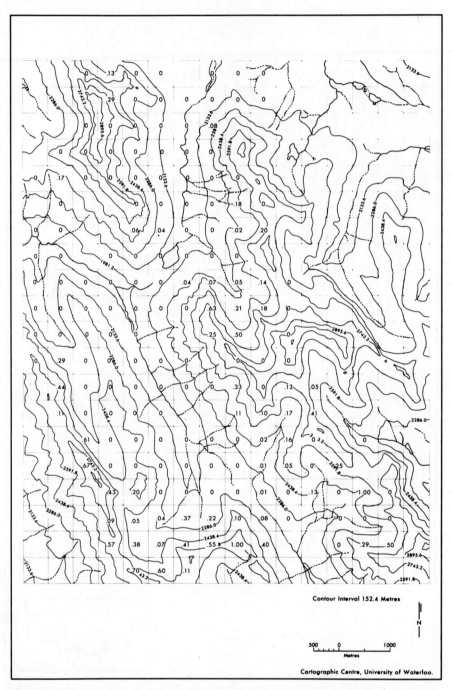

FIGURE 13 *Spatial distribution of rockfall/rockslide probability (i.e., number of hours with one or more rockfalls/total number of hours of observation for each grid cell).*

286

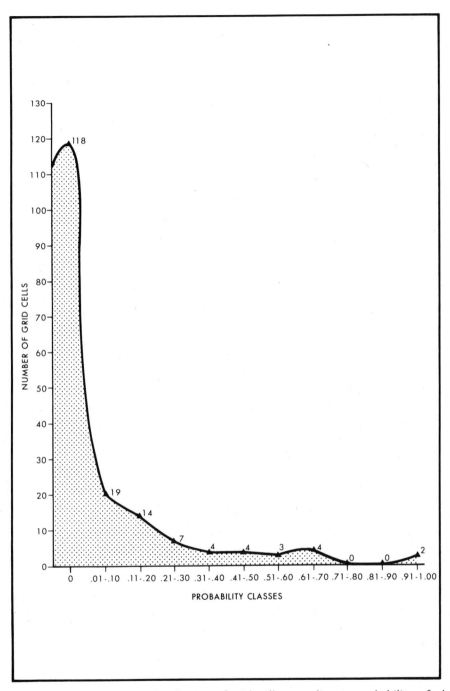

FIGURE 14 *Frequency distribution of grid cells according to probability of observing one or more rockfalls/rockslides in 1 hr of observation.*

are generally northwest- to northeast-facing. In Figure 13 the same data are displayed as the probability of observing in any given cell one or more rockfalls in an hour of observation. Again, high probabilities are concentrated primarily along the northwest to southeast lineations that mark the distribution of major free faces. The spatial distribution of rockfall occurrence is summarized in the frequency distribution (Fig. 14), which portrays a clustered pattern. In general, the majority of grid cells had zero rockfall/rockslide probability, with significantly fewer cells having probabilities ranging from 0.01 to 1.0.

Remnants of high-magnitude rockfalls and rockslides are scattered throughout the study area (Fig. 5). All are associated with free faces, seven deposits are beneath scarp slopes, and one is beneath a dip slope. Given this association, it is probable that the seven high-magnitude events were initiated as rockfalls and the one as a rockslide.

INTERPRETATIONS

Interpretation of the study results takes two forms. First, a tentative summarization of frequency and magnitude characteristics of rockfalls and rockslides is made for the study area. Second, associations between rockfall/rockslide occurrence and various environmental factors such as geology, topography, and microclimate are described.

Frequency and Magnitude Characteristics

One can only estimate the frequency-magnitude characteristics of rockfalls and rockslides from the data that have been obtained. In addition, without the benefit of absolute dates for the occurrence of some of the high-magnitude events in the area, the derived frequency-magnitude curve must be treated as very tentative at this time. Figure 15 illustrates the relationships between frequency and magnitude of rockfalls and rockslides in the study area. Ranges of frequency and magnitude are plotted rather than absolute values. As would be expected, frequency of occurrence decreases as magnitude increases. Also shown in Figure 15 is a partial frequency-magnitude curve based on observed events in the Lake Louise area (see Gardner, 1970a) and an estimated frequency for very high magnitude events in the Alps (Gretener, 1967). This added information also fits a negative exponential relationship between frequency and magnitude. The estimate from the Alps appears high in relation to the curve derived from the study area. Given that very high magnitude events have occurred through postglacial and in recent time elsewhere in the Rockies, a frequency-magnitude curve for an area of comparable size to that considered by Gretener probably would bring the Alps estimate into line.

If one accepts in principle the relationships shown in Figure 15, it is clear that low frequency/high magnitude events individually are responsible for the transport of significantly greater volumes of material than high frequency/low magnitude rockfalls and rockslides in postglacial time. However, several qualifications must be placed on this statement. An examination of Figure 5 suggests that talus and other debris slopes are extensive and important landforms in the study area—certainly more extensive and of greater volume than high-magnitude rockfall and rockslide deposits. Present rates of rockfall and rockslide, given by the inventory, are probably insufficient to explain the debris accumulations in the study area during postglacial time. This observation is not unique. Rather, it is supported by a similar observation by Gray (1972) from data gathered in the Yukon Territory. It is also known that other agents and processes, including snow avalanches,

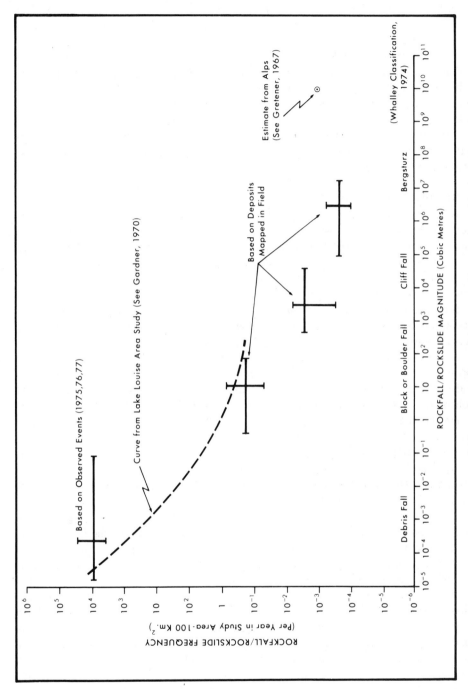

FIGURE 15 *Tentative relationship between rockfall/rockslide frequency and magnitude for the study area.*

289

debris flows, and surface runoff, contribute material to alpine debris slopes (Luckman, 1971; Rapp, 1960a, 1960b; Gardner, 1970b). Present rates of low-magnitude rockfalls and rockslides may not reflect rates throughout postglacial time. Given that high-magnitude rockfall and rockslide appear to have been more frequent in the past, the same may hold for low-magnitude events. In addition, some talus and debris slopes may include morainic debris which predates deglaciation. Other landforms in the study area also point to higher rates of rockfall and rockslide in the past. Features such as protalus ramparts with vegetated or lichen-covered surfaces appear at the base of some slopes, and some rock glaciers suggest high rates of debris accumulation in the past.

Thus, the inventory data and the lack of recent deposits from high-magnitude events in the study area indicate that under existing environmental conditions, the rockfall/rockslide system is dominated by high frequency/low magnitude events.

Some Environmental Associations

As spatial and temporal patterns of rockfall/rockslide occurrence emerge from the data, possible associations with various environmental factors can be suggested. The associations provide a basis for discussing causal mechanisms in rockfalls and rockslides. From this study and previous work, topography, geological structure, lithology, slope aspect, snow and ice, freezing and thawing temperatures, and water are the important factors in rockfall and rockslide occurrence.

Steep slopes are a topographic prerequisite for rockfall and rockslide. Areas of high rockfall occurrence in the study area as shown in Figures 12 and 13 correspond to free faces as shown in Figure 5. The association is particularly clear in the southwestern quarter of the study area, where a long, linear free-face system extends for about 4 km in a northwest-to-southeast direction and a 1-km free face extends in a west-to-east direction. Grid cells along these free-face systems exhibit high rockfall/rockslide frequencies and probabilities based on inventory data. In addition, at the base of the longer free-face system are the remnants of the rockfalls of greatest magnitude mapped in the study area (Fig. 5).

The topographic conditions are structurally and lithologically controlled. Throughout the study area free faces are associated with the major thrust faults and the Mississippian-Permian and Devonian limestones and dolomites (the Rundle and Palliser Formations) (see Fig. 2). In addition, where thrust faults are not present, slopes composed of the cliff-forming limestones and dolomites have been steepened to free faces by glacial erosion. This phenomenon is illustrated in the well-developed cirques around Mt. Rae (Fig. 1). In the southwestern quarter of the study area, the long free face is located along the Lewis thrust fault in a glacially eroded valley and is composed of Rundle limestones and dolomites in which well-developed bedding and extensive jointing is evident. The west-east free face noted above is not related to a thrust fault and is developed in Kootenay shales, siltstones, and coal (Fig. 2). The slope represents a southern cirque wall and has been subject to glacial erosion. Detailed examination of the thinly bedded shales, siltstones, and coal indicates that numerous tight folds are present, and the presence of numerous slickensided surfaces reveals further structural disturbances. Elsewhere in the study area, the Kootenay Formation appears as the most easily weathered rock group, and forms low hills, passes, and valley bottoms. Thus, in the southwestern quarter of the study area, high rockfall occurrence is associated with areas of major thrust faulting, minor folding and faulting, glacially steepened slopes, and, in come cases, rocks that are easily weathered.

About 75% of all observed rockfalls occurred on northwest-, north-, and northeast-facing slopes. In part, this reflects the orientation or aspect of free faces in the study area (Fig. 5). However, it also reflects the environment and microclimate of northerly aspects. Previous work (Gardner, 1971) has shown an association between high rockfall frequency and northerly facing slopes in the Lake Louise area. Free faces with this exposure in the Canadian Rocky Mountains retain snow and ice patches well into the summer, and in some years and in some locations, throughout the summer. Gardner (1969) also showed that northerly exposures in the presence of snow exhibited significantly more freeze and thaw cycles than did southerly exposures. Freezing and thawing temperatures in the presence of snow and/or ice are contributing factors in low-magnitude rockfall occurrence. An association with freezing and thawing temperatures is also suggested by the diurnal pattern of rockfall occurrence shown in Figure 11, wherein rockfall frequency reaches a maximum during the warm temperatures of midday. Many individual observations in the field support the idea. Most rockfalls, in addition to occurring on northerly exposures, occurred on warm, calm, and sunny days. Rockfall frequency on cool and cloudy days was generally low. The presence of melting snow and ice patches was frequently observed with high rockfall frequencies. This was particularly true in June and again in September (Fig. 10).

In summary, then, high rockfall frequencies were associated with free faces of structural and erosional origin, easily weathered rocks, rocks that have been disturbed by folding and faulting, and melting snow or ice.

CONCLUSIONS

Rockfalls and rockslides operate discontinuously in time and space. Data gathered in this study suggest that frequency of occurrence varies within several time scales from post/glacial time to the diurnal scale. The occurrences of observed low-magnitude events were clustered in time and space, with relatively few hours and a few locations accounting for the majority of observations. At the diurnal scale, maximum frequencies were recorded in the hours between 1100 and 1500. It is suggested that low-magnitude rockfall frequency may have been greater than at present in other periods of postglacial time on the basis of large volumes of material accumulated in related landforms such as talus slopes. No high-magnitude rockfalls and rockslides have occurred recently in the study area, although eight deposits created by such events have been mapped. Their spatial distribution shows a concentration beneath scarp slopes of Rundle limestones and dolomites in glacially eroded valleys and cirques. The nature of the deposits suggest a possible greater frequency of occurrence during or shortly after deglaciation.

The idea of geomorphic threshold provides a useful tool for conceptualizing the frequency, magnitude, and distribution of rockfalls and rockslides. Specifically, threshold values cannot be defined without more detailed analysis of the rock masses involved and more data to describe the operation of potential causal factors such as freeze-thaw and moisture conditions. The discontinuous temporal and spatial characteristics of rockfalls and rockslides in the study area indicate that special conditions, if not specific thresholds, are involved in the processes. Take the case of a low-magnitude rockfall, for example. The special conditions are those found in locations where the probability of occurrence is greatest—a free face of limestones, dolomites, and/or thinly bedded shales and coals: a northerly exposure with remnant snow and ice patches; and a time about midday on a warm, sunny day. Many rockfalls are of a secondary nature, meaning that their ultimate release downslope comes as the rocks lie in a weathered and loose state in a niche or on a ledge of the free face. The release is a relatively sudden process triggered by the melting of ice by which the rock has been attached to other material; the stress of moving water, snow, other rocks, and wind on the rock; and the formation of ice and related expansion or heaving of the rock and surrounding material. All the triggers are produced by the action of extrinsic factors and each occurrence signals that a force or resistance threshold has been exceeded. The resulting disequilibrium is manifest as a rockfall or rockslide. Further research could usefully examine the trigger factors in detail and specify thresholds of the number and intensity of freeze-thaw cycles and the volume and velocity of moving water, snow, and so on, most likely to initiate movement.

The threshold concept is also of use in thinking about high-magnitude events. High free faces or long, steep slopes are prerequisite for the rockfalls and rockslides described in this chapter. For large masses of rock to be dislocated, the presence of extensive lines of weakness, as along joint systems and bedding surfaces, are also necessary. The fact that the rockfalls and rockslides are highly discrete, occurring suddenly after long periods of quiescence, speaks of a delicate equilibrium between force and resistance which is upset by a trigger factor that increases the force or decreases the resistance. The occurrence may also represent the culmination of a long buildup of forces and/or decline in resistance to movement. In the former case, the trigger factor could be defined in terms of a threshold value at which point it destroys the equilibrium. The geomorphic system is adjusting in response to external changes or extrinsic thresholds. In the latter case the geomorphic system catastrophically adjusts in response to internal conditions when intrinsic thresholds are exceeded. The suggestion of a high frequency of high-magnitude rockfalls on deglaciation rests on the idea that glacial erosion may have steepened valley-side slopes. As long as it was present, the glacier ice would provide some lateral support for the slopes. When the ice receded, the support would be removed. This alone may have been sufficient to cause failure on the steepened slopes. Or, the glacial unloading and related groundwater changes may cause adjustments within the rock mass in the form of opening joints and other lines of weakness. Moisture penetrating the joints and bedding planes may than act as a trigger by decreasing resistance to movement.

Rockfalls and rockslides are rapid adjustments to disequilibrium in the geomorphic systems of steep mountain slopes. Disequilibrium is brought about by transitory stresses created by microclimatic, hydrological, and geological factors, or it may result from long-term stresses and decreasing resistance created by major environmental changes such as deglaciation and intrinsic changes in the rock mass. In each case the occurrence of a rockfall or rockslide signals that a threshold intrinsic or extrinsic to the geomorphic system has been exceeded.

ACKNOWLEDGMENTS

The author wishes to acknowledge the financial support of the National Research Council of Canada, Grant A9152, and the field assistance of Peter Kershaw, Linda Kershaw, Pam Tarquin, Dan Smith, Bill Mitchell, and Joe Deslodges.

REFERENCES CITED

Albritton, C. C., ed., 1967, Uniformity and simplicity: Geol. Soc. Amer. Spec. Paper 89.

Bjerrum, L., and Jorstad, F. A., 1968, Stability of rock slopes in Norway: Norwegian Geotechn. Inst. Publ. n. 79, p. 1–11.

Browning, J. M., 1973, Catastrophic rock slides, Mt. Huascaran, north-central Peru, May 31, 1970: Amer. Assoc. Petrol Geol. Bull., v. 57, p. 1335–1341.

Corbel, J., 1959, Vitesse de l'erosion: A. Geomorphol., v. 3, p. 1–28.

Cruden, D. M., 1976, Major rock slides in the Rockies: Can. Geotech. J., v. 13, p. 8–20.

———, and Krahn, J., 1973, A re-examination of the geology of the Frank slide: Can. Geotech. J., v. 10, p. 581–591.

Dishaw, H. E., 1967, Massive landslides: Photogram. Eng., v. 33, p. 603–608.

Duffy, P. J. B., 1967, The valley of the rocks: Can. Alpine J., v. 50, p. 88–90.

Dury, G. H., 1973, Magnitude frequency analysis and channel morphology: *in* M., ed. Morisawa, Fluvial geomorphology: Binghamton, State Univ. New York, Publ. Geomorphol., p. 91–122.

Eisbacher, G. H., 1971, Natural slope failure, northeastern Skeena Mountains: Can. Geotech. J., v. 8, p. 384–390.

———, 1977, Rockslides in the MacKenzie Mtns., District of MacKenzie: Geol. Survey Canada Paper 77-1A, p. 235–241.

Gage, M., 1970, The tempo of geomorphic change: J. Geol., v. 78, p. 619–625.

Gardner, J., 1968, Debris slope form and processes in the Lake Louise district, a high mountain area: unpubl. Ph.D. thesis, McGill University, Montreal, 263 p.

———, 1969, Snowpatches: their influence on mountain wall temperatures and the geomorphic implications: Geografiska Ann., v. 51 (A), p. 114–120.

———, 1970a, Rockfall: a geomorphic process in high mountain terrain: Albertan Geographer, v. 6, p. 15–20.

———, 1970b, Geomorphic significance of avalanches in the Lake Louise area, Alberta, Canada: Arctic Alpine Res., v. 2, p. 135–144.

———, 1971, A note on rockfalls and northfaces in the Lake Louise area: Amer. Alpine J., v. 17, p. 317–318.

———, 1977a, Some geomorphic effects of a catastrophic flood on the Grand River, Ontario: Can. J. Earth Sci., v. 14, p. 2294–2300.

———, 1977b, High magnitude rockfall/rockslide: frequency and geomorphic significance in the Highwood Pass area, Alberta: Great Plains–Rocky Mountain Div. Assoc. Amer. Geographers J., v. 6, p. 228–238.

Gould, S. J., 1965, Is uniformitarianism necessary?- Amer. J. Sci., v. 263, p. 223–228.

Gray, J. T., 1972, Debris accretion on talus slopes in the central Yukon Territory: *in* Slaymaker, O., and McPherson, H. J., eds. Mountain geomorphology: Vancouver, B.C., Tantalus Research Ltd., p. 75–84.

Gretener, P. E., 1967, The significance of the rare event in geology: Amer. Assoc. Petroleum Geol. Bull., v. 51, p. 2197–2206.

Harris, S. A., and Howell, J., 1977, Chateau Lake Louise moraines—evidence for a new Holocene glacial event in southwest Alberta: Bull. Can. Petrol. Geol., v. 25, p. 441–455.

Harrison, J. V., and N. C. Falcon, 1938, An ancient landslip at Saidmarreh in S.W. Iran: J. Geol., v. 46, p. 296–309.

Heim, A., 1882, Der Bergsturz von Elm: Deutsch. Geol. Gesell. Z., v. 34, p. 74–115.

——, 1932, Bergsturz and Menschenleben, Zurich, Fretz's Wasmath Verlag, 218 p.

Hewitt, K., 1972, The mountain environment and geomorphic processes: in Slaymaker, O., and McPherson, H. J., eds. Mountain geomorphology: Vancouver, B.C., Tantalus Research Ltd., p. 17–36.

Howe, E., 1909, Landslides in the San Juan Mtns., Colorado: U.S. Geol. Survey Prof. Paper 67, 58 p.

Hsu, K. J., 1975, Catastrophic debris streams (sturzstroms) generated by rockfalls: Geol. Soc. Amer. Bull., v. 86, p. 129–140.

Kent, P. E., 1966, The transport mechanism in catastrophic rockfalls: J. Geol., v. 74, p. 79–83.

Ladd, G. E., 1935, Landslides, subsidences and rockfalls: Proc. Amer. Railway Eng. Assoc., v. 36, p. 1091–1162.

Leopold, L. B., Wolman, M. G., and Miller, J. P., 1964, Fluvial processes in geomorphology: San Francisco, W. H. Freeman, 522 p.

Locat, J., and Cruden, D. M., 1977, Major landslides on the eastern slopes of the southern Canadian Rockies: in Humphrey, C. B., ed., Proc. 15th Ann. Symp. Eng. Geol. Soils, Pocatello, Idaho State University, p. 179–197.

Luckman, B. H., 1971, The role of snow avalanches in the evolution of alpine talus slopes: Slopes, form and process: Inst. Brit. Geog. Spec. Publ. 3, p. 93–110.

——, 1976, Rockfalls and rockfall inventory data: some observations from Suprise Valley, Jasper National Park, Canada: Earth Surface Processes, v. 1, p. 287–298.

Mathews, W. H., and McTaggart, K. C., 1969, The Hope landslide, British Columbia: Proc. Geol. Soc. Canada, v. 20, p. 65–75.

McConnell, R. G., and Brock, R. W., 1904, Report on the great landslide at Frank, Alberta, Canada: Ann. Rep., Canada Dept. Interior, pt. 8, 17 p.

Mudge, M. R., 1965, Rockfall-avalanche and rockslide avalanche deposits at Sawtooth Ridge, Montana: Geol. Soc. Amer. Bull., v. 76, p. 1003–1014.

Muller, L., 1964, The rockslide in the Vaiont Valley: Felsmechanik and Ingenieurgeologie, v. 2, p. 148-212.

Ollerenshaw, N. C., 1975, Bedrock geology map: Kananaskis Lakes Sheet, 1:250,000 Geological Survey of Canada, Ottawa.

Osborn, G., 1975, Neoglacial deposits in the Lake Louise area, Banff National Park, Alberta: Geol. Soc. Amer. Programs with Abstracts, v. 7, p. 635–636.

——, and Jackson, L. E., 1974, Physical environment: the mountain environment and urban society: Environmental Sciences Centre, Calgary, University of Calgary, p. 53–150.

Rapp, A., 1960a, Recent development of mountain slopes in Karkevagge and surroundings, northern Scandinavia: Geografiska Ann., v. 42(A), p. 66–200.

——, 1960b, Talus slopes and mountain walls at Tempelfjorden, Spitzbergen: Norsk Polarinstitutt Skrifter, n. 119, 96 p.

Schumm, S. A., 1973, Geomorphic thresholds and complex response of drainage basins: in Morisawa, M., ed. Fluvial geomorphology: Binghamton, State Univ. New York, Publ. Geomorphol., p. 299–310.

——, and Chorley, R. J., 1964, The fall of threatening rock: Amer. J. Sci., v. 262, p. 1041–1054.

——, and Lichty, R. W., 1963, Channel widening and floodplain construction along Cimarron River in southwestern Kansas: U.S. Geol. Survey Prof. Paper 352-D, p. 71–88.

Shreve, R. C., 1966, Sherman landslide: Science, v. 154, p. 1639–1643.

——, 1968, Leakage and fluidization in air layer lubricated avalanches: Geol. Soc. Amer. Bull., v. 79, p. 653–658.

Simpson, H. E., 1973, Map showing areas of potential rockfalls in the Golden Quadrangle, Jefferson County, Colorado: U.S. Geol. Survey Map I-761-C.

Stock, R., 1968, Morphology and development of talus slopes at Ekalugad Fjord, Baffin Island, N.W.T.: unpublished B.A. thesis, Dept. Geography, London, Univ. Western Ontario, 84 p.

Thornes, J. B., and Brunsden, D., 1977, Geomorphology and time: London, Methuen, 208 p.

U.S. Geological Survey, 1964, The Hegben Lake, Montana, earthquake of August 17, 1959: U.S. Geol. Survey Prof. Paper 435.

Whalley, W. B., 1974, The mechanics of high-magnitude, low frequency rock failure and its importance in a mountainous area: Reading, Univ. Reading Geog. Papers, 27, 48 p.

Wolman, M. G., and Eiler, J. P., 1958, Reconnaissance study of erosion and deposition produced by the flood of August 1955 in Connecticut: Amer. Geophys. Union Trans., v. 39, p. 1–14.

——, and Miller, J. P., 1960, Magnitude and frequency of forces in geomorphic processes: J. Geol., v. 68, p. 54–74.

14

THRESHOLDS IN GLACIAL GEOMORPHOLOGY

Cuchlaine A. M. King

INTRODUCTION

A geomorphological threshold may be defined as a state above which one set of processes operate while below another set operates. Thus the passing of a threshold indicates a change of process. The point at which the threshold is passed depends on the variables that control the action of the processes involved. These variables may act continuously or intermittently. Temperature, for example, is a continuous variable, while precipitation is a discontinuous one. Many of the variables have a considerable range of values, the extreme values usually occurring more rarely than the moderate ones.

The effectiveness of a threshold is closely related to the feedback processes involved in the interactions between the variables and the processes. A threshold will be particularly effective when it is associated with a positive, self-generating, feedback relationship. Positive feedback is particularly important in glacial geomorphology; thus thresholds play a very significant role in glacial geomorphology. Nevertheless, positive feedback relationships cannot continue their self-enhancing operation indefinitely. They will eventually sow the seeds of their own destruction, as changes cannot continue permanently in one direction. There comes a point where another threshold is crossed and the condition either becomes stable or a reverse change is initiated. When the limit of development has been reached under some circumstances a reversal of the trends will be set in motion, leading to a reverse positive feedback situation being established. Again, the changes cannot continue in this reverse direction indefinitely, so a cyclic pattern is developed.

Thus in considering the effects of thresholds in glacial geomorphology, it is necessary to bear in mind the effects of feedback, both positive and negative, and to consider the cyclic element of the glacial processes with which these feedback relationships are often connected. The various glacial cycles operate on a wide range of time scales and tempos of activity.

ICE SHEET GROWTH AND DECAY

One of the most conspicuous cycles associated with glacial activity is the alternating advance and decay of the major ice sheets of the northern hemisphere. The two major, land-based ice sheets of North America and northern Europe have advanced and retreated with great rapidity and reasonable regularity during the last few hundred thousands of years at least. Their major fluctuations are in marked contrast with the relative stability of the great southern hemisphere Antarctic ice sheet, which has a circumpolar situation and whose extent is limited by the area of the land on which it rests. There is no such limit to the southern spread of the northern hemisphere ice sheets along a significant extent of their margins. They are inherently unstable and grow with great speed, melting and disappearing with even greater speed.

The whole process of growth may take only a few thousand years, while their collapse and decay may be accomplished in only two to three thousand years. The cycle involved in the growth and decay of these ice sheets is thus asymmetrical and the period of glaciation is considerably longer than that of deglaciation or interglacial conditions. The former period is of the order of 90,000 years, compared with about 10,000 years for the interglacial periods. There are shorter and less intense cycles of advance and retreat superimposed on the major oscillations of the ice masses.

The inherent instability of the glacial system is very important from the geomorphological point of view. It means that the activity of glaciers is continuously varying according to whether they are advancing or retreating, and there is rarely time for conditions to reach an ultimate state in which stability is achieved. The glacial and interglacial cycles are short on the geological time scale, and this continuous switch from one state to another means that the geomorphological processes are nearly always acting on a landscape that is out of phase and out of equilibrium with the currently acting processes. Glaciers are modifying a fluvial landscape, or rivers and periglacial processes are modifying a glaciated landscape. This accounts for the relatively great effectiveness of glacial processes, as geomorphological activity is always greatest when a new system is put in operation before equilibrium is established.

The activity of thresholds is particularly important in the cyclic swing from glacial to nonglacial conditions in the areas alternately covered by the expanding Laurentide and Scandinavian ice sheets and deglaciated as they retreat. A great deal of valuable work has been done recently on the causes of the build up and decay of these great ice sheets. At its maximum the Laurentide ice sheet was very similar in size to the Antarctic ice sheet, which contains 96% of all the ice on earth at present. The Scandinavian ice sheet was about half this size, but still it covered a very large area of northern Europe. Thresholds and trigger mechanisms affect the growth and collapse of these unstable ice masses.

The buildup of the ice sheets may be related to changes in the oceanic circulation which at times allows relatively warm water and air to penetrate the Canadian Arctic. This increases the precipitation on the upland plateau areas of Baffin Island, Labrador-Ungava, and Keewatin. At a critical point the snow banks that survive the summer melting cover a large enough area to increase the albedo sufficiently to set up a positive feedback system, whereby the snowfall increases as less melts, resulting in lowering temperatures. In time, however, the low temperature reduces the moisture-holding capacity of the air enough to reduce the precipitation, and the snow may decrease, and with it the buildup of ice. The area of maximum accumulation then spreads farther south, where conditions are not so extreme, and thus the Baffin Island ice center is followed by the growth of the Labrador-Ungava center. Finally, the Keewatin center expands until the three merge to form the one large Laurentide ice sheet.

The collapse of the system is also the result of trigger action. The weight of ice depresses the crust isostatically with a time lag, resulting in a relative rise of sea level in the glaciated area. The ice floats, water can penetrate beneath it, and it rapidly breaks up as a result. It has been suggested that the ice of northern Europe retreated first about

15,000 to 17,000 years ago in the Spitzbergen area. When this ice was returned as water to the sea, the rise in level triggered off the melting of the Scandinavian ice mass about 10,000 to 12,000 years ago. The subsequent rise of sea level then triggered the collapse of the Laurentide ice sheet, then centered over Foxe Basin, as water penetrated through the Hudson Strait, causing flotation and collapse of the Laurentide ice sheet between 8000 and 6000 years ago. The further rise of sea level that ensued caused the western Antarctic ice sheet to start collapsing about 5000 years ago. This is a good example of successive thresholds being passed, triggering off the collapse of the ice sheets successively as the critical water depth allowed flotation. In this theory the climatic modifications are the result and not the cause of the fluctuations in the ice sheets. If also draws attention to the many variables that have to be considered in dealing with the complex matter of glacial systems and their growth and decay. The problem of which is the initial cause and which the effect is not easy to solve in such a complex system.

Thresholds are nevertheless important, and it is of interest to note that the situation of the Baffin Island plateau surface during the Little Ice Age 300 to 500 years ago was such that a threshold that would have led to rapid glaciation was almost reached. There is evidence that the surface area covered by shallow snow patches increased rapidly during this period and nearly reached the state that would lead to almost instantaneous glaciation, had the threshold been crossed (Andrews et al., 1976).

INITIATION OF ICE FLOW

One of the most important thresholds from the geomorphological point of view of glacier activity is that at which accumulating snow and ice becomes thick enough to flow. The threshold is not dependent solely on the thickness of the snow and ice, because flow is related by Glen's flow law to the relationship between the stress and the strain rate. The stress depends both on the thickness of the ice and on the temperature. The density of the ice is another variable. Flow takes place more readily as the temperature rises toward the melting point of the ice; thus very cold ice will be less mobile than ice that is not as cold. The threshold at which flow starts may be passed either by an increase in thickness of the ice, an increase in density, or an increase in temperature. The thickness at which temperate ice, which is at its pressure melting point, will start to flow is about 30 m.

The relief will play a part, as well as the nature of the bedrock surface on which the ice is resting, so again it is not possible to specify an exact thickness even for a constant temperature. The threshold between moving and immobile ice is a very important one from the point of view of glacial geomorphology. Ice can be one of the most erosive geomorphological agents or it can protect the surface over which it lies (Linton, 1963). The difference depends on the speed of movement to a great extent. The speed, in turn, depends on supply and relief. Areas undergoing severe erosion often lie adjacent to those under a protective ice cover that are not being altered at all when ice is channeled along troughs with a critical orientation, as the through valleys of the Appalachian Plateau. This strong contrast gave rise in the early days of glacial geomorphology to the two opposing schools, the glacial protectionists and those that advocated effective glacial erosion. Both are correct for specific areas and conditions, the difference being whether the threshold for movement has been exceeded or not.

TYPE OF FLOW

One threshold determines whether ice will flow or remain immobile, but there are also other important thresholds involved in the type of ice flow, which in turn reflect in the nature of the landforms produced by the flowing ice. Ice can be in one of two states at its base. It can be cold ice that has a temperature below the pressure melting point and is frozen to its bed, or it can be at the pressure melting point so that water can exist between the ice and its bed.

The threshold between these two states is important from many points of view. It has been invoked, for example, by Wilson (1964) in his theory of the initiation of glacial periods. He argues that an abrupt change in conditions is brought about when the Antarctic ice sheet reaches a threshold that initiates much more rapid flow as the ice reaches pressure melting point at its base. The condition that often leads to the passing of this threshold is an increase in ice thickness; this in turn leads to more rapid flow, and the generation of heat by friction further enhances the speed of flow by positive feedback, until the more rapid flow thins the ice and the process is reversed. This is a good example of the cyclic situation referred to in the introduction.

Cold ice flow and temperate ice flow have also been differentiated in terms of their respective geomorphological results both in erosion and deposition. The presence of water at the base of a temperate glacier is the factor that is mainly responsible for the contrasts produced.

Other important ice flow thresholds are those that differentiate between normal laminar flow and extending or compressing flow. In laminar flow the flow lines are parallel to the bed of the glacier, the slope of which is uniform in gradient, and the longitudinal strain rate is constant down the glacier. If, however, the longitudinal strain rate is not constant, it can be either positive or negative. When it is positive, the glacier is stretching at the surface and flow is extending; when it is negative, the glacier is compressing at the surface and the flow is compressive. The two possible types of flow are associated with the development of planes of weakness or shear zones within the body of the ice. These can have significant effects on the response of the glacier bed to the movement of the ice, in terms of the patterns of erosion or deposition. The types of flow in turn depend on the nature of the glacier bed. Again a feedback relationship exists between the morphology and the process which has very important repercussions on the landforms produced by glacier flow. In fact, the nature of the bed provides the threshold that determines the type of flow. A convex bed is associated with extending flow and a concave bed with compressive flow.

Surges

Another aspect of glacier flow that is even more closely associated with thresholds is that of glacier surging. This phenomenon occurs only in a small number of glaciers, but affects a wide range of glacier types of differing size, mass balance, and temperature regime. Glaciers that surge do so more or less regularly, the glacier building up until the threshold is passed and a surge takes place, with velocities often several orders of magnitude greater than normal. The glacier then tends to become stagnant, with its lower reaches raised in elevation and its upper part lowered. Gradually the instability is built up again as the upper part rises and the lower part ablates, until another surge takes place.

The causes of glacier surging are not fully understood and it is likely that different glaciers surge for different reasons. One cause of surging that may apply to at least some glaciers is the threshold that divides cold ice from temperate ice. When the ice reaches pressure melting point a surge may be set off. The stress is built up beyond the normal yield stress of 1 bar before the surge takes place, possibly as a result of the mass of stagnant ice forming a dam. This theory does not however, explain the first surge.

Surges may be the explanation of a number of geomorphic features that are difficult to explain in other ways. They give rise to characteristic moraine forms in valley glaciers and may be responsible for certain features associated with ice sheets. The great bend of the Susquehanna River has been tentatively attributed to the action of a surge, and more recently a surge has been suggested to account for anomalies of glacial deposits on the east coast of England (Boulton et al., 1977).

Surging seems to affect glaciers of all sizes, from small valley glaciers to ice caps and major ice sheets. Only a minority of glaciers are affected, however, and Meier and Post (1969) state that only 204 North American glaciers out of more than 10,000 surge. The time interval between surges is about 15 to more than 100 years, and flow rates during surges range from 150 m/yr to more than 6 km/yr, with horizontal displacements of less than 1 km to more than 11 km at the snout. Surges occur in a wide range of climates, but all surging glaciers surge repeatedly. The long profile of the glacier increases in gradient just before the surge. Some very rapid surges could be accounted for by flotation, but there is probably no single cause of surging, although they all probably involve some form of threshold. Three possibilities have been suggested: (1) stress instability, (2) water film instability, and (3) temperature instability, the second and third being connected in some instances. Surges can occur in both temperate and cold glaciers.

The instability of the long profile induces surge conditions. In quiescent phases the upper ice reservoir thickens, and when the basal shear stress, given by $\rho g h \sin \alpha$, where ρ is the density, g the gravity, h the ice thickness, and α the surface gradient, reaches the critical threshold value the glacier surges. The condition of rapid flow propagates very quickly up and down glacier from the point of initiation. The large longitudinal stresses set up in this way help to maintain the total shear stress and the ice and its bed become decoupled. The active phase comes to an end when the total bed shear stress reaches another low threshold value.

The surge of Bruarjökull exemplifies one type of surge. This glacier, on the north side of Vatnajökull in Iceland, surged in 1625, 1720s, 1810, 1890, and 1963. In the latter year a bulge was forming near the glacier margin by the end of August, by October 14 the ice was very crevassed, and by November 14 the snout had advanced 2 to 3 km and its front was 20 to 50 m high. By January l, 1964, the front was advancing at 1 m/hr in a rhythmic way but was not forming moraine. It only moved a farther 200 to 300 m forward subsequently, and then started the gradual lowering and frontal decay that usually follow surges.

Weertman (1969) and Lliboutry (1968) support the thermal instability cause of surging, as it affects the amount or presence of meltwater at the glacier base. The change from cold to warm conditions at the glacier base could initiate a surge as the bed becomes lubricated with water.

A model for surging applicable to cold glaciers has been suggested by Shoemaker (1976). In cold glaciers the ice is often frozen to the bed at the upper and lower levels with a temperate area in the middle where the ice is thick and moving faster. The mean temperature must increase downglacier. The surge is the result of a positionally dependent temperature distribution. A thickening of the glacier in excess of the constant-state conditions tends to induce instability. With a linear increase of accumulation rate upslope, an oscillation of the region of thickening up and down slope becomes unstable. The theory does not require basal sliding, although it can be included. Friction should decrease downslope if temperature increases in this direction. In this theory mass instability is induced by the temperature gradient.

So far in the discussion of the cause of surging the threshold that triggers the surge is self-induced, in that it depends on an instability developed within the glacier system, due to temperature change, accumulation change, or to a slow buildup without external change until the critical limit is reached. Some surges, however, may be induced by an external trigger, causing the exceedence of a threshold. Such a trigger could be the result of an earthquake or sudden, abnormal avalanching of ice or snow into the glacier.

It has been argued that surging of the Antarctic ice sheet as a result of reaching a critical threshold thickness resulting in the basal temperature rising to pressure melting point and causing enhanced flow could trigger a northern hemisphere glaciation. It has also been argued that the northern hemisphere might lead the southern, in that a fall of sea level associated with northern ice advance could cause the Antarctic ice sheet grounding line to advance seaward, thus allowing the ice sheet to thicken and perhaps again reach its pressure melting point and hence surge forward even farther.

One of the geomorphic characteristics of surging glaciers is their very deformed lateral and medial moraines. The extension of lobes of ice beyond the normal smooth margin of an ice sheet may also mark the position of surges, and if moraines form around the margin their form will bear witness to the presence of surging ice in the vicinity.

JÖKULHLAUP

The Icelandic term "jökulhlaup" describes the occasional sudden emptying of a glacial lake. The discharge from Grimsvotn on Vatnajökull in Iceland is one of the best known examples of the phenomenon, although many other ice-dammed lakes also empty suddenly at times (Fig. 1). The process as described by Bjornsson (1974) and Nye (1976) illustrates very well the operation of a glacial threshold.

The variables that are important in the process are the subglacial topography and the glacier surface form, which is very important. During a period of about 5 to 6 years the lake level rises about 100 m, the glacier is lifted off a subglacial ridge and water escapes subglacially through a 50-km-long tunnel beneath Skeiðararjökull. When the water level has fallen about 100 m, the tunnel becomes sealed by plastic deformation before the water level is lowered to the subglacial rim. Another factor in the jökulhlaup is the geothermal heat, which greatly aids the melting process, providing the water to fill the subglacial lake. The drainage area of the lake is about 300 km^2, while the lake itself covers between 30 and 40 km^2. About three-fourths of the water is provided by geothermal melting, so this is one of the main causes of the outburst. The surface slope of the ice is changed to allow the water to drain into the Grimsvotn depression.

Jökulhlaups have occurred about every 5 to 6 years lately in 1941, 1945, 1948, 1954, 1960, 1965 and 1972, the next being expected in 1977-78. The localized source of geothermal heat forms a depression in the surface of the ice cap and allows water as well as the ice to drain into the largely subglacial basin. A jökulhlaup occurs within a few months of the water level in the basin reaching the critical threshold height. Most of the water leaves the glacier by three or four main streams, amounting to a total of 3 to 3.5 km^3 during the last few jökulhlaups. In former times about twice as much water was discharged during a jökulhlaup, and the water level change before the outburst was 150 to 200 m. Such jökulhlaups took place in 1903, 1913, 1922, 1934, and 1938, the interval being rather longer than recently.

The relief below the water is of importance in explaining the jökulhlaup. The base of the Grimsvotn depression is at about 1000 m with a lower level to west and south. The ice is 600 m thick above the saddle of the basin and 220 m thick over the subglacial lake of Grimsvotn, which permanently holds 2 to 3 km^3. The ice thickness increases again southeast of Grimsvotn to 1000 m at the head of the Skeiðararjökull valley. The relief is such that water accumulates in Grimsvotn until the level has risen about 100 m, when the catastrophic flood occurs.

The ice overburden pressure is $p_i = \rho_i g H_i$, the subglacial hydrostatic pressure due to the lake water is $P = \rho_w g(h_w - h_b)$. H_i is the ice thickness and h_w is the water level in Grimsvotn and h_b is the elevation of the glacier bottom; ρ_i and ρ_w represent the density of ice and water, respectively. The value of P_i is greater than P, so the ice seal remains effective and the water level rises. The difference between p_i and P represents the potential barrier around the lake, and it has a specific width and threshold value. When the lake rises to a critical level, the threshold is exceeded and the water can begin to escape. The width of the barrier is least to the southeast of the lake, being 5 km when the level is lowest. The erosion by the escaping meltwater could well create channels beneath the ice, and these would effect the precise level at which the lake reached the threshold value.

Once the seal is broken, water can escape from the lake along the entire length of the 50-km channel to Skeiðararsandur, assisted also by the generation of heat by friction to form and enlarge subglacial tunnels through which the water flows. The flood stops when the water level falls about 100 m but before it reaches the level of the subglacial ridge. This is when p_i exceeds P, and the pressure of the ice rapidly closes the tunnels by plastic deformation. The rapidity of the flood, which only takes about 10 days to build up to its maximum and then returns to normal within about 1 day, is partly due to the generation of heat by friction as the water flows through and enlarges the tunnels. The subglacial flow system is unstable, according to Nye's analysis, if the water escapes through tunnels. If it escaped along veins, the driving force would be toward and not away from the lake in a system of veins.

The geomorphic effects of jökulhlaups include the cutting of the tunnel floor and the deposition of large amounts of debris on the outer sandur, while the inner part is often trenched by the flood waters. The new road built since the last jökulhlaup over Skeiðar-arsandur will be vulnerable to the next one. An artificial spillway to lower the lake level is the best means of preventing any further jökulhlaups occurring, but it would need to be 1 km long at least.

Meltwater Features

Jökulhlaups involve abnormal, extreme, short-lived discharges of glacial meltwaters. The water can achieve a great deal of geomorphic work in a short time, both in erosion and deposition. Even under more normal conditions, however, glacial meltwater can create impressive features, both of erosion and deposition. There is a subtle threshold

(a)

(b)

(c)

304

(d)

(e)

FIGURE 1 *Series of photographs taken showing the 1954 jökulhlaup at Skaptafellsheidi in south Iceland. The flood took 10 days to reach maximum flow, from July 8 to 18. By July 20, the flow had returned to normal. About 3.5 km³ of water were discharged during this threshold-exceeding event. (a) Normal flow just prior to jökulhlaup. (b) beginning of jökulhlaup. (c) and (d) Maximum flow during the flood. (e) Flood waters are starting to wane. Notice the icebergs in the upper right of the photograph.*

that separates the two processes, so that their morphological results can often be found in close association. Discussion will be limited to those features formed beneath the ice.

In the discussion of the jökulhlaup, it was shown that Nye has established that the escaping water must flow in tunnels rather than veins within the ice. The water flow must be concentrated as a result, and provided its velocity and turbulence are above the critical threshold, deposition will be inhibited, and erosion take place. The sandur in front of Skeiðararjökull demonstrates the effectiveness of this process. Large channels have been cut through the deposits of the sandur, often with kettle holes forming where lumps of ice torn from the tunnel walls became embedded in the channel floor. One such channel was 200 m wide and about 6 m deep, and must have been cut in 2 or 3 days.

Closed channels have been recorded in the floor of the North Sea. These deep channels are 10 to 50 km long, 1 to 4 km wide, and have a relief of between 20 and 150 m, but mainly in the range 50 to 100 m. They have been equated with the tunnel valleys of Denmark. Some of the North Sea depressions are filled with sediment, while others are relief features. They are probably similar to the buried tunnel valleys discussed by Woodland (1970) from East Anglia. A few of the East Anglian channels are more than 100 m deep and extend below sea level. Some of the larger tunnel valleys, including those in Denmark, are now thought to be the result of glacial erosion, although others may have been formed by erosion of migrating streams, which would account for their considerable width. At any one time only part of the channel would be occupied by flowing meltwater, the rest being filled with ice. During the last stages of occupation, when the discharge was reduced below the threshold necessary to carry the heavy load of debris, deposition would replace erosion. Thus an esker could be formed in the last position of the stream in the final tunnel. This would account for the presence of depositional eskers within a large erosional channel.

The undulating nature of the long profile of some of the tunnel valleys supports their subglacial origin. Water under hydrostatic pressure in a tunnel beneath the ice can flow uphill and, provided that the ice is not too thick, the tunnel can reach to the floor of the glacier to erode a depression within it.

The maximum depth to which subglacial streams can penetrate beneath the ice surface depends on several variables. The amount of meltwater and its temperature are critical because they determine the rate of melting of the tunnel. The water pressure offsets the tendency of the ice pressure to close the tunnel. There is a critical threshold between the hydrostatic pressure of the water and the pressure of the overburden of ice. A tunnel will stay open provided that it remains water-filled, and that the hydrostatic pressure exceeds the weight of ice tending to close the tunnel, because water is denser than ice. An empty tunnel soon closes, and this is more likely to occur in winter when meltwater is not plentiful.

Water normally penetrates into temperate ice to depths of about 100 to 150 m, rarely to 200 to 300 m, and very infrequently to depths greater than 300 m, except where the ice is stagnant and very crevassed. These conditions only apply to temperate glaciers, as temperatures are too low for meltwater to occur in cold glaciers, except under very unusual conditions. Thus the penetration of water to the bed of the ice mass is a function of the critical threshold between water and ice characteristics. Only when water can penetrate to the bed can it produce permanent features of either erosion or deposition. Which of the two processes will operate also depends on the thresholds of stream competence and capacity; when this threshold is passed the stream will deposit.

Glaciofluvial erosion is likely to occur as the ice moves slower, and glacial erosion is reduced. The meltwater load is lessened, but the volume of water may for a time increase as melting accelerates, allowing more efficient water erosion. Deposition will occur in the final stage when both water and sediment supply are limited. Owing to the variability of flow, however, deposition can at least occur intermittently. It is likely that the deposition of an esker can take place rapidly under suitable conditions, often within the channel formed by the more erosive conditions preceding the final stages of ice dissolution. Eskers are most likely to form in the final stagnant phase, as exemplified in the dead ice areas of Baffin Island.

FORMS OF EROSION RELATED TO THRESHOLDS

Ice works in several different ways to erode the surface over which it is flowing. The features produced by this erosion also differ according to the type of process acting. They can be divided into small-scale features and larger ones. The small-scale features illustrate the significance of the details of the action of the erosional processes and thus provide a useful starting point. The theory associated with these processes has been considered by Boulton (1974), and, as it invokes the operation of thresholds, the main points will be briefly reiterated.

Sliding over its bed by a glacier is an important aspect of erosional processes. It is related to the presence of cavities between the ice and the bed. In the absence of cavities, the processes of regelation and plastic flow must be considered. The processes of erosion include abrasion, crushing, and plucking.

Abrasion is achieved by the dragging of particles over the bed of the glacier as it flows. The particles vary in size, and the load is often bimodal, with large blocks and finer debris predominating. The larger particles, in the clast mode, play the greater part in abrasion. The abrasion rate can be related to the effective normal pressure for any one ice velocity (Fig. 2). The rate rises to a maximum with increasing pressure, then falls off rapidly as a threshold is passed, and finally deposition takes place. The threshold rises with increasing velocity, being about 10 bars for a velocity of 5 m/yr to 30 bars for a velocity of 100 m/yr. There is thus a critical condition for lodgement of particles, as shown in Figure 2a and b. The effective normal pressure depends on the density and thickness of the ice, and on the water pressure at the glacier bed, so that subglacial conditions play an important part in the process. The permeability of the bed plays an important part in the effective water pressure.

The small-scale features formed by abrasion include streamlined forms. As ice moves over bed irregularities the effective normal pressure is modified, first falling, then rising, then falling again. The pattern of asymmetry of the forms produced varies according to whether the abrasion rate is greater on the up or downglacier side of the hump. At effective normal pressures below the critical threshold a cavity occurs on the downglacier side and erosion is most effective on the upglacier side, the reverse occurs at pressures above the threshold value.

The process of subglacial crushing is related to failure of the solid rock beneath the glacier bed, and the formation of cavities is important in the process. A rock will fail under certain specific conditions. The optimum condition is the threshold where cavities are just supressed. Increasing ice velocity makes the crushing process more effective,

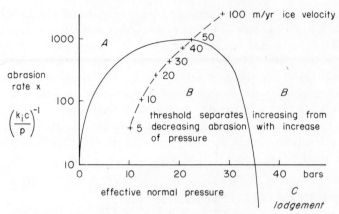

FIGURE 2a   *Relationship between abrasion rate, effective normal pressure, and ice velocity in determining the threshold between increasing and decreasing abrasion rate, and between abrasion and lodgement, for 50-m/yr ice velocity.*

and there is a threshold velocity below which the process cannot take place whatever the ice thickness; this is about 3 m/yr. Above this speed the form of the bedrock hummocks affects the critical velocity and thickness.

Plucking occurs under some conditions when the glacier bed is hard. It is the process whereby the material produced by abrasion and crushing is plucked from the bed by the tractive force of the ice. Plucking alone is required where ice is moving over unlithified

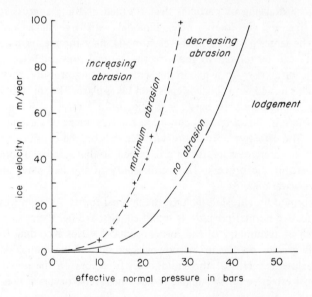

FIGURE 2b   *Threshold values between increasing and decreasing abrasion, and between decreasing abrasion and lodgement. (Partly after G. S. Boulton, 1974.)*

sediments. The threshold for plucking to occur is the overcoming of the frictional drag between the particle and the bed. Plucking takes place on the lee side of rock hummocks where cavities occur most readily. The absence of cavities causes a rapid rise of frictional resistance, and hence the threshold rises and plucking is inhibited. Where the glacier movement is directed away from the rock surface, plucking should be most effective.

The details of the glacier bed, the presence of cavities between the ice and rock, and the nature of the bed material all play a very important part in the processes of glacial erosion and the entrainment of material into the ice. There are thresholds involved in the effectiveness of the processes considered, causing their operation to be variable both in space and time. The bed form is modified to fit the variations in the ice thickness and velocity. Thus small irregularities in the bed tend to become larger by positive feedback processes and interactions.

A well-developed cirque is one of the most distinctive of glacial landforms. There is, however, a continuum of forms of which the typical cirque is a member. At one end of the continuum are snow patch nivation hollows, while at the other end are cirques where erosion has gone so far that either several have joined or their backwalls have been considerably reduced in steepness and curvature. The cirque in its optimum stage of development has a deep rock basin and a steep, curved back wall. Although there is a continuum of forms, there is a threshold between the processes that produce the nivation hollows and those that produce the typical glacial cirque. This threshold is the point at which the snow accumulation in the nivation hollow becomes sufficiently thick to flow and thus becomes a glacier.

A cirque glacier is a thin, usually steeply sloping ice mass confined within the cirque hollow. Its flow is generally slow, accelerating to a maximum at the equilibrium line and then decreasing toward the snout. The most detailed study of flow in a cirque glacier is that reported by Lewis (1960), in which it was shown that the ice was moving almost as a rigid body rotating about a circular arc that closely fitted the cirque floor. This pattern of movement, however, does not account for the reversed slope of the bed that is characteristic of the best developed cirques, nor does it require the measured upward movement of the ice below the equilibrium line. The upward movement compensates for the loss of mass in the ablation zone, and counteracts the downward movement found in the accumulation zone. Thus there is some deformation of the ice as well as the rotational slipping that takes place over the glacier bed. The rotational slipping is associated with flow over the bed, the speed of which was 90% of the velocity recorded at the surface at the point of maximum ice thickness near the equilibrium line.

The ice thickness and recorded velocity were such, however, that the threshold for quarrying, which involves both crushing and plucking, would not be exceeded. Thus the most likely process operating at the glacier bed to produce the hollow form is abrasion. Tools used for this process could be supplied by the thaw–freeze processes that take place around the margins of the cirque. The ice thickness and velocity both reach their maxima at about the equilibrium line, so it is at this point that optimum conditions for erosion would occur, as cavities are likely to occur throughout such a glacier. The abrasion could lead to preferential erosion and the development of a hollow, provided conditions were such that the threshold for abrasion covered only a limited stretch of the glacier bed around the zone of maximum thickness and flow. The system will be self-generating up to a point, because as the hollow develops, the ice at this point gets thicker and its erosive capacity increases, by positive feedback.

The same type of positive feedback relationship can be considered in relation to the characteristics of a glaciated valley long profile. This is typically irregular, with rock basins and riegels within the upper parts of the valley, and sometimes fjords develop toward the mouth of the valley where it descends below sea level. The compressing flow characteristic of concavities in the bed produces a positive feedback cycle. The slip planes that slope tangentially upward from the glacier bed in this type of flow allow material to be carried up into the ice, thus exposing the bed to further erosion where the ice is thickest. There is, however, a threshold beyond which the self-generating, overdeepening process can no longer continue, as discussed by Nye and Martin (1968). They suggest that the pattern of slip planes is such that the ice tends to modify the bedrock floor until its form fits closely to the pattern of planes. The ice can then rotate smoothly across the hollow produced by the excessive erosion. At this stage its capacity for further overdeepening will be limited because the upward ice movement will no longer occur effectively. Material can no longer be plucked from the bed and entrained into the ice, so that the bed is protected from further abrasion or quarrying.

The degree to which overdeepening can take place depends on the relationship between the thickness of the ice and its velocity. As Boulton has shown, these variables critically affect the effectiveness of abrasion and the other erosional processes. Thus where the ice is very thick it must have a high velocity if it is to erode effectively. The combination of great thickness and high velocity provide the optimum conditions for a high abrasion rate, within a narrow critical zone, because, as the effective normal pressure increases still further, there is a rapid drop in the abrasion rate, which at a threshold value changes to deposition. This relationship also helps to account for the restricted and local nature of glacial overdeepening.

Because abrasion rates are higher for thick, fast-flowing ice, within the necessary threshold range of these variables, active glaciers will be associated with the creation of rock basins and fjords. Thus the distribution of these features can be explained in terms of glacier activity. Fjords are found mainly on the west coasts of continents in high latitudes, including Norway, Alaska, British Columbia, Patagonia, southwest New Zealand, and Greenland. These are the areas where precipitation is high, slopes steep, and ice volumes large with thick and active glaciers.

The term "threshold" is used in this chapter mainly to refer to the crossing of a critical instability point to initiate the operation of a different set of processes. In discussing fjords and glacial rock bars in the valley longitudinal profile, the term can be applied in its more literal sense, as a step to be crossed in a doorway. At the mouth of a fjord a shallow threshold is very often present, separating a deep basin upstream from the open sea at the mouth of the valley. Farther upstream the long profile of a glacial valley is usually characterized by a succession of rock bars or riegels that form a series of thresholds along the valley floor.

The largest glacial trough that would become a fjord if the ice were to melt is the Lambert glacier trough in the Antarctic. It is 50 km wide and 3400 m deep, extending to more than 2000 m below sea level. The profiles of Inugsuin fjord in east Baffin Island illustrates the essential characteristics of fjords well. It is 120 km long and has a maximum depth of 634 m in a basin where the mountains rise to their highest elevations of nearly 1700 m on either side of the fjord, giving a total relief of well over 2000 m. Near the seaward margin of the fjord is its threshold, which reaches within about 30 m of the surface, giving a basin about 600 m deep. The Sogne fjord in Norway reaches a depth of 1308 m below sea level, but at its threshold it is only 3 km wide and has a depth less than 200 m. There are many islands scattered above the surface at the threshold. There are other thresholds and basins upstream, which continue as a sequence into the glacial troughs draining from the Jostedalsbreen ice cap.

There is no difference in the processes that give rise to fjords and those that create the rock basins and thresholds of a normal glacial trough. The basins occur where conditions for effective glacial erosion are operative and the thresholds were glacial erosion is least efficient. The formation of the threshold at the seaward end of fjords is not easy to explain. Flotation may play a part as the ice can erode until it floats and thus could start the erosion of a basin, which by positive feedback would tend to grow deeper. As the ice thickens, the erosion could increase, while over the threshold the ice velocity might exceed that necessary for erosion as the ice is thinner over the threshold, thus again tending to maintain its height and form.

The formation of the typical U-shape cross profile of a glacial valley can also be related to threshold values that allow the most effective abrasion and other erosion processes where cavities form. The thickness and speed of flow of most valley glaciers is such that cavitation is likely to occur beneath the ice. This increases the likelihood for reaching the critical conditions for erosion, by crushing and plucking. If the valley cross profile were originally V-shaped, the effective normal stress would increase linearly toward the floor of the valley. The maximum pressure at the bottom could fall within any of the three critical zones. If it all lies within the zone of increasing abrasion rate with effective pressure, the abrasion will be at a maximum at the base, widening the lower part of the valley and producing a U-shaped form. If the zone of maximum erosion by abrasion is reached and passed, maximum erosion will occur at some point toward but not at the bottom of the valley. This situation will lead to a reduction of erosion at the bottom of the valley and a widening to form a flatter-bottomed U-shape cross profile. If the zone of deposition is reached as the threshold of effective pressure is passed, the bottom of the valley will be filled with deposits, while erosion takes place farther up the hillslope, again producing a U-shaped form. Thus whatever pattern occurs, a U-shape form is likely to be created.

FORMS OF DEPOSITION

It has been shown that there is a threshold that divides abrasion from deposition, when the effective normal pressure increases beyond a critical value for any single ice velocity. This has been termed the critical lodgement index by Boulton (1974). The variables that determine the lodgement of a particle on the bed is given by a complex equation, which includes the effective pressure, the velocity of the ice and of the particle, a friction coefficient, the cross-sectional area of the particle, as well as variables associated with the state of the ice, the particle, and the bed. A small increase in effective pressure will greatly increase the tendency for lodgement, and this is more important than the ice velocity. A large flow-law exponent, n, will cause more ready lodgement. The temperature also affects the process. The particle shape affects the ease with which it will lodge; thus plate-shaped particles will lodge more readily than spheroidal ones, which will tend to roll. The bed variables also affect the ease of lodgement. The permeability affects the normal effective pressure to a considerable degree. Thus it seems likely that there are many different variables that influence the threshold that divides deposition from erosion beneath a glacier.

The transporting power of a glacier decreases as the bed roughness increases, leading to deposition when the bed is rough. If the sliding velocity increases, the transporting power will increase. If the effective pressure increases, the transporting power will decrease in the zone of traction. The effective pressure is sensitive to the permeability of the glacier bed. Where water can penetrate readily into the bed, and transmissibility is high, lodgement should occur readily, except in the marginal zone where the ice is thin and moving slowly. Lodgement is likely to occur on upglacier flanks of irregularities. Till with a high percentage of clasts will not lodge so readily as will a fine-grained till with a large clay mode. The lodgement index is likely to decrease away from the source of the load as the size of the particles is reduced by comminution. Thus the rock type will also affect the amount of lodgement likely to take place, being less on hard igneous rocks and greater on rocks that readily break into small pieces. Many variables are involved in deposition as in erosion.

The tendency for tills to be bimodal in size distribution, with a large mode in the boulder or cobble group and a small mode in the clay or sand group has already been commented upon. The fabric of the till, which is concerned with the preferred orientation of the elongated particles, is also often bimodal in form. One mode lies in the direction of ice flow, whereas there is often a second mode transverse to this direction. The transverse mode is sometimes related to a specific pattern of flow or to the form of the surface over which the ice is moving. Thus transverse fabrics sometimes occur where ice is being forced through a narrow channel or is moving upslope. The length of time the fabric has been forming also influences the pattern, as does the size and shape of the stones. Thus many different threshold phenomena may be involved in explaining any one fabric orientation pattern. Owing to the complexity of the controlling variables, it is only to be expected that fabrics tend to be variable, and it is generally agreed that more than one fabric is needed to provide a reliable estimate of past ice flow directions.

The processes of glacial deposition depend to a considerable extent on the thermal characteristics of the glacier bed. Boulton (1972) has examined the situation in four different conditions. In the first there is net basal melting, the second is a zone in which melting and freezing balance, the third is where freezing at the glacier base of meltwater is enough to maintain the ice at pressure melting point, and the fourth zone is one where the glacier is frozen to its bed.

In the first zone the glacier slips over its bed and material is incorporated partly by regelation. The load is mainly carried near the glacier bed in the lowest meter of ice or is deposited by lodgement as subglacial till. Deposition will first occur behind obstacles and then more generally. The second zone is similar to the first, but if meltwater is not plentiful lodgement may be greater.

In the third zone net erosion is likely to occur as material is incorporated into the ice by freezing, and plucking can occur. Deposition is likely to be supraglacial derived from englacial material by melting.

In the fourth state material tends to move up into the ice along shear planes and no relative movement occurs at the bed. Active ice can only deposit till subglacially. These processes are dependent on critical temperature states which provide the thresholds that divide the different zones.

Because of the wide range of conditions and the large number of variables that affect the processes of glacial deposition, there are also many different features of glacial deposition. Discussion will be limited to those that are directly deposited by the ice. These include drumlins and a wide range of features usually referred to as moraines of various types. Drumlins and associated features are streamlined forms that must have been shaped by moving ice, while many types of moraine have been formed in contact with slow-moving or stagnant ice.

Features formed by moving ice into streamlined forms range from purely erosional to entirely depositional forms. The erosional forms include roches moutonnées while many drumlins are purely depositional, although some contain a rock or till core and many include glaciofluvial material. Drumlins have given rise to much speculation, and many explanations have been put forward to account for their formation.

The most successful theory so far put forward is that due to Smalley and Unwin (1968). This theory is particularly relevant to a consideration of thresholds, as it depends on critical pressure conditions to account for the formation and distribution of drumlins. For drumlins to form the material at the base of the ice must have been continuously deformed by shearing. The material must consist of boulders in a clay and water system. The large particles in the material form a dilatant system, which is a property of granular masses. The till lies between the glacier ice and the bed and when flow in this layer is obstructed, part of the layer forms into an obstruction while the rest flows around it, thus streamlining the obstruction.

When a granular mass is disturbed it expands, and the dilatant material becomes more resistant to shear stress. It is this high resistance to shear stress that causes drumlins. The load on the till first leads to deformation by expansion until a critical load is reached. There is increasing resistance to deformation from a load of zero to the critical threshold value. Further deformation causes collapse and the required deforming load is reduced to a lower value. Additional deformation can now take place with lower loads. If the loads are greater than the higher critical value, the material is moved by the glacier, but if the load falls to less than the lower critical value, deformation is not possible. Drumlins must form in the zone where the stress or load range is between the first and second values, that between the minimum load necessary to initiate dilatance and the minimum load necessary to maintain dilatancy. Thus there are thresholds at both sides of the conditions suitable to the deformation of the till and the formation of drumlins by the flowing ice. This stress range will depend mainly on the thickness of the ice. Thus where the ice is very thick, no drumlins can form, and where it is too thin, their formation is also prevented. This would account for the occurrence of drumlins in certain well-defined belts where the stress range is within the necessary limits. These limits will vary to a certain extent according to the nature of the till, which will affect its dilatant properties. The large rock fragments in the till are responsible for its dilatant character. The collapse of the till layer within the critical zone gives rise to the obstructions that cause the flow to mold the moving till sheet into drumlin form. The presence or absence of water within the clay matrix may also affect the resistance of the material and hence the formation of drumlins. Once an initial obstruction is present, the drumlin is formed by the addition of layers of till. Often, stratified material is also included in the drumlin, indicating the presence of water where drumlins form. This may be produced by the regelation process. The width of a drumlin zone will depend on the zone over which the necessary conditions occur. In northern Ireland and south of the Great

Lakes of America the belt is wide. In the former this is probably due to the wide extent of fairly thin ice sheets generated in the local highland areas.

The form of drumlins and features of streamlined origin vary from the very elongated fluted ridges to features of ridge form but oriented perpendicular to ice flow. The latter are sometimes called Rogen moraines. The elongation of drumlins themselves varies according to the constancy of ice flow, being most elongated where the ice is flowing fastest and most consistently, the elongation indicating accurately the direction of flow of the ice forming the drumlins. This pattern is clearly seen in the drumlin field of the Solway lowland in northern England. The nature of the pattern probably depends on the characteristics of the subglacial material, the till, and the ice flow conditions. Thus a wide range of forms is possible provided that the critical thresholds for formation are fulfilled. The relatively small extent of streamlined forms shows that these thresholds are specific and limited in their occurrence.

Various attempts have been made to establish zones of types of depositional land-scape associated with glaciated terrain (Sugden and John, 1976; Boulton, 1972; Clayton and Moran, 1974; Flint, 1971). That suggested by Sugden and John consists of six depositional zones outside the zone of erosion. Next to the erosion zone in the sequence comes the zone of fluted ground moraine, a transitional zone to the active zone in which the dominant forces are streamlining and till lodgement. In the outermost part of this zone are drumlinized ridges and in the inner zone drumlins occur. The next zone is another transitional one in which drumlins, eskers, and localized Rogen moraines are found. This leads into the zone of wastage, where deposition from stagnant ice is domi-nant. Again there are two zones, the innermost consisting of disintegration features and the outermost of end moraines at the limit of glacial advance.

Clayton and Moran identify four zones, an inner suite, characterized by longitudinal shear marks, a transitional suite in which washboard moraines, eskers, circular disintegra-tion ridges, and partly filled meltwater channels occur. The third zone includes the mar-ginal suite, in which ice-walled lake plains, transverse compressional features, and simple hummocks are found. The final zone is the fringe suite, which consists of alluvial plains or sandar and meltwater channels.

Both suites have an essential element in common, in that the streamlined features occur in the inner zone where the ice is more active and the features characteristic of stagnant ice occur in the marginal zone. The separation of the features into zones illus-trates the importance of threshold conditions under which each type of feature can form. The relationship can be considered in connection with the pattern of occurrence of drumlins and Rogen moraines.

In Jämptland, where the two features occur in association, the drumlins tend to occur on the convex slopes, where the ice moves with extending flow, while the Rogen moraines are found on the concave slopes of the valleys, where the ice may also be mov-ing uphill. The pattern is shown in Figure 3. Where the ice is extending, the shear planes are directed downward toward the glacier bed, allowing effective movement over the bed, lodgement, and streamlining of the till or deposits into drumlin form by active ice. On the other hand, in the zone of compressive flow the shear planes pass upward from the glacier bed and allow material to be carried from the bed up into the ice. This type of flow has been associated with the formation of Rogen moraines. It would account for the pattern found by Lundqvist (1969), in which drumlins are found nearest the ice source, while progressively downglacier truncated drumlins are replaced by crescentic ridges and

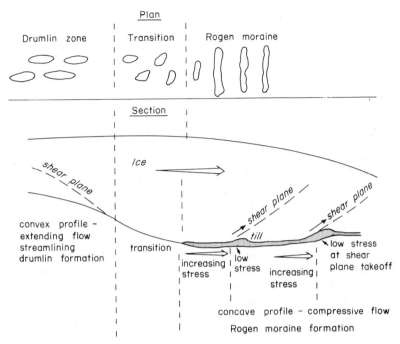

FIGURE 3 *Diagram to illustrate the possible connections between drumlins and Rogen moraines based on Lundqvist's theories.*

in turn by Rogen ridges, which lie transverse to the ice flow direction. Rogen moraines have been described by Moran (1971), who showed that they included sheared bedrock, which could have been carried up into the ice along the shear planes developed in the zone of compressive flow. The pieces of sheared bedrock become distributed in an imbricate pattern on the concave bedrock and are often buried by till to form the Rogen ridge. Thus the term "ice-thrust" ridge, which has also been used to describe these features, would appear appropriate from the genetic point of view.

The Rogen ridges are fairly large features that give a generally ribbed appearance to the landscape. Individual ridges may be over 1 km long, 10 to 30 m high, and spaced between 100 and 300 m apart. They are usually gently arcuate with a concave upglacier profile. They are often composed of stony till with large boulders on the surface, which probably originated as ablation moraine.

The ridges show a preferred till fabric oriented transverse to the ridge crest, which is, therefore, parallel to the ice flow. This could be explained if the material had moved up along thrust planes in the ice. Sometimes the ridge crests have been fluted or drumlinized. In order to account for the fluting of the crests, a subglacial origin seems more likely than a marginal one. It also implies that the ice was active when the ridges were formed. They could be accounted for if the moving ice passed up along a shear plane, at which point the stress on the surface would be reduced below the threshold value to allow deposition of the ridge. The next ridge would then mark the position of the next shear plane when the same threshold would again be reached and deposition occur by pressure melting and lodgement of the till (see Fig. 3). Where both the transverse features and the streamlined ones occur together a common origin seems likely. This could occur where a critical range of debris load, basal ice temperatures, and type of ice movement occurred together. Other threshold values could give rise to either one or the other process occurring alone. The fact that they often are found in close association suggests that subtle thresholds exist in this part of the glacial system.

The formation of moraines depends on a wide range of different processes that produce many different forms. Some of these depend on thresholds for the necessary conditions to be fulfilled. As an example of moraines associated with cold glaciers, the Thule-Baffin moraines will be considered. These are ice-cored moraines that have restricted occurrence, forming on the margins of some cold ice masses. The moraines of the Barnes ice cap of Baffin Island are of this type and have been subject to detailed investigation (Weertman, 1961; Hooke, 1973). They have also been referred to as shear moraines, but shearing is no longer thought to be an essential element in their formation. They have an ice core covered by thin till. The problem is to account for the presence of the till on the ice surface, as there are no outcrops or surface debris above the marginal moraines. The most likely process is related to temperature changes within the ice. When the basal temperature rises to the pressure melting point, subglacial material would be incorporated into the basal ice, which on subsequent refreezing would then be transported to the snout, where it would eventually melt out at the surface aided by compression and thrusting. Thus the process requires the ice to pass through the freezing threshold at least twice. The compressive strain toward the margin of the ice mass would account for upward movement of the debris laden ice near the snout. Compression also leads to foliation within the marginal ice, giving steep upglacier dips, especially if the flow is obstructed by earlier moraines or the accumulation of wind-driven snow around the ice front. The effectiveness of a frontal dam is seen in the Svalbard glaciers, where it results in shearing and thrusting of the ice behind it. The margin of the Barnes ice cap also illustrates this point (Hooke, 1973). According to his theory a wedge of snow and superimposed ice forms at the margin of the ice cap, gradually extending its limit outward as it builds up. The ice behind the growing obstruction moves increasingly slowly toward the margin, but with an increasing upward element of flow. This ice slowly overrides the wedge of snow and superimposed ice to form a zone of increasing height. Where the debris amount within the ice is sufficiently great enough, material will cover the ice to prevent or slow down ablation and an ice-cored ridge will form. The compressive flow beneath the ridge causes its height and gradient to increase. Up-ice from the growing ridge the flow will become less compressive, the vertical velocity will decrease, and a trough will tend to form. Owing to the lower flow velocity, the trough will get deeper as ablation takes place more effectively with the thinner debris cover and the ice is not replaced by flow. In this way an isolated debris-covered ridge can form a moraine at the glacier margin. The process requires that the amount of debris within the ice is above the threshold necessary to provide a cover thick enough to prevent or reduce ablation below the amount of upward ice movement. In this way the ridge can gain in height with time, until it becomes separated from the active part of the ice cap. Moraines could form in this way when the glacier mass budget is balanced, although a preliminary advance may be required to initiate the process, while a negative balance would facilitate the formation of the trough behind the growing ridge. Recycling of debris may play an important part in the building of the moraines, whereby material slumps down the steepening ridge face to be covered by snow and superimposed ice in subsequent seasons. This material may later reemerge and again provide a protective cover. The process is a slow one and it would take between 40 and 90 years to form one ridge, including recycled material. Without this material it could take several hundred years, which is the time needed to transport all the material near the ice margin at any one time to the edge of the ice cap, for a moraine ridge to be formed.

Another very different type of moraine that also requires the operation of a threshold is the cross-valley moraine, which is similar to the de Geer moraine. These moraines are long narrow ridges parallel to former glacier fronts and may be formed annually. The cross-valley moraines are associated with former lakes and were formed below the water level of the lake. Their distribution is thus related to the conditions under which lakes could form as glaciers retreated. They are common on the west side of Baffin Island, in parts of Scandinavia and Iceland, and around the shores of Hudson Bay. The spacing of the moraines varies between one every 320 m to one every 35 m.

In Baffin Island they are composed of sandy till. The orientation of the stones within the till has a preferred direction perpendicular to the length of the ridge. The fabrics are stronger on the proximal side of the ridge and more random on the distal side. The characteristics of the moraines where they are closely spaced supports the view of Andrews that they are formed at the margin of a retreating ice mass ending in a proglacial lake. They probably form annually. It is argued that they form when water-soaked till is squeezed into ridges at the margin of the glacier. The influx of melt water into the frontal ice during the spring melt could act as the threshold or trigger that results in the structural collapse of the till, as the pore water pressure increases. The ice would then settle into the till and this would be forced toward the margin, where it would be squeezed into a ridge. The squeezing would produce the strong proximal fabrics, while the weak distal fabrics occur as the till slumps over the crest of the moraine ridge. The threshold in this instance is due to the annual melting that leads to a critical increase in the pore water pressure and collapse of the frontal ice under certain conditions. Similar features have been noted in Iceland, where they have formed subaerially.

PATTERNS OF EROSION AND DEPOSITION

The patterns of erosion and deposition by ice sheets have been influenced by the alternation of growth and decay of the ice (Fig. 4) Thus an area will first be affected by advancing ice, it will later be covered by thick ice at the maximum, and finally will be affected by the conditions of glacial retreat. It is not always easy to differentiate the features formed under these different conditions, and earlier ones may be modified later as conditions change. The thresholds that determine whether the ice is erosive or depositional are critical to the study of large-scale glacial patterns. The model proposed by Boulton et al. (1977) for Britain illustrates some of these points.

Increased glacial erosion will occur as ice velocity increases provided that other variables are suitable and the threshold for maximum erosion is not passed. Increased effective pressure will increase frictional resistance and decrease erosion. The ice velocity and ice temperature are important in determining erosion rates, erosion being less effective where the ice is cold. Effective erosion would not occur near the center of an ice sheet, owing to low velocities. Where such areas show marked erosion, this probably occurred later when the ice was thinner and moving faster, owing to high local ablation gradients. Effective erosion occurs near the margins of the ice sheet where flow is fast, the ice is still fairly thick, and the basal ice is temperate.

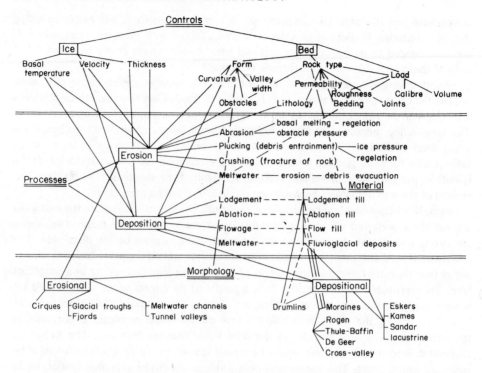

FIGURE 4   *Some elements of the glacial system to stress the complex interrelations that determine the value of threshold phenomena.*

Patterns of glacial deposition include drumlin belts and melt out supraglacial deposits. The former occur where the ice was flowing effectively and lodgement till thick, and the latter occur where the ice became stagnant. Subglacial lodgement till is dominant where the ice was temperate, while the supraglacial deposits occur in cold ice where most of the debris is carried englacially and later melts out. Drumlins usually occur in the lodgement zone till fairly near the margins of glaciers. The availability of lodgement till depends on the sliding velocity of the ice. It occurs where the sliding velocity is less than the threshold value of 50 to 100 m/yr, depending on the ice thickness and other controls. At speeds above the threshold the bed is swept clear. Velocities below the threshold on the upglacier side of the fast-flowing marginal zones may have allowed the deposition of lodgement till and the formation of drumlins. The absence of drumlins along the east coast of England, apart from the Tweed valley, may be due to high velocities in this zone. Surging has been suggested to account for these high velocities. A final phase of deposition can follow one of erosion as the ice sheet retreats and decelerates.

Clayton and Moran (1974) have applied a somewhat similar argument to the glacial features of North America, recognizing four distinct zones in which processes and conditions differ. These are the preadvance, the subglacial, the supraglacial, and the postglacial. Each is controlled by specific processes and can be differentiated by the features produced. Thresholds separate the different processes that dominate the four zones.

CONCLUSIONS

Critical conditions are necessary to explain the origin of many glacial features. Triggers or thresholds control the nature of processes and their effect, both in space and time. The most important variables affecting the thresholds include:

1. The thickness of the ice.
2. The speed of flow, which is partly dependent on the thickness.
3. The temperature of the ice, which also affects its flow.
4. The nature of the glacier bed, particularly its permeability, which is related to the effective ice pressure on the bed.
5. The load of the glacier, which can be subglacial, englacial, and supraglacial.

All these variables change both over time and spatially; thus the resulting landscape revealed by deglaciation is highly complex, and variable.

An attempt has been made in Figure 4 to indicate the complexity of the glacial system by drawing attention to some of the many links between the three main elements of the system. These are the processes, the material on which they are acting and which they produce, and the morphology that they are modifying. The thresholds that have been considered in the paper are related to the controls that determine the operation of the processes. These are essentially the character of the ice and of the surface over which it is contact. Any one of the variables associated with these controls can affect the thresholds that determine the effect of the ice. In most situations all the controls exert a combined influence on the thresholds. It is, therefore, not possible to set specific limits on any one variable, except within a large range, because they are interdependent. Thus the threshold for lodgement to occur depends on the ice thickness, basal temperature, and velocity as well as on the characteristics of the bed and the material carried by the ice. The system illustrated in the figure has been much simplified and many relevant forms have been omitted; in general, only those features specifically discussed have been included and small forms have been omitted.

In some areas evidence of effective erosion may be found alongside unaltered ground or thick deposition, while evidence of flow in streamlined forms may be partially hidden under evidence of stagnation in disintegration deposits. A good example of selective glacial erosion is seen in the through valleys of the Finger Lakes area of New York state in the glaciated Appalachian Plateau and in the Cairngorms of eastern Scotland. In both these areas powerful, thick, and fast-flowing ice streams from distant sources eroded deep troughs through rising ground across their paths. On the adjacent high ground slow-moving, thin ice protected the land surface. The orientation of preexisting valleys guided the ice flow along discrete paths. This factor is an important one in the distribution of through valleys and deeply glaciated troughs. The juxtaposition of diverse features can often be explained by the operation of thresholds, giving abrupt switches from one state to another, and from one process to another both in space and time. Some specific examples of these abrupt discontinuities have been outlined. As more is learned of the controls that affect the processes, the pattern of glaciated landforms will be more fully understood. The critical or threshold variables can be recorded and related to the processes and the materials on which they are acting to produce the landforms. There are many complex feedback processes involved in these interrelationships.

REFERENCES CITED

Andrews, J. T., Davis, P. T., and Wright, C., 1976, Little Ice Age permanent snowcover in the eastern Canadian Arctic: extent mapped from Landsat-1 satellite imagery: Geogr. Ann., v. 58A n. 1, 2, p. 71–81.

Barry, R. G., Andrews, J. T., and Mahaffy, M. A., 1975, Continental ice sheets: conditions for growth: Science, v. 190, p. 979–981.

Bjornsson, H., 1974, Explanation of jökulhlaups from Grimsvotn, Vatnajökull, Iceland: Jökull Ar., v. 24, p. 1–26.

Boulton, G. S., 1972, The role of thermal regime in glacial sedimentation: Polar Geomorphology: Inst. Brit. Geog. Spec. Publ. 4, p. 1–19.

——, 1974, Processes and patterns of glacial erosion: in Coates, D. R., ed., Glacial geomorphology: SUNY Publ. Geomorph., p. 41–87.

——, 1975, Processes and patterns of subglacial sedimentation: a theoretical approach: in Wright, H. E., and Mosley, F., eds., Ice ages: ancient and modern: Geol. J. Spec. Issue, Liverpool.

——, Jones, A. S., Clayton, K. M., and Kenning, M. J., 1977, A British ice sheet model and patterns of glacial erosion and deposition in Britain: in Shotton, F. W., ed., New York, Oxford University Press, p. 231–246.

Clayton, L., and Moran, S. R., 1974, A glacial process-form model: in Coates, D. R., ed., Glacial geomorphology: SUNY Binghamton Publ. Geomorph., p. 89–119.

Donovan, D. T., 1973, The geology and origin of the Silver Pit and other closed basins in the North Sea: Proc. Yorks. Geol. Soc., v. 39, n. 2, p. 267–293.

Flinn, D., 1973, The topography of the sea floor around Orkney and Shetland in the northern North Sea: J. Geol. Soc. London, v. 129, p. 39–59.

Flint, R. F., 1971, Glacial and Quaternary Geology: New York, Wiley, 892 p.

Hooke, R. LeB., 1973, Flow near the margin of the Barnes ice cap and the development of ice-cored moraines: Bull. Geol. Soc. Amer., v. 84, p. 3929–3948.

Lewis, W. V. ed., 1960, Norwegian cirque glaciers: R. Geog. Soc. Res. Ser., v. 4, 104 p.

Linton, D. L., 1963, The forms of glacial erosion: Trans. Inst. Brit. Geog., v. 33, p. 1–28.

Lliboutry, L., 1968, General theory of subglacial cavitation and sliding of temperate glaciers: J. Glaciol., v. 7, p. 21–58.

Lundqvist, J., 1969, Problems of the so-called Rogen moraine: Sv. Geol. Unders., v. C648, p. 1–32.

Meier, M. F., and Post, A., 1969, What are glacier surges?: Can. J. Earth Sci., v. 6, n. 4, p. 807–817.

Moran, S., 1971, Glaciotectonic structure in drift: in Goldthwait, R. P., ed., Till, a symposium: Columbus, Ohio State Univ. Press, p. 127–148.

Nye, J. F., 1976, Water flow in glaciers: jökulhlaups, tunnels and veins: J. Glaciol., v. 17, n. 76, p. 181–207.

——, and Martin, P. C. S., 1968, Glacial erosion: Assemblee Gen. de Berne, Commn. Snow and Ice (1967) Int. Assoc. Sci. Hydrol. Publ. 79, p. 78–86.

Palmer, A. C., 1972, A kinematic wave model of glacier surges: J. Glaciol. v. 11, n. 61, p. 65–72.

Shoemaker, E. M., 1976, Temperature-gradient induced mass-instability theory of glacier surges: J. Glaciol., v. 17, n. 77, p. 433–445.

Smalley, I. J., and Unwin, D. J., 1968, The formation and shape of drumlins and their distribution and orientation in drumlin fields: J. Glaciol., v. 7, p. 377–90.

Sugden, D. E., and John, B. S., 1976, Glaciers and landscape: London, Arnold, 376 p.

Weertman, J., 1961, Mechanism for the formation of inner moraines found near the edge of cold ice caps and ice sheets: J. Glaciol., v. 3, p. 965–978.

——, 1969, Water lubrication mechanism of glacier surges: Can. J. Earth Sci., v. 6, n. 4, p. 929–942.

Wilson, A. T., 1964, Origin of the ice ages: an ice shelf theory for Pleistocene glaciation: Nature, v. 201, n. 4915, p. 147–149.

Woodland, A. W., 1970, The buried tunnel valleys of East Anglia: Proc. Yorks. Geol. Soc., v. 37, p. 521–578.

15

RIVER ICE PROCESSES: THRESHOLDS AND GEOMORPHOLOGIC EFFECTS IN NORTHERN AND MOUNTAIN RIVERS

Derald G. Smith

INTRODUCTION

In winter, rivers located in northern and mountain environments are severely affected in cold weather and subsequent accumulations of river ice. Several types of ice accumulations can occur, depending on channel characteristics, weather conditions, and volume of river discharge. Thresholds are indicated by dramatic changes in the processes of ice accumulation or destruction, which can in turn result in significant geomorphologic changes to river channels and floodplains.

The main objective of this chapter is to show how the concept of a geomorphic threshold can be applied to river ice processes and to a better understanding of the ensuing geomorphologic effects. A second objective is to create greater awareness of the various river ice processes that are common to northern rivers. A final objective is to present some new results from an ongoing study on river ice breakups and drives in western Canada.

A limited number of published studies plus personal research and observations will provide the information and field examples of specific river ice processes and effects. Much of the discussion on anchor ice and channel icings which come from the literature will be supplemented by the author's observations in Idaho and Montana. Most of the discussion on ice breakup, drives, and jams comes from my research on rivers in Alberta carried out between 1974 and 1978.

CONCEPT OF THRESHOLD

The concept of geomorphic threshold as used in this chapter refers to a dramatic change in the rate or character of landform response in a given fluvial system. A geomorphic threshold occurs when a landform response changes from a linear to a nonlinear rate (Fig. 1). For example, if the slope of a graded single-channel river is increased, the channel banks will eventually unravel at some slope threshold and flow will form a braided network. On the other hand, if river velocity is increased and flow changes from laminar to turbulent and there is still no movement of bed sediment, then neither has the landform changed nor has a geomorphic threshold taken place but instead a hydraulic threshold has occurred.

The concept of geomorphic thresholds has not been applied per se to river ice processes, although most researchers of geomorphic processes have no doubt been aware

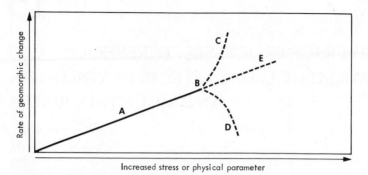

FIGURE 1 *Geomorphic threshold (B) occurs when the rate of. landform change per unit of stress or physical parameter changes from a linear rate (A) to nonlinear rates (C, D).*

of the threshold idea, and have occasionally used equivalent terms, such as tipping point, triggering effect, boundary condition, chain reaction, dramatic transition, and dynamic response. In fluvial hydraulics most of the dimensionless numbers (i.e., Froude and Reynolds) define threshold conditions or transitions in fluid behavior. The concept of threshold has been around for a long time, but has not been vigorously applied to geomorphology processes and their effects on landforms.

Progress in the study of geomorphologic effects of river ice has been slow because most studies consist of observations with few data. Civil engineers, concerned primarily with river structures, have collected most of the observations and data so far; consequently, geomorphic aspects have been a secondary concern. Both engineers and geomorphologists who study river ice quickly realize that research is very difficult and costly. Poor weather conditions, limited access to rivers, lack of mobility, and the unpredictable nature of river ice accounts for most of the field problems and the limited number of good process studies available.

Four distinct geomorphic thresholds involving different river ice processes commonly occur in northern and mountain rivers. Each threshold marks a sudden landform response or change. In each case the dominant threshold parameter appears to be a characteristic of channel morphology. However, secondary parameters, such as the strength of ice, a sudden increase of flow discharge, and temperature changes, are often important.

The first geomorphic threshold applies to river ice drives, with the *threshold of channel width* as the dominant physical parameter. The landform response is an enlargement of channel area by a factor of about three times (Smith, 1979). Numerous personal observations of rivers in Alberta during breakup suggest that the width determines whether the ice will move as a drive and scour the channel or not move and melt in place.

The second threshold involves river ice jams during a breakup of an ice drive. A *threshold of irregular channel morphology* consisting of sharp meander bends, channel constrictions, and islands are proposed as the dominant physical parameters. The major landform response of flooding is overbank deposition and scour. A precise determination of the threshold for ice jams is very difficult to make, because of the influence of important secondary parameters, such as volume of flow, discharge, and ice strength. Despite a lack of precise data, we know that jams often recur at the same channel locations from year to year.

The third threshold involves anchor ice which forms on river beds. The *channel slope threshold* is proposed as the dominant parameter which influences the formation of anchor ice or a surface ice cover. The landform response is the removal of bed sediment frozen to the anchor ice when it lifts and floats down river. Secondary parameters such as weather conditions and water temperature are also important. Slope, however, is necessary to initiate anchor ice in the first place. Without it, secondary parameters would be insignificant.

The fourth threshold involves channel icings, a condition whereby all the winter flow freezes and accumulates in-channel and sometimes spreads onto the floodplain. A *flow depth threshold* appears to be the dominant parameter that delimits whether an icing or a surface ice cover will form. The geomorphic effect from an icing includes channel relocation and flooding in spring. Icings are common in braided river systems, where winter flow in channels is very shallow. Channel icing can begin either as anchor ice or as surface ice cover before part or all of the channel flow is frozen and stored as ice until spring.

The morphologic thresholds are presented intentionally without numerical values, because the value of a threshold parameter fluctuates from year to year for any given river system, depending primarily on flow discharge, atmospheric conditions, and ice strength. In reality, a range of threshold values exist. For example, a threshold range is particularly apparent at bankfull stage. Because bankfull varies along a reach of channel, a range of bankfull stages must exist. Since the bankfull variation is often expressed as an average value, most people interpret it as a critical or threshold value. As with bankfull discharge, other geomorphic thresholds, more often than not, occur as a range of values over time and space.

FORMATION OF FRAZIL ICE

Frazil ice is the foremost building block in the accumulation of river ice. In the fall, water becomes supercooled and frazil ice begins to form at about $-0.01°C$, as measured at Goldstream Creek near Fairbanks, Alaska (Gilfilian et al, 1972). Frazil particles are hexagonal, disk-shaped plates (0.1 to 0.5 mm in diameter) (Williams, 1959 and Michel, 1966). In a turbulent stream frazil can form through the entire vertical column of river flow (Fig. 2). In very slow moving rivers, or in lakes without mixing, however, frazil occurs only on the surface.

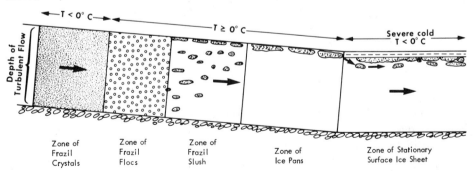

FIGURE 2 *Schematic diagram showing various ice formation stages from frazil ice crystals to a surface ice cover.*

The formation of floes of frazil slush and ice pans are the next stage in the development of stationary ice sheets that straddle rivers in late fall (Fig. 2). As frazil crystals float to the surface, they floc into soft, mushy-like masses of slush. The frazil slush masses then float downriver, bumping into each other and forming rim-like edges on each slush mass. With continued cooling the slush masses come together, enlarge, and form ice pans. If rapids or strong riffles are encountered, the slush breaks up and disintegrates, but re-forms again in pools and quiet water. In its movement downriver slush floes will lodge along the shoreline, freeze, and contribute to the growth of border ice. As the area of open water is reduced, so is the production of frazil. During extreme cold conditions slush and ice pans press together, freeze, and come to a stop, forming a continuous stationary ice cover. Frazil slush and pans will sometimes come to rest against the ice cover, freeze, and cause the cover to advance upriver. At other times the incoming slush and ice pans will flow underneath the ice cover, accumulate, and freeze in place, as shown in Figure 2.

A. Frazil ice deposition

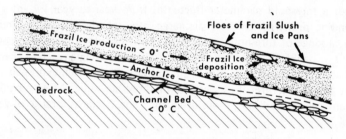

B. Sediment transport via rafting of anchor ice.

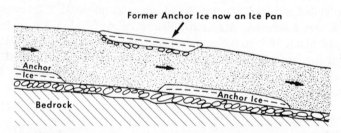

FIGURE 3 Schematic diagram showing the accumulation (A) and removal (B) of anchor ice.

Stationary ice covers on rivers tend to form below specific flow velocities. Many years of observation by researchers have shown that slush floes cannot develop an ice cover at velocities that exceed 0.7 to 0.8 m/sec (2.3 to 2.6 ft/sec) (Estifeev, 1958). Kivisild (1959) stated that the Froude number of flow in front of an ice cover is a valid criterion to determine whether the cover will progress upriver or remain stationary as ice floes pass underneath it. From measurements in many rivers, he concludes that the limiting value for the Froude number is about 0.8.

Frazil crystals are also the building blocks for anchor ice, a second major type of river ice accumulation (Fig. 3). In supercooled river water, bed material also becomes supercooled. In rivers with strong turbulent mixing, frazil crystals accumulate in low-flow areas and cling onto projections (boulders and weeds) on the river bed. Adhesion is rare on flat muddy beds, where heat transfer occurs from the mud to the water (Williams, 1959). In contrast to gravel bed rivers, ice covers form rather quickly over muddy bed rivers which have decreased velocity and turbulence.

In summary, while frazil ice may seem insignificant in a discussion of geomorphic thresholds, it is the basic material essential in the formation of river ice.

CHANNEL WIDTH THRESHOLD AND ICE DRIVES

An ice drive, run, or floe is the "flushing out" or removal process of surface ice from a channel (Fig. 4). Some ice drives are very spectacular while others are uneventful, de-

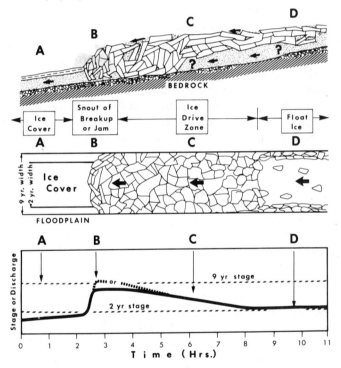

FIGURE 4 *Idealized diagram showing the stages of ice break-up and drive over distance and time of a typical "large" northern river.*

pending on the strength of the ice cover, duration, and magnitude of runoff which initiates a breakup and drive. If a period of persistent warm spring weather should occur, or an early spring rain, a rapid rise in discharge and flood wave will often cause a sudden and powerful breakup and drive. If cool spring weather persists, the surface ice cover may slowly melt in place or become so weak that it will turn into slush and float downriver without incident.

In rivers of the northern plains, the occurrence of ice drives is directly influenced by channel width. In such rivers, a threshold of bankfull channel width apparently divides rivers into two groups: those that have ice drives and those that do not. Of course, secondary factors, such as channel morphology, river discharge, velocity, and ice strength, will all affect the occurrence of ice drives.

The geomorphologic effects of ice drives include channel enlargement, formation of a distinctive channel-cross-section geometry, deposition of cobbles and boulders on upper channel margins, and erosion of cut banks at meander bends. All four effects will be discussed in greater detail near the end of this section.

The rivers that have the most dramatic ice drives and associated geomorphologic effects are those in which breakup begins in the upper part of the watershed and moves downriver. In North America two groups of rivers have such ice-drive regimes. The most obvious group consists of northward-flowing rivers, with the Yukon, Mackenzie, Liard, and Nelson as the most prominent Western examples (Mackay, 1963; Newbury, 1968). The second group of rivers are those with their headwaters in the chinook belt of the high plains of B.C., Alberta, and Montana. Severe springtime chinooks (warm westerly winds) can suddenly raise temperatures and cause rapid snow-melt runoff to initiate a breakup and drive. In most other rivers, ice breakup occurs in segments, starting at the downstream end of a river. In general, south-flowing rivers follow this breakup pattern.

Results from an ongoing study of 20 rivers in Alberta, Canada, (Table 1) indicates that a threshold of channel width separates rivers into those that have ice drives and those that do not. Data on the hydraulic channel geometry at bankfull were plotted to demonstrate the threshold of channel width (Fig. 5). A comparison of width, depth, and velocity in Figure 5 indicates that there is a significant break in width between "small" (6 to 150 m^3/sec bankfull) and "large" (991 to 14,160 m^3/sec bankfull) rivers. The break occurs between 54 and 107 m width.

Field observations of rivers during spring breakup can help to interpret this graphical break in channel width. In the Medicine and Beaver Rivers, two of the three largest small rivers, no ice drives were observed following breakup over a period of 4 years. In Figure 6, the ice cover in the Medicine River was broken into large pans but had not moved any significant distance downriver and was melting in place. The same situation existed in the Beaver River, a larger river, but with more meanders. Neither river had riparian trees with scars as evidence of previous ice drives. However, for "large rivers," ice drives were commonplace having been observed, photographed (Fig. 7), and well documented in newspapers. Nearly every tree along the bankfull channel boundary has been severely scarred by moving ice (Fig. 8). Willows on the channel banks all exhibited multiple scars, repeatedly "raked" by ice drives. Apparently, channel widths were sufficient to allow ice movements and a sustained ice drive.

The two obvious clusters of small and large rivers in Figure 5 might indicate that two populations were intentionally selected for study and the "middle-size rivers" were neglected. The absence of medium rivers is explained, for the most part, by the enlargement of channels caused by bank and bed scour from ice drives. In a recent study, Smith (1979) found that channel-cross-sectional areas at bankfull for ice-drive rivers in Alberta are about three times larger than for non-ice-drive rivers. The enlarged channels were compared at bankfull discharge and discharge at the 1.6-year return period, a return period suggested by Dury (1970) as the average return period for bankfull. All of Dury's (1970) rivers, however, were selected from Australia, England, and the United States.

TABLE 1 *Data on Bankfull Stage, Discharge, and Return Periods and Associated Hydraulic Geometry for 20 Small and Large Rivers in Alberta, Canada*

River and location	Bankfull stage (method)	Bankfull Q at or near a station (m³/sec)	Bankfull return period (yr)	Mean bankfull channel geometry			
				W (m)	D (m)	A (m²)	V (m/sec)
Small Rivers							
1. Beaver—nr. Cold Lake	8.90	132.0	2.02	53.6	3.20	171.9	0.77
2. Big Knife—nr. Gadsby	2.32	6.5	1.50	10.4	1.62	16.7	0.38
3. Little Paddle—nr. Mythp.[a]	2.07	29.5	2.80	23.5	1.40	33.0	0.88
4. Little Red Deer—nr. mouth[b]	2.12	79.0	1.96	44.5	1.30	57.6	1.38
5. Manyberries—Bodin Fm.	2.13	17.2	2.35	21.9	0.70	15.4	1.12
6. Medicine—nr. Eckville	4.90	80.7	2.19	34.1	2.62	90.0	0.90
7. Mistaya—nr. Waterfoul Lk.[a]	1.74	27.8	1.24	36.9	1.13	40.9	0.67
8. Paddle—nr. Rkft. Br.[a]	3.32	46.2	2.98	22.9	2.04	46.6	1.01
9. Prairie—nr. Rocky Hs.[a]	1.52	62.3	3.60	31.1	1.22	37.6	1.65
10. Swan—nr. Kinuso[a]	4.69	149.0	1.38	42.1	3.78	158.6	0.94
11. Vermillion—nr. Mannville[a]	2.01	7.9	1.66	18.6	1.10	20.4	0.40
			Average return period = 2.15				
Large Rivers							
12. North Sask—Edmonton	8.84	2,676.2	8.65	182.9	7.16	1309.9	2.04
13. Oldman—Brockit	3.66	999.2	6.25	106.7	3.53	376.3	2.64
14. Oldman—Lethbridge	4.88	1,529.3	8.32	201.2	4.02	808.2	1.89
15. Peace—Peace R.	14.63	14,160.0	8.73	438.9	9.45	4134.1	3.41
16. Red Deer—Red Deer	4.27	948.7	7.64	140.2	2.90	407.4	2.33
17. Red Deer—Drumheller	4.82	1,047.8	8.46	131.7	3.57	469.2	2.23
18. Red Deer—Bindloss	3.66	1,132.8	10.88	131.1	4.36	571.3	1.98
19. Smoky—Watino	7.71	5,522.4	13.50	274.3	5.76	1583.9	3.47
20. South Sask.—Med. Hat	5.49	2,662.1	9.37	167.6	5.64	947.6	2.80
			Average return period 9.09				

[a]Data from Kellerhals et al. (1972).
[b]Bankfull at low-level bench.

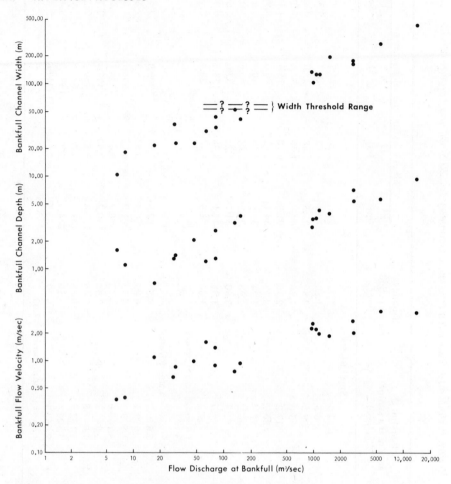

FIGURE 5 *Hydraulic geometry from Table 1, showing the range of width threshold separating rivers without ice drives from rivers with ice drives.*

Smith calculated that bankfull discharge of the enlarged channels was about 4.7 times larger than for "normal" rivers with 1.6-year return periods.

If it is assumed that all the large rivers with enlarged channels in Table 1 and Figure 5 have been formed by ice drives, it is possible to narrow down the range of threshold width. Dividing the smallest large river (949 m³/sec) by 4.7, the enlargement discharge factor, a bankfull flow of 202 m³/sec would approximate a near non-ice-drive river. Therefore, for Alberta, a width threshold should occur somewhere between 45 and 60 m. This threshold range should be taken with caution until it is confirmed by additional field research. Both the Beaver and Swan Rivers are very close to the width threshold, beyond which dramatic channel enlargement could occur.

The apparent absence of medium-size rivers in Alberta is also explained by the fact that only river channels at bankfull are considered. The smaller "large"-size rivers in Figure 5 would be medium-size if their channels at bankfull had not been enlarged. In terms of average annual flow they are medium-sized, but they become large when con-

(a)

(b)

FIGURE 6 *Comparison of summer flow (a) with spring break-up but no ice drive (b) in the same channel reach in the Medicine River. Views looking downstream.*

sidering channels at bankfull. In essence, channel enlargement has resulted in the hydraulic geometry of the large rivers in Figure 5 to be shifted to the right.

The average return period of 16.7 years in Smith's 1979 study of bankfull discharge in Alberta rivers is now regarded as too high. The study did not include enough small rivers, which have small return periods and the inclusion of rivers with abnormally high return periods (e.g., 45 years), which may be attributed to man-induced upstream flow regulation or misinterpretation of a terrace for a floodplain.

Some large regulated rivers have little or no chance of having a drive. If a steady flow can be maintained during the spring time, a breakup and drive is very unlikely; such a condition exists on the Bow River at Calgary and South Saskatchewan at Saskatoon.

(a)

(b)

(c)

FIGURE 7 *Time-lapse photography of a meander in the Red Deer River taken about 15 minutes apart of prebreakup (a), breakup (b), and ice drive (c). Views looking downstream.*

FIGURE 8 *Two cottonwood trees on a bank of the North Saskatchewan River, Edmonton, severely affected by ice-drive abrasion.*

On the other hand, in nonregulated rivers, drives are almost an annual event, as in the Red Deer River north of Calgary. Most drives go unnoticed because being at the right place at the right time to observe a breakup and drive is a problem. Sometimes breakups occur at night or during rain and snow storms. Most researchers do not have the time or resources to wait and travel quickly to breakup events. Thus future research of breakups will probably be very slow.

Channel enlargement by ice drives is not an instantaneous event, but is essentially a maintenance activity of a previously ice shaped channel. If ice drives should cease because of upriver regulated flow, the cross-sectional area of a channel would decrease by sedimentation and vegetation encroachment, as is presently happening on the Bow River at Calgary. The response time for channel-size reduction is slow, requiring perhaps several hundred years.

The morphology of enlarged channels shaped by ice-drive rivers is somewhat different, often consisting of an inner channel within a larger channel (Fig. 9). Bray (1972) observed a low-level bench associated with the present river. He found that the movement of bed sediment occurred for all reaches with a discharge in excess of the 2-year summer flood stage equal to the low-level bench, concluding that this represented dominant channel-shaping dishcarge. The 2-year flood corresponds closely with Dury's (1970) conclusion that the 1.6-year flood is the dominant discharge which shapes channel geometry.

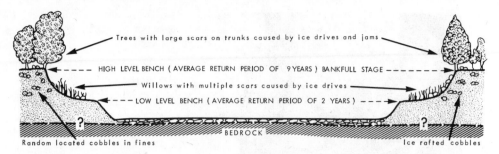

FIGURE 9 *Idealized channel cross section common in large rivers of Alberta. The low-level bench corresponds to the 2-year flood by Bray (1972).*

Research in progress by the author partially confirms Bray's (1972) observations that the low-level bench occurs frequently and corresponds to the 2-year summer flood. This compares closely to the small rivers in Figure 5, which have an average return period of 2.15 years. Bray also related his high-level bench with the 10-year summer flood. An example of both benches along the Little Red Deer River is shown in Figure 10.

The channel geometry, with two bench levels shown in Figure 9 and 10, is believed to be shaped and maintained by two different channel processes: summer floods, which maintain the low bench, and ice drives, which preserve the higher bench. The high-level bench or floodplain corresponds to the 9.09-year bankfull summer flood return period shown in Figure 11 for the large rivers listed in Table 1. Summer flood return periods were calculated for small and large rivers shown in Figures 5 and 11. Bankfull discharge return periods for ice drive flows were not calculated for a number of reasons: (1) ice jamming is frequent during drives; (2) ice content in ice-drive flows is unknown; (3)

FIGURE 10 *Right bank of the Little Red Deer River, showing both the lower and upper benches. Photo was taken in September during very low flow. The trees lining the upper bench are all ice-scarred.*

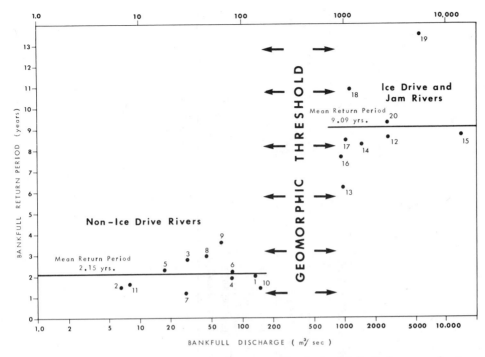

FIGURE 11 *Using the annual series of summer floods, bankfull discharges and return periods were calculated for small and large rivers in Alberta.*

cold water temperatures and ice blocks greatly increase viscosity, reducing velocity; (4) stage recorders are often disrupted during ice drives; (5) ice drives may only have a duration of a few hours; and (6) measurement of mean velocity of an ice drive is not possible due to moving ice blocks. Data in Table 2 demonstrated the discrepancy in using ice-jam-flow data. However, return periods for bankfull stage can be calculated and perhaps should be used for study of northern rivers.

Abrasion of channel banks by ice drives probably ceases early during the recession of river stage, causing ice blocks to become grounded on the lower bench (Fig. 9 and 10). In Figure 12A and B grounded ice blocks form a gorge-like border on the lower bench, forcing the remaining drive to continue within the 2-year inner channel. In many parts of the channel bed, gravel and cobble armoring is often stripped, gouged, and pushed into mounds, exposing bedrock in places. Both channel incision and widening are suggested as the major geomorphic effects resulting from ice-drive processes. Ice drives are complex, dangerous, and most safely studied by photography, as pictorially summarized in Figure 8.

Sediment deposition on floodplains is a minor effect of river ice drives. McPherson (1966) observed that ice had pushed sediment onto the floodplain to form a low discontinuous ridge some 30 to 50 cm high bordering the channel. Henoch (1973) observed ice-pushed ridges up to 17.3 m above the Liard River at summer base flow. From personal observations of large rivers in B.C., Alberta, Saskatchewan, and Montana, gravel, cobbles, and boulders often have been rafted and pushed onto floodplains. Usually, the coarse sediments are deposited in mounds or cobble clusters.

TABLE 2 Comparison of Ice-Jam-Caused Floods with Summer Floods in the Red Deer River, Red Deer, Alberta, Canada[a]

Ice-jam floods			Summer floods		
Date	Peak stage[b]	Daily discharge[c]	Date	Peak stage[b]	Daily discharge[c]
April 8, 1913	5.45	113.3	June 28, 1915	5.81	1585.8
April 7, 1925	5.32	300.2	June 2, 1923	4.42	985.4
April 4, 1943	6.69	274.4	June 4, 1929	4.34	968.4
April 7, 1955	5.61	339.8	June 24, 1952	4.52	1073.2
April 8, 1958	4.68	88.1	August 26, 1954	5.09	1206.9

[a]Haden et al. (1960); modified from Burton (1965).
[b]Meters (m) on the gaging staff.
[c]Cubic meters per second (m³/sec).

(a)

(b)

FIGURE 12 *(A) Ice blocks stranded on the low-level bench during the recession of an ice drive in the Red Deer River, 1978. (B) Close-up view of ice blocks and inner channel boundary located directly in front of the pumping station on the floodplain in photograph (A). Views looking upriver.*

IRREGULAR CHANNEL MORPHOLOGY THRESHOLD AND ICE JAMS

Ice jams which commonly occur during ice drives in large northern rivers are influenced by a complex of channel morphologic factors, such as sharp meander bends, channel constrictions, and islands. Jams of varying severity generally occur at the same channel locations from year to year in a given river. For example, at the Ramparts, a narrow bedrock constriction on the lower Mackenzie River near the Arctic Circle, annual ice jams here shoved debris up to 14 m above summer flow stages (MacKay and Mackay, 1973).

Two volumes that summarize much of the literature and ideas about ice jams are Bolsenga's (1968) voluminous review and Williams' (1973a) edited Seminar on Ice Jams in Canada. In spite of all the published works, there is still a great need for more research, particularly observations and measurements of jam characteristics. As Gerard (1975) points out, research on jams is very expensive, time consuming, and frustrating for the limited information that is obtained.

Ice jams are defined here as drives, floes, or runs of moving ice which have come to a stop, with ice clogged in a channel obstruction. Two kinds of ice jams are common in northern rivers, early winter jams and spring jams. Winter jams, caused by frazil slush and floating anchor ice, have little geomorphic impact and therefore will not be discussed. On the contrary, spring jams result from thick ice blocks of broken surface ice and often have major geomorphic impact.

Determination of a numerical threshold value for irregular channel morphology of ice jams is a difficult but not an impossible task. An approach attempted by Williams (1973b) was to measure and determine the average channel width of the Mackenzie River, then locate the sites where channel width was less than 60% of mean width. He found a good correlation between known jam sites and constricted channel reaches. The same approach could be tried for sinuosity or meander curvature for jams in sharp meander bends. Finally, the position of channel islands with respect to the thalweg or main channel could be related to frequent jam sites.

The geomorphic effects of ice jams are threefold: ice scour of channel bed and banks, flooding, and reducing sinuosity by meander cutoffs. The effects of bed and bank scour have already been discussed in the previous section on ice drives. Flooding and accompanying overbank sediment deposition can be particularly dramatic for ice-jam floods. Ice-jam floods have an added surprise; they occur without warning and the water levels often rise quickly. For example, at Red Deer, Alberta, the ice-jam flood of 1943 occurred at night when the river ice rose more than 4.6 m (15 ft) in less than 1 hour at about 4 A.M., April 4 (Table 2; Burton, 1965). But this ice-jam flood was small compared to others. A spectacular ice-jam flood occurred in 1875 at Ft. McMurray on the Athabasca River, Alberta (Moberly and Cameron, 1929, p. 151). "In less than an hour the water rose 17.5 m [57 ft], flooding the whole flat [floodplain], mowing down trees some 1 m [3 ft] in diameter, like grass." Finally, Henery (1965) describes a jam on the Yukon River at Ruby, Alaska, in 1930 which rose to a height of 20 m (65 ft) above the winter ice cover, and extended 24 km across the Yukon Valley. The striking features of ice-jam floods are the extreme high stages attained in a short time, often occurring without warning.

Ice jams are not easily broken and generally must break themselves by hydrostatic pressure from the backwater upriver. In 1945, a similar ice jam to the one at Ruby occurred near Galena Air Base on the Yukon River and was bombed with the hope of breaking the jam and releasing the flow. Approximately 18,208 kg (40,150 lb) of bombs were dropped onto the jam without success (Henery, 1965). Four 2000-kg (4000-lb) bombs were among the bomb sizes dropped. The result appeared to have a reverse effect, by shaking loose anchor ice and augmenting the jam with more ice. Ice jams seem to have a great ability to absorb explosive shock without breakage.

An additional geomorphic feature induced by ice jams is the cutting off of meanders occupied by a jam. This process is hypothesized by Williams (1973b) but has not been confirmed by field observations.

CHANNEL SLOPE THRESHOLD AND ANCHOR ICE

Anchor ice, common to mountainous terrain, is primarily related to steep channel slopes. While slope is paramount, several secondary parameters are important, such as sufficient discharge and cold air temperatures. In plains rivers anchor ice commonly occurs during the early freeze-up period, but does not occur after a stationary ice cover has formed. However, in most steep mountain rivers with turbulent mixing, ice covers rarely form. Thus a slope threshold exists whereby rivers will form either anchor ice or a surface ice cover.

Anchor ice is a hardy, laminated ice accumulation which deposits as frazil or grows as crystals on the bed of turbulent streams (Fig. 3), usually in mountain settings. Most observers believe that frazil crystals within the turbulent column of flow adhere to the bed material, while others believe that ice crystals grow from the bed upward. Little research has been carried out on anchor ice.

The thickness of anchor ice can vary from several centimeters to extremes of several meters. On the Bow River near Calgary, Alberta, a thickness of anchor ice raised the river flow in the channel by almost 1 m (anonymous, 1952). In the Madison River in southwestern Montana, during a period of severe cold weather, anchor ice thickened to several meters (Stevens, 1940) and eventually formed a channel icing.

Excessive flow depth does not seem to restrict the formation of anchor ice. In the St. Lawrence River, anchor ice was observed under 6.2 m of water and in the Neva River in Russia under 20 m of water (Bolsenga, 1968).

Atmospheric conditions are very important in the diurnal accumulation and destruction of anchor ice. Under mild winter conditions anchor ice will form at night but will detach itself and float to the surface as sunlight penetrates and warms the water the next day. During a period of severe cold weather, anchor ice will continue to thicken during both the day and night. For example, in the upper Salmon River in Idaho, the Madison River in Montana, the upper Bow in Banff Park, Alberta, and the Kicking Horse River in B.C., continuous thickening has occurred which in some cases has forced the river to flow out of its channel.

Sediment transport by rafting of anchor ice is the dominant geomorphic effect. The volume and grain size distribution of sediment transported downriver is unknown. In the Salmon River in Idaho, and Kicking Horse River in B.C., fist-sized cobbles were observed frozen onto float ice, former anchor ice. However, sand and pebbles were most commonly observed. Both Gilfilian et al. (1972) and Michel (1966) observed sediment in slush floes. The former paper mentions that some of the floating anchor ice is redistributed and refrozen under a stationary ice cover.

FLOW DEPTH THRESHOLD AND CHANNEL ICINGS

Channel icings are influenced chiefly by flow depth in winter and occur commonly in mountain braided rivers (Fig. 13). The threshold of flow depth is very difficult to isolate since other factors also affect icings, such as weather conditions, channel morphology, and incoming flow discharge. For example, whether a braided channel consists of 2 or 20 channels has a considerable effect on the rate of freezeup for a given discharge. If the discharge from the 2 or 20 braids was concentrated into one deep channel, the chances of an icing would be greatly reduced.

FIGURE 13 *Channel icing covering braided channels of the North Saskatchewan River, Banff Park. View looking downriver.*

The term "icing" in the English language is synonymous with the terms "aufeis" in German and "naled" in Russian. A general definition for river icing in all three languages is the complete freezing of river flow, followed by progressive surface accumulations of ice, layer by layer, a process similar to the buildup of lava flows. Icings are fed by either groundwater springs or channel flow that becomes blocked and is forced to the surface. The growth of a river icing can exceed the channel and extend laterally onto adjacent floodplains.

Several papers have been written on channel icings. Three comprehensive literature reviews written by Carey (1970, 1973) and Anisimova (1973) and three good good field studies by Eager and Pryor (1945), Kane (1975), and Sloan et al. (1976) have examined different aspects of icings. A number of other miscellaneous papers on icings deal with engineering and highway maintainence problems.

River icings exhibit considerable variability in size, shape, and thickness. Kane and Slaughter (1972) in Alaska measured a channel icing that varied from 2 to 3 m thick and which amounted to about 4% of the total yearly discharge. Carlson and Kane (1973) concluded that the thickness of a channel icing may exceed the normal thickness of river surface ice by two to three times. They and others have observed that, when a channel icing has reached bankfull stage, lateral expansion onto the floodplain accommodates most of the remaining ice accumulation. Berkin and Maliy (1969) determined that the runoff contributions from icings comprised 11 to 30% of the spring runoff and 6 to 17% of the annual runoff.

Several geomorphologic effects can result from channel icings. When spring thaw begins, runoff if forced to flow wherever favorable topographic depressions allow it. Thus, if an icing extends onto adjacent floodplains, flood stages can be considerably higher for a given discharge than they would be without an icing. Besides flooding and obvious overbank deposition of sediment, a second major effect is new channel incision into floodplains and relocation of channels, a process observed by Stevens (1940) on the Madison River in Montana. The same process has been observed on the North Saskatchewan River downriver form the Icefield Parkway bridge shown in Figure 13. The process of new channel incision and relocation may be an important mechanism for maintaining braided channels at slopes below the threshold values proposed by Leopold and Wolman (1957) and Schumm and Khan (1972). This has not been field-tested.

CONCLUSIONS

The concept of geomorphic threshold as used in this paper refers to a parameter or stress level beyond which the rate of landform response changes suddenly. Or, put differently, a threshold occurs when the rate of landform change accelerates or decelerates with respect to constant increases or decreases of stress or parameter change (Fig. 1). The thresholds of channel characteristics in this study, summarized in Table 3, are presented as a first attempt to classify river ice processes and related thresholds.

TABLE 3 *Summary of River Ice Processes, Thresholds, and Geomorphic Effects*

River ice process	Dominant threshold parameter	Region and morphology of river	Major geomorphic effects
Ice drive	Width	Northern meander rivers	Enlarged channels, bed scour
Ice jam	Sharp meanders, islands, constrictions	Northern meander rivers	Flooding, overbank deposition
Anchor ice	Slope	Mountain steep-slope rivers	Sediment transport by rafting
Channel icing	Depth	Mountain braided and small meander rivers	Flooding, channel erosion and abandonment

A variety of river ice processes are very important in shaping fluvial landscapes in cold winter environments. Beyond the threshold of bankfull channel width (54 to 60 m), ice drives can occur, resulting in channel enlargement by bank scour and incision. Sharp meander bends, channel constrictions, and islands are morphologic thresholds for ice jam formation, which can result in severe rapid flooding without warning. Channel slope in mountain rivers is believed to be the dominant threshold parameter for anchor ice. Excessive slope and related high flow velocity usually prevent ice covers but are conducive to the formation of anchor ice. Flow depth is regarded as the threshold parameter for channel icings which result in flooding and channel relocations. Additional research is necessary to isolate the numerical ranges of threshold parameters and to aid in the basic understanding of processes, river planning, and forecasts.

ACKNOWLEDGMENTS

Research on river ice drives and jams was supported by the National Research Council of Canada and Alberta Research Council. Thanks are also extended to the Water Survey of Canada, for supplying hydrometric data. Special thanks go to George Ballard, Robert Gerard, and Jack Vitek for their critical review of the manuscript.

REFERENCES CITED

Anisimova, N. P., et al., 1973, Groundwater in the cryolithosphere: CRREL Draft Trans. 437, 172 p.

Anonymous, 1952, Royal commission inquiry into ice jamming and winter flooding on the Bow River: unpublished report, 59 p.

Berkin, N. S., and Maliy, V. A., 1969, Accumulation in naleds of winter runoff in small rivers of Pribaykal region and eastern Sayan: *in* Siberian Naleds, Naledi Sibirii, trans. from Russian: USSR Academy of Science, Siberian Branch, CRREL Draft Trans. 399, p. 246–251.

Bolsenga, S. J., 1968, River ice jams: a literature review: U.S. Army, Corps of Engineers Detroit, Mich., Res. Rept. 5-5, 567 p.

Bray, D. I., 1972, Generalized regime-type analysis of Alberta rivers: unpublished Ph.D. thesis, University of Alberta, Edmonton, Alberta, Canada.

Burton, I., 1965, Flood-damage reduction in Canada: Geograph. Bull. Canada, v. 7, 161–185.

Carey, K. L., 1970, Icing occurrence, control and prevention: an annotated bibliography, CRREL Spec. Rept. 151, 56 p.

——, 1973, Icings developed from surface water and groundwater: CRREL Monogr. 111-D3, 65 p.

Carlson, R. F., and Kane, D. L., 1973, Hydraulic influences on aufeis growth: First Canadian Hydrotech. Conf., Edmonton, Alberta.

Dury, G. H., 1970, Meandering valleys and underfit streams: *in* Dury, G. H., ed., Rivers and river terraces, Geographical Reading Series: London, Macmillan Press, p. 264–275.

Eager, W. L., and Pryor, W. T., 1945, Ice formation on the Alaska highway: Public Roads, v. 24, n. 3, p. 55–74, 82.

Estifeev, A. M., 1958, Frazil control at power plants: (in Russian), Moscow, Gasenirgoizdat.

Gerard, R., 1975, Preliminary observations of spring ice jams in Alberta: Proc. Third Int. Symp. Ice Problems, Hanover, N.H., p. 261–277.

Gilfilian, R. E., Kline, W. L., Osterkamp, T. E., and Benson, C. S., 1972, Ice formation in a small Alaskan stream: UNESCO, Int. Symp. Role of Snow and Ice in Hydrology.

Haden, Davis and Brown (Alberta) Ltd., H. G. Acres and Co., 1960, Red Deer River Flood Stages and Industrial Land Development: an unpublished report commissioned by the city of Red Deer, Alberta, 12 p.

Henery, W. K., 1965, Ice Jam floods of the Yukon River: Weatherwise, v. 81, p. 80–85.

Henock, W. E. S., 1973, Data on height, frequency of floods, ice jamming and climate from three-ring studies, part 1: Hydrologic aspects of northern pipeline development: Tech. Rept., Information Canada, p. 157–177.

Kane, D. L., 1975, Hydraulic mechanism of aufeis growth: unpublished Ph.D. thesis, University of Minnesota, 107 p.

——, and Slaughter, C. W., 1972, Seasonal regime and hydrological significance of stream icings in central Alaska: Int. Symp. Role of Snow and Ice in Hydrology,

UNESCO-5, Properties and Processes of River and Lake Ice.

Kellerhals, R., Neill, C. R., and Bray, D. I., 1972, Hydraulic and geomorphic characteristics of rivers in Alberta: Research Council of Alberta, River Eng. Surface Hydrol. Rept. 72-1, Edmonton, Alberta, 52 p.

Kivisild, H. R., 1959, Hanging ice dams: Proc. 8th Congr. AIRH.

Leopold, L. B., and Wolman, M. G., 1957, River channel patterns: braided, meandering and straight: U.S. Geol. Survey Prof. Paper 282-B.

MacKay, D. K., and Mackay, J. R., 1973, Locations of spring ice jamming on the Mackenzie River, N.W.T.: *in* MacKay, D. K., ed., Hydrologic aspects of northern pipeline development: Tech. Rept., Information Canada, 73-3, p. 237–257.

Mackay, J. R., 1963, Progress of breakup and freeze-up along the Mackenzie River: Geograph. Bull. Canada 19, p. 103–111.

McPherson, H. H., 1966, Morphology and fluvial processes of the lower Red Deer River Valley, Alberta: unpublished Ph.D. thesis, McGill University.

Michel, B., 1966, Morphology of frazil ice: *in* Oura, H., ed., Physics of snow and ice: The institute of Low Temperature Science, Hokkaido University, Sapporo, Japan.

Moberly, H. J., and Cameron, W. B., 1929, When fur was king: Toronto, Dent, J. M., p. 151.

Newbury, R. W., 1968, The Nelson River: a study of subarctic river processes: unpublished Ph.D. thesis, Johns Hopkins University.

Schumm, S. A., and Khan, H. R., 1972, Experimental study of channel patterns: Geol. Soc. Amer. Bull., v. 83, p. 1755–1770.

Sloan, C. E., Zenone, C., and Mayo, L. R., 1976, Icings along the Trans-Alaska pipeline route: U.S. Geol. Survey Prof. Paper 979, 31 p.

Smith, D. G., 1979, Effects of channel enlargement by river ice processes on bankfull discharge in Alberta, Canada: Water Resources Res., v. 15, p. 469–475.

Stevens, J. C., 1940, Winter overflow from ice-gorging on shallow streams: Trans. Amer. Geophys. Union, v. 21, p. 973–978.

Williams, G. P., 1959, Frazil ice: a review of its properties with a selected bibliography: Eng. J., November, p. 55–60.

——, 1973a, Characteristics of ice jams: *in* Williams, G. P., ed., Seminar on ice jams in Canada, Univ. Alberta, Edmonton, N.R.C. Tech. Memo. 107, p. 17–29.

——, 1973b, Seminar on ice jams in Canada: Univ. Alberta, Edmonton, N.R.C. Tech. Memo. 107, 182 p.

16

THRESHOLD AND LIMIT EFFECTS IN KARST GEOMORPHOLOGY

D. C. Ford

INTRODUCTION

Introducing the symposium on Geomorphic Thresholds at the American Association of Geographers meetings, 1976, R. L. Frederking and J. D. Vitek wrote:

"Threshold" and "threshold concept" are employed as self-explanatory terms to denote existence of zones or critical conditions at which change in topographic form or process occurs.

These definitions are taken as sufficient for the following discussion.
Sweeting (1972, p. 5) defined the karst geomorphic system in these terms:

The sinking of water and its circulation underground is the essence of the karst process. This process is dominated by a chemical (solutional) activity, and true karst landforms result largely from the action of one erosive process, namely solution.

True karst forms are distinguished from pseudokarst features (such as thermokarst) by the necessity of rock solution. True forms may be excavated entirely by aqueous solution, or other processes may contribute largely to their dimensions; but where the latter applies, solution plays an essential precursor or "trigger" role, as in the case of collapse dolines. The common karst rocks are limestone and dolomite, gypsum and salt. Limestone karstlands predominate and many specialists study no others.

A working classification of true karst landforms is given in Table 1. Karren (lapiés) are solutional indentations developed upon bare rock or beneath soil. There is a wide range of form: individuals may be as small as 1 centimeter. Topographically closed depressions, termed dolines, sinkholes, and so on are *the* diagnostic surface karst feature. They are the most abundant form among the intermediate and large-scale groups. Again, morphology is extremely varied. Residual hills of roughly conical or tower form between densely packed dolines or upon corrosional plains are also particularly important at the intermediate scale. Poljes are larger depressions with flat floors that have often been planed by corrosion; ideally, the depressions are topographically closed and drained underground but there are also "open" poljes drained by surface channels. Dry valleys and gorges have been created by the fluvial suite of processes and are relict as a consequence of abstraction of their waters underground. Cave systems vary considerably in scale and morphology.

TABLE 1 *Classification of Karst Landforms*

Locus	Scale	Principal landforms	Dependency
Earth's surface	*Small*: greatest dimensions (e.g., length) commonly <10 m	Karren—solutional pits, channels, and grooves	Varies; types dependent and independent of groundwater systems exist
	Intermediate: greatest dimension commonly 10 to 1000 m	Closed depressions—dolines and compound forms (uvalas); Upstanding residuals—karst cones and towers	Dependent; the existence of efficient systems of underground water discharge is essential
	Large: greatest dimension commonly >1000 m	Poljes, dry valleys and gorges	
Underground	*All scales*	Cave systems	Not applicable

Development of most intermediate and large features depends upon establishment of efficient groundwater circulation through solutional channels (cave systems). This is also true for many karren types, especially the subsoil ones.

Examination of the karst system may offer four distinctive contributions to our general understanding of thresholds in geomorphology. First, attention is focused upon thresholds in chemical reactions and resistances. Most previous work has considered physical, especially mechanical, thresholds. Second, it is concerned with a few types of rocks, where other systems must consider all rocks. Rock characteristics are investigated in more detail in karst studies. Third, in most geographical regions, more water is in storage and transit underground than is at the surface. By definition, karst research is particularly concerned with the role of groundwater. Finally, concepts of climatic control have been investigated more thoroughly in karst geomorphology than in other branches of the subject, offering a special opportunity to investigate climatic thresholds of morphogenesis.

These four points are considered in turn in this chapter. Because there has been little explicit discussion of thresholds in karst geomorphology (although much is implicit) my intent is to identify and define them as far as is possible, leaving the reader to place them in general frameworks suggested elsewhere. It transpires that discrete thresholds are rather difficult to find in many parts of the karst system, but certain *limits* are clearly defined. These are mentioned, also.

SOLUTION PROCESSES AND MINERAL SOLUBILITY

Calcite, dolomite, gypsum, and halite are the principal minerals of the karst rocks. All are soluble in pure water by the process of molecular dissociation- for example, for halite,

$$NaCl \xrightleftharpoons{H_2O} Na^+ + Cl^- \tag{1a}$$

and for calcite,

$$CaCO_3 \xrightleftharpoons{H_2O} Ca^{2+} + CO_3^{2-} \tag{1b}$$

These reactions proceed forward (to the right) until a solution is saturated with the relevant ions, and the rate of back reaction equals the forward reaction. Mineral and solution are then at equilibrium. Equilibrium concentration of ions (in other words, mineral solubility) constitutes a well-defined limit to each reaction. Limits are temperature- and pressure-dependent. Equilibrium solubilities for the karst minerals are given in Table 2.

Natural waters are not pure, being carbonated to an extent and containing other "foreign" ions in at least trace amounts. This is of little significance in the cases of gypsum and salt, which are already sufficiently soluble, but carbonation greatly increases the solubility of calcite and dolomite by the well-known sequence of reactions that may be written

TABLE 2 *Solubility of Selected Minerals*

A. Solubility in pure water, at $25°C$ and 1 bar

Halite	368,000 mg/l
Gypsum	2,410 mg/l
Calcite	14 mg/l
Silica	5 mg/l

B. Approximate range of solubility in natural
waters, 0 to $30°C$ and 1 bar

Calcite	14–400 mg/l
Silicate minerals	<5–90 mg/l

$$CO_2(gas) + H_2O \rightleftharpoons H_2CO_3 \tag{2a}$$

$$CaCO_3 \overset{H_2O}{\rightleftharpoons} Ca^{2+} + CO_3^{2-} \tag{2b}$$

$$H_2CO_3 + CO_3^{2-} \rightleftharpoons 2HCO_3^- \tag{2c}$$

These reactions are dominant in the solution of limestone and dolomite, as Table 2 indicates.

It follows from equations 1 and 2 that there are no chemical thresholds in the solution processes themselves, only chemical limits. The requirements are merely that air, water, and rock be brought into contact. There is no additional requirement that some minimum velocity (shearing force) be attained in the water, unlike the case for corrasion and transport of clastic debris in a fluvial system. Solution occurs in static water and throughout the known range of natural water velocities. This partly explains why karst processes work more uniformly over the climatic year in areas where there is strong wet–dry or warm–cold seasonality than do the dominant processes of other systems (Drake and Ford, 1974).

Next, consider the chemical evolutionary paths that waters take between dissolving their first molecule and attaining saturation. Such paths, especially those of groundwater, are much studied at present. Contrasting waters, such as snowmelt on bare rock and seepage from an acid bog, will follow different routes across a saturation diagram (Harmon and Drake, 1973). But analysis of published paths for *individual* waters indicates no change of gradient that can be associated with change in a landform or the creation of a new form.

Mixing of different waters (merging of two evolutionary paths) may have a threshold result. Where both waters are at or close to their individual saturation limits, mixing may boost the combined limit above the arithmetic average of the mixture—the "mischungs-korrosion effect" (Bögli, 1964). Where mixing is point-located, as when water flowing in a crack joins that in a larger pipe, it may trigger development of distinctive corrosion recesses termed "blind pockets." These can attain depths of tens of meters in water-filled caverns. Mischungskorrosion is one of a series of chemical effects currently being investigated that serve to boost or depress the saturation limits of given karst waters (see Picknett et al., 1976). Others in the series may induce threshold situations, but at present it appears that their effect will be only to increase or reduce rates of development.

There being little threshold effect discernible in the aqueous processes that dissolve karst rocks, solubility of their minerals should next be considered. A most important point is that a majority of "non-karst" minerals are also slightly soluble in pure water and somewhat more so in natural waters. Silica has been included in Table 2 as an example. Its solubility in pure water is scarcely less than that of calcite when both are compared to gypsum and salt. A wide variety of limestone karst develops in alpine tundras, where the saturation limits for calcite are often below 80 mg/l (Ford, 1971a). This value falls within the solubility range of many silicate minerals in natural waters (Table 2), indicating that there can be no exact solubility threshold for karstification that places the "karst" minerals on one side of a line and all others upon the other.

True karst forms *do* develop upon silicate and other "non-karst" rocks. Beck (1977) has studied karren on granodiorites in Puerto Rico. The chemical reactions were hydrolysis and carbonation of plagioclase, yielding 30 to 35 mg/l total dissolved solids. The waters were from root meshes and had a pH of only 4.8. Many other instances of silicate karren are reported, although without this chemical detail. At the intermediate and large scales, a karst of spectacular shaft dolines and caves in massive, Precambrian sandstones and quartzites is being explored in southeastern Venezuela (Szczerban and Urbani, 1974).

Most outcrops of silicate rocks do not display karst features. The problem may be resolved by noting that the solubility of silicate minerals is less than 25 mg/l in most natural waters that they encounter, whereas the solubility of calcite is greater than this in almost all natural waters. Approximately 25 mg/l appears to be the effective solubility threshold of karstification on this planet, with the rider that it is modulated by rock control and competition factors (below). Rocks displaying that minimal solubility have a "sporting chance" of hosting karst forms.

ROCK CONTROL

Bulk chemical purity, petrography, lithology, structural features, and so on, of the karst rocks are here included under the general heading "rock control." This is of central importance in karst studies. It profoundly influences rock solubilities and determines, at the initiation of karst groundwater circulation, the critical properties of porosity and permeability. At the very least, it appears to confuse all search for distinctive climatic thresholds of morphogenesis. The topic is too large for more than a few aspects to be touched upon here.

First, consider bulk chemical purity within a formation or other unit that would be classified wholly as a limestone or gypsum, etc.: there are no "non-karst" interbeds. By definition, a limestone is a rock consisting of $\geqslant 50\%$ $CaCo_3$ + $MgCO_3$ by wt/vol. It may be absolutely pure, 100% $CaCO_3$. A fully karstic landform assemblage (a holokarst) does not normally develop unless bulk purity is 70 to 80% or greater. Limestones of lesser purity may display a low density of small dolines within an otherwise fluvial landscape, as do some rocks containing less than 50% carbonate minerals (e.g., calcareous sandstones). Such development has been termed "parakarstic" (Birot, 1954). The great volume of insoluble residue produced by dissolution of, say, a 60% carbonate limestone serves to largely choke development of the karstic groundwater system essential for the growth of a holokarst. Similarly, clay and hydroxide residues inhibit karstification of most silicate rocks. Many dolomite rocks exhibit marked differential solubility at the granular scale so that, in addition to solution by vertically descending water, the fluvial processes of differential weathering and washout of residues operate to permit only para-karst, even where carbonate purity exceeds 95%. Dolomite grus is particularly well known in Hungary (Jakucs, 1977).

There are many stratigraphic sequences in which carbonates of high purity predominate but there are significant interbeds or greater units of marl, shale, and so on. Their effect upon propagation of karst was recognized in the earliest and most fundamental classifications of karstland typology. Grund (1914) proposed the term "halbkarst." Cvijic (1924) suggested "merokarst," which together with "holokarst" are the two basic types. A merokarst terrain is one of trunk rivers (not necessarily allogenic) flowing in fluvial valleys through the predominantly carbonate sequence, with the interfluves intensively karstified [e.g., the Quercy region of France (Avias, 1972)]. Röglic (1960) proposed "fluviokarst" for both parakarst and merokarst terrains.

It appears infeasible to develop a numerical index for the limit of fluviokarst (i.e., threshold of holokarst) development. The real stratigraphic sequences and other factors are too varied. For example, in the Front Ranges of the southern Canadian Rockies, thick, pure, and massive Devonian and Mississippian carbonates are separated by 100 m of interbedded cherts and thin limestones, underlain by 10+ m of carbonaceous shale. These units induce fluviokarst conditions except at Crowsnest Pass, where groundwater has breached them to combine both carbonate formations into one deep holokarst.

Lithologic typology, as in the limestone classifications of Folk (1959) or Dunham (1962), is important. Sweeting and Sweeting (1969) show that micritic beds are more soluble than sparite-rich beds among Carboniferous limestones of Yorkshire, England. Here solubility will not differ in a chemist's sense, in that standard powders prepared from both would dissolve at the same rate in the laboratory; but in the field, the greater facial area exposed by the smaller micrite grains permits significantly greater solution in a given time. On a larger scale, Sweeting (1972) finds the occurrence of karst towers in Belize, Jamaica, and Sulawesi (Indonesia) to be limited to the outcrop of granular dolomitic limestones.

The measure of textural homogeneity is particularly significant. A limestone may assay 99.9% $CaCO_3$ and yet be heterogeneous e.g., lithoclasts in a micrite groundmass partially replaced by sparite. At the scale of the hand specimen, increasing homogeneity appears to be associated with an increasing number of different karren types that may be hosted, as well as increasing regularity of form for a given type. Linear karren types cannot develop on very heterogeneous rocks such as conglomerates, many reefs, and so on. High homogeneity is needed for growth of solutional scallops and rillenkarren, while trittkarren, a mini-cirque form, appear to be confined to porcellaneous beds or fine-grained marbles. Threshold homogeneities have not been quantified for many karren types, but it would be feasible to do so.

At a larger scale, textural homogeneity may be a major control of effective fracture density. Here, fine-grained marble might appear as the low-density end member in a continuum extending to reef rock or a cemented conglomerate at the high-density, heterogeneous end.

Sweeting (1972) argues that the mechanical strength of rock as defined in the laboratory (e.g., with core slices) is important. Specifically, she suggests that chalks of southern England lack the strength to support large groundwater conduits (caves) or doline sidewalls and therefore karst landform expression is generally lacking. The compressive strength of chalk is generally below 5000 psi, whereas that of other limestones ranges from 8000 to 30,000 psi in published tables (Chilingar et al., 1967, p. 362–363). This implies a threshold compressive strength between 4000 and 8000 psi. Others would attribute the lack of karst forms in the British chalk to its high porosity. The White Cliffs of Dover are chalk. However, strength of this kind does play a part. There are large collapse components in the morphology of most gypsum landforms of intermediate or greater scale. Like chalk, it is weak.

Density of fissures penetrable by water (bedding planes and fractures such as joints and faults) is the fundamental control of cavern patterns and affects most aspects of surface karst morphology. Massive rocks (i.e., thickly bedded and therefore with low joint frequency as a normal consequence) are everywhere favored. In the extensive tower karst of South China, presence of massive strata with strong, vertical (deep joint) permeability is the foremost explanatory factor (Silar, 1965; Balazs, 1971). Smith and Atkinson (1976) find that the famous "cockpit" (or polygonal) karst of Jamaica is confined to outcrops of massive limestones. However, real upper limits apply to this trend, as is explained later in terms of interaction with hydraulic gradient.

Fracture density determines the mechanical strength of limestones and dolomites subject to karstification in a more real sense than do the compressive tests described above. Unfortunately, this cannot be tested and precisely quantified in the laboratory. Thin-bedded rocks with high joint frequency may not develop karst forms other than the smallest karren. Enterable caves, if they exist, are dominated by a morphology of roof and wall collapse. Dolines have a limited range of form, variants of a rubble funnel. The boundary separating high and low fracture density has not been explicitly stated. Hydraulic gradient factors intervene to ensure that it cannot be a precise figure but, rather, a transition range. But the threshold of the desirable "massiveness" in bedding can be set at about 40 cm between each parting plane penetrable by groundwater.

Greater structural features, such as folds, are also significant. Deep phreatic cave systems show preferred development where strata dip steeply. But there is no threshold; they also occur in flat-lying rocks. More important is the case where surface slope is concordant with steep stratal dip. In the Canadian Rockies there are many limestone dip slopes of 30 to 50°. Here, with high potential energy available to surface processes, karst processes may not be able to compete effectively and fluvial forms result despite chemical purity, lithology, fissure density, and so on, that would elsewhere favor karstification. This is to suggest a "threshold of effective competition," which is further commented upon in the Conclusions.

Finally, we may consider limits where karst rocks are overlain by others (i.e., they do not crop out). "Interstratal" karsts may develop where the overburden is of consolidated rock, "mantled" karsts where it is unconsolidated. Interstratal karst, expressed as collapse dolines or geological organs in the cover strata, is abundant. From limestone interstrata, collapse landforms have propagated to the surface through as much as 100 m of overlying quartzitic sandstone (Thomas, 1954). From salt, collapse and subsidence forms have propagated through 1000 m of consolidated cover in Saskatchewan.

In contrast, there are instances of allochthonous mantle burying karst rocks. In lowland Canada great areas of limestone, dolomite, and gypsum are overlain by Wisconsinan glacial drift. This is normally rich in carbonate clasts. In Ontario we find that where a carbonate-rich till is deeper than 1.2 to 1.8 m, underlying carbonate bedrock has rarely suffered solutional damage; that is, karstification has been stalled throughout the 13,000 years or more since the ice receded because waters have exhausted their solvent potential upon carbonates in the mantle (Pluhar and Ford, 1970). But over the more soluble gypsum, large sinkholes have propagated through as much as 60 m of drift during postglacial times (Wigley et al., 1973).

TEXTURAL KINETIC CONTROL

Before proceeding to the next major topic, it is of interest to consider two small karren features whose genesis and form is precisely determined by kinetic factors operating in a narrow range of rock textures. Solution kinetics have to do with physical controls of solutional processes, such as diffusion of ions to and from the rock–water interface.

Solutional scallops (Fig. 1) are shallow, asymmetric scoops that are common features of caves and river channels in limestone, gypsum, or salt. They may also form on glacier ice and firn, where the mechanism is sublimation into flowing air. They colonize all available surface area to form patterns of packed, overlapping sequences.

Curl (1966) argued in a theoretical analysis that scallops develop when the static, chemically saturated, fluid boundary layer normally present between rock and flowing water becomes detached from the interface, permitting enhanced diffusion rates (i.e, enhanced erosion) under the detached portion. Detachment requires points of exceptional roughness in the solid surface, such as small projections or holes. It was predicted to occur at a Reynolds number of 22,000. Mean scallop length would be inversely proportional to mean velocity in the body of the fluid. In a physical simulation at three dimensions, Goodchild produced stable patterns of solutional scallops on plaster of paris in a flume, confirming the predictions (Goodchild and Ford, 1971). The kinetic threshold of scallop formation is $N_R \propto 22,000$.

Rillenkarren (Fig. 2) are patterns of solutional channels on inclined, bare rock surfaces. They appear to outrage the basic principle of Horton's formulation for drainage channel patterns (1945) for they propagate at the *crest* of a slope, washing out downslope into a "belt of no (channeled) erosion," and are packed with great regularity rather than being spaced apart by unchanneled interfluves of regular width.

Glew (1975) conducted a simulation with artificial rain falling upon inclined slabs of plaster of paris, and produced rillenkarren of outstanding realism. He determined that rill cross sections approximated the quadratic parabola in form, rather than the semicircular cross section of an ideal river channel which they appear to possess when casually inspected in the field. The unique property of the quadratic parabola is that it focuses parallel forces through a point—focuses parallel raindrops at the trough base in this instance. Packed rillenkarren are, therefore, the expected form where the dominant erosion process is solution by droplet impact onto inclined surfaces, because there is maximum energy conservation. The rills are not channels *sensu stricto*, despite their appearance. They are eliminated downslope when water accumulated and flowing in their troughs attains a critical depth, which prohibits the focused splash from penetrating directly to the rock. Glew found this depth to be 0.15 mm for rainfall at 35 to 40 mm/hr and droplet size appropriate to that intensity. Rill length is proportional to slope in a

FIGURE 2 *Rillenkarren upon a landslide block in Surprise Valley, Jasper National Park, Alberta. Scale is in inches.*

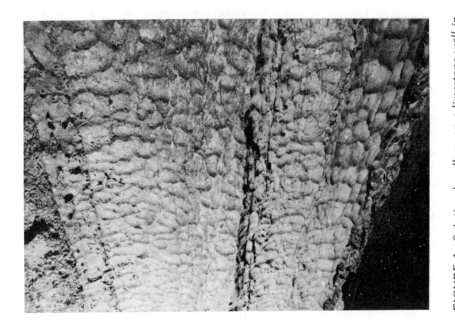

FIGURE 1 *Solutional scallops on a limestone wall in Bonnechère Caves, Ontario.*

log-linear manner. Short rills appear at threshold gradients of 8 to 12°, and lengthen as slope is increased to 70°. At higher gradients, they are irregularly replaced by patterns of scallops. There may be thresholds of droplet size and rainfall intensity for propagation of rillenkarren, but these have not been determined experimentally. The limit is the critical depth of the water film.

Scallops and rillenkarren are limited to soluble rocks of medium and fine grain and a high measure of textural uniformity. Presence of strong, point inhomogeneities spaced more closely than the kinetic length of a scallop or rill will prohibit their full extension and, if there are several such inhomogeneities per kinetic length, they will not develop at all. Scallop and rill are examples of karst forms strictly governed by textural–kinetic relationships. To coin more jargon, they may be termed "uphikinetic" forms (Greek; $\nu\phi\eta$ = texture).

POROSITY, PERMEABILITY, AND KARST GROUNDWATER CIRCULATION

An efficient underground plumbing system to discharge karst water is essential for development of most intermediate and large surface forms. Caves of explorable dimensions are the largest parts of the plumbing, although they constitute only a small proportion of all that is present beneath a karst surface. Characteristics of porosity and permeability determine whether karst circulation will develop, and the geometric properties of that circulation.

Primary porosity refers to interstices at grain contacts, and so on, within a rock. Primary permeability is the measure of flow through these interstices, if they interconnect. Secondary porosity is that contained in bedding planes, faults, and joints. It is generally accepted that rocks displaying good karst development are dominated by secondary permeability. At the onset of karstification, they function as "fracture aquifers," which are then modified by solutional enlargement to become "conduit-flow aquifers."

The minimum diameter of rock pores is about 1 micron, but Bocker (1969) and others find that a threshold diameter of 10 microns is necessary if solvent water is to modify the medium to any significant extent. Flow must be laminar below diameters of 5 to 16 mm, the threshold for turbulence. White and Longyear (1962) suggested that at onset of turbulence there would be an "hydraulic jump" in solution rate, with profound effect upon evolution of the circulation system thereafter; but revision of the kinetics (Curl, 1968) indicates this is not so. However, at least the turbulent minimum diameter must be attained if a conduit is to be capable of absorbing waters supplied by a doline of average dimension (i.e., this is a requirement if such a doline is to evolve).

Laminar flow in fractures and pipes is approximately governed by the Hagen–Poisseuille equation, and turbulent flow by the D'Arcy–Weisbach equation (see Thrailkill, 1968). However, at the scale of detail required for morphological modelling in karst rock, it proves most difficult to apply realistic values to several terms in each of these equations. General approximation is adequate for much work in water supply management, but the geomorphologist falls back upon empiricism.

Smith et al. (1976) have compiled an excellent diagrammatic summary of porosity in selected limestones, and fitted it with approximate permeability contours (Fig. 3). Data are principally from British rocks. I have added a boundary line to distinguish those

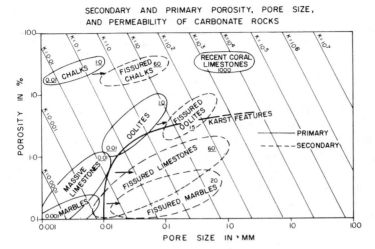

SECONDARY AND PRIMARY POROSITY, PORE SIZE,
AND PERMEABILITY OF CARBONATE ROCKS

FIGURE 3 *Primary and secondary porosity, pore size, and permeability of carbonate rocks. Values of K, the theoretical permeability, are in meters/day and derived $Q/a = K(h/L)$, where Q is the discharge, a the cross section, h the head, and L the length of system. The values are based on the assumption that the rock behaves as a bundle of straight, parallel capillary tubes (Smith et al., 1976). Double underlining indicates total permeability; single underlining, primary permeability. (Modified from Smith et al., 1976, p. 183.)*

conditions that appear to favor development of enterable cave systems, intermediate and large surface forms, from those that do not. The line originates at Bocker's effective pore threshold diameter and, thereafter, is a limit rather than a threshold. The principle it indicates is that where there is comparatively high primary porosity, solutional attack is widely diffused, producing a "diffuse-flow aquifer" rather than conduit flow. Much of the solvent attack is expended at the soil base upon abundant pores that are about or below the effective threshold diameter, especially in chalk. Point-centered forms such as dolines, or linear cave systems, cannot be produced where solution is diffused at a scale of centimeters or less. With high values of primary porosity, permeability itself becomes of little significance. The permeability range of British chalks and oolitic limestones, which are largely devoid of karst landforms, is concordant with that of fissured massive limestones, which display excellent karst. The primary porosities are very different.

The boundary line in Figure 3 is most tentative. There is, in reality, a transition range that it would be inappropriate to draw on a diagram of this configuration. The Jurassic oolites of Britain have particularly porous cement and should not be supposed typical of all oolitic limestones. Chalk rocks in Poland support doline karst (Maruszcsak, 1966) and small natural caves are known in chalks of France (Chabert and Maingonat, 1977). Their primary porosities may be lower than those of Figure 3.

In the many limestones able to host a wide range of karst landforms the "fissure frequency" (density of secondary permeable bedding planes and fractures) varies considerably. The real variation is a continuum between low and high values. But Ford and

Ewers (1978) have shown that with respect to cave systems that develop beneath a vadose groundwater zone, the continuum reduces to four basic states which produce caves with different relationships to a water table (Fig. 4). For example, where fissure frequency is very low, waters must normally penetrate to great depth before they can discover an outlet, = state 1. Where frequency is high, most waters circulate close to the water table and generate a cave there, = state 4. Fissure frequencies greater than the threshold of state 4 merely produce an ever-flatter water table cave. Precise frequency values for each state cannot be assigned as yet.

This "states model" is generalized in Table 3. There exists a lowest state, O, where fissure frequency is too little for cave systems to develop. This is common in many outcrops of marble of local scale. Proceeding through states higher than state 4, a situation similar to that of much of the British chalk is approached. Values of secondary porosity become very high and the solvent attack is too diffuse to produce karst forms. This limit is represented by state n.

TABLE 3 *States of Fissure Frequency Related to Genesis of Cave Systems*

State	Type of cave system	Minimum hydraulic gradient (dh/dl) required to *initiate* karst water circulation
0	No caves	Not applicable
1	Bathyphreatic	$\geqslant 0.0285$
2	Multiloop phreatic	c. 0.025
3	Mixed, phreatic, and water table elements	c. 0.015
4	Ideal water table cave	$\leqslant 0.0125$
Higher states	Ideal water table cave	Flatter than state 4
N	No caves	Not known

At present it does not appear that there is close concordance between the states model for caves and the genesis and density of intermediate or large surface features. Doline and cockpit karsts are known to overlie systems of states 2 to 4. By definition, there are no dolines, and so on, at state 0, and the threshold for solution doline genesis probably lies somewhere within the frequency range encompassed by state 1. Assemblages of densely packed dolines (polygonal karst) are probably eliminated at a state little higher than 4.

Topographic relief may exercise threshold control when karst groundwater circulation is being initiated. This is indicated by the suggestion of initiating hydraulic gradients in Table 3. Values quoted are tentative, from my field experience. If a rock possesses, for example, a frequency of state 2 but the local relief between potential sink and spring positions is so low that the initiating value (dh/dl $\geqslant 0.025$) cannot be attained, the groundwater system will not develop and a fluvial system will prevail instead. This explains why many karsts originate at an escarpment edge, where dl is minimized, and then propagate back into plateaus.

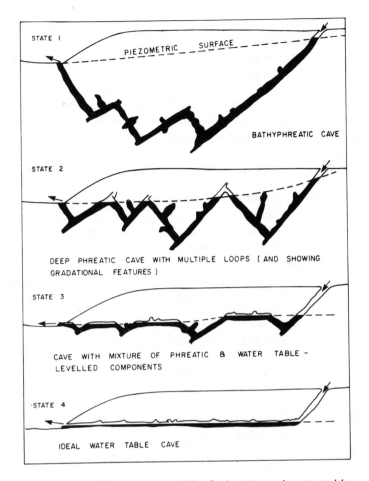

FIGURE 4 *"Four-state model" of phreatic and water table cave system genesis. Natural variation of the frequency of fissures penetrable by groundwater may be reduced to four significantly different states, each supporting a cave system possessing a different relationship to a stable water table.*

Topographic control applies *only* at initiation, when fractures are tightly closed. Once an efficient groundwater system is established, even drastic reduction of hydraulic gradient may not significantly affect it. The Yucatan sinkhole plain, Mexico, is an instance. Postglacial rise of sea level raised boundary springs there by perhaps 80 m. The plain retains efficient holokarst drainage although the hydraulic gradient over many kilometers is only 0.00007 (W. E. Back, personal communication).

ARE THERE CLIMATIC THRESHOLDS OF KARST MORPHOGENESIS?

Climamorphic precepts have developed to a greater extent in karst studies than in other branches of geomorphology. Given the dominance of a single aqueous process with laboratory-established temperature dependencies and the limitation of karst to a few rock types, it is reasonable that possible climatic control should be especially emphasized. Debate has centered upon limestone landforms. Lehmann (1954) asserted that each climatic region displays a clima-specific karst morphology. He listed six types: (1) polar and periglacial, (2) high alpine, (3) cool oceanic (west European), (4) mediterranean, (5) dry desert, and (6) humid tropical, indicating that others remained to be found and that there might be scope for subdivision. Criticism from those who suspected that rock control factors accounted for much of this apparent clima-specificity mounted in the 1960s (Verstappen, 1964; Panos and Stelcl, 1968) so that in her major discussion of 1972, Sweeting recognized but four types—glacial/nival, temperate, humid tropical, arid/semiarid.

Climatic control of karst development in an absolute fashion is evident. The modern climate of Mars lies below the threshold for karstification of our earthly karst rocks, where as that of Venus far exceeds the limits of stability. Similarly, salt rock is unstable at the surface in most of the earth's climates, being limited to patches of karren and micro-doline topography in the driest places.

Sweeting (1972) suggested a rainfall threshold of 250 to 300 mm/yr for limestone karst, but many dolines are reported in the Hamada of Guir, Moroccan Sahara, where precipitation is 50 to 100 mm (Castellani and Dragoni, 1977). There is interbasinal karst water circulation in Nevada (Maxey, 1966). Karst specialists have the same problem seeking a precipitation threshold in deserts as do other geomorphologists there. Water is the principal agent, despite its paucity.

There is a thermal threshold, in the sense that deep continuous permafrost will limit new karst forms to the karren zone. Where karst was established before permafrost, matters may be more complex. Karst water circulation is vigorous in regions of discontinuous permafrost (Brook, 1976). Thaw–freeze processes may suppress karst forms on comparatively thin-bedded rocks but on massive strata scoured by glacial ice, karst attack itself may be the threshold requirement for significant thaw–freeze action (Ford, 1971b). Distinctive morphology of glacial/nival regions relates to solutional impact on glacial topography or glacier impact on karst topography. These are not clima-specific characteristics.

There do not appear to be significant climatic threshold effects in cave erosional morphology. At the surface, the characteristic temperate karst assemblage is of dispersed dolines, with dry valleys or poljes in some areas. This assemblage is also common in the humid tropics, but in addition, there are polygonal karsts, often with cones or towers at interfluves, and the dispersed tower assemblage on alluviated plains. The search for climatic control largely reduces to seeking threshold conditions for polygonal and tower development. Polygonal karst is reported in temperate China (Sweeting, 1972) and New Zealand (Gunn, 1977). There are karst towers in Poland, Czechoslovakia, and, most important, at latitude 61°N in the Mackenzie Mountains, interior N.W.T. (Brook and Ford, 1976). But there is no doubt that most occurrences of these types are in the tropics, although it has not been established that they predominate over other types there.

Smith and Atkinson (1976) have tackled this problem by a most thorough study of 231 published sets of data for mean annual carbonate solute concentrations, extrapolated erosion rate (m^3/1000 yr), and surface versus underground solutional proportionality.

Although statistically significant differences occur in some parts of the analysis, their physical significance is rightly doubted and the authors conclude that

> The traditional latitudinal divisions of climate do not reflect a corresponding division of processes and erosion rate. The principal climatic parameter . . . is the mean annual runoff . . . (p. 401).

Verstappen (1964) suggested a precipitation threshold of 1500 mm/yr for polygonal karst, and Balazs (1971) 1200 mm/yr and $T = 18°C$ for the tower karsts of south China. These precipitation thresholds are exceeded on many temperate highland karsts. Karst towers of the Mackenzie Mountains, noted above, have developed where precipitation is about 350 mm/yr (much falling as snow) and $T = -3°C$. Strata there are massive, with strong vertical permeability, and there is a history of regional alluviation, nonclimatic factors that Balazs cites as important in the south China tower karst.

Sweeting (1972) argues that it is *intensity* of tropical rainfall that is important, as this yields flash flooding and gullying of dolines, plus intensity of evaporation which leads to case hardening of rock, permitting steep doline and tower side slopes to stand up. But recent workers have not observed gully flow during rains, and case hardening is absent from a well-developed polygonal karst in Belize (Miller, in litt.).

To conclude, it appears that the strong climatic threshold effects implicit in Lehmann's morphoclimatic scheme (1954) do not exist. Rock control, multiphase and polygenetic effects have been misinterpreted. But when these other factors have been equated, it is probable that particular polygonal and tower assemblages of strong relief will be found to be located predominantly in the humid tropics. The climatic effects, if any, producing these quasi-singularities have not been established by analysis of current data. Much more data collection in tropical areas may resolve the matter: data are sparse there at present. But it is possible that a proposal of Corbel (1959) holds much of the truth: the distinctive tropical relief is a product of lengthier exposure to denudation than is common in temperate areas (where there have been greater thicknesses of cover strata to be stripped), coupled with a less variable climatic history, in which case, the "threshold" is a complex entity and cannot be established from modern climatic and chemical runoff data.

CONCLUSIONS

Considering karst morphogenesis, there are no thresholds in the aqueous solution processes, but there are well-defined limits of mineral solubility. Rock control is profound, at all scales from hand specimen to massif. It ensures that most other kinds of threshold or limit can be but imprecisely stated. It determines much distribution and variation of morphology. It is the primary determinant of whether or not significant groundwater circulation will be established, and its type. In turn, certain potentially quantifiable circulation types are threshold necessities for development of intermediate or large surface forms. Two forms amenable to laboratory investigation, scallops and rillenkarren, are known to be controlled by kinetic thresholds and limits within a frame of rock textural control. Further investigations will probably discover more such relationships. Karst is the system best suited to morphoclimatic differentiation, but climatic thresholds and limits prove to be elusive, perhaps illusory, except in the broadest terms.

Because karst is "rock-specific" and the other geomorphic systems are not, there is, at first exposure of a karst rock to denudation, a situation of competition between karst processes and those of the zonal system (e.g., fluvial). There is a threshold of effective competition which, if exceeded, results in other landforms or a polygenetic mix. This concept has many different aspects. On the slightly soluble silicate rocks, the contest is most often resolved in favor of competing fluvial or pedimentation processes, and so on, but there is a great silicate karst in Venezuela. On dolomites and mixed formations, there may be a drawn result (i.e., a fluviokarst). Glaciers, including those with water at the sole, are normally overwhelming competitors, but in certain circumstances karstification is accelerated beneath them. The greater the hydraulic gradient at inception, the swifter karst develops: but if high hydraulic gradient is coupled with steep surface slope, surficial processes may win.

Time is an important component of the equations of competition. An efficient karst water circulation must be established against high friction underground, where overground channel flow, and even wash, faces proportionately much less friction. Much time is required to create the underground circulation system, during which fluvial processes may carve a competing system of channels to regional base level, confining karst expression to a few interfluves. Certain fluviokarsts on rock of holokarstic composition may be of this "time-limit-elapsed" type. The time elapsed since the last glaciation of lowland Canada has been insufficient to establish karst circulation on most carbonate outcrops, although circulation has advanced from the crest of the dolomite Niagara Escarpment at 180 m/1000 yr in optimal situations, and at 1 km/1000 yr in some exceptional limestones in Quebec. But, once established, karst circulation systems are difficult to erase. Caves are the longest-lived (least-time-limited) elements in our landscapes.

REFERENCES CITED

Avias, J., 1972, Karst of France: in Herak, M., and Stringfield, V. T., eds., Karst: Amsterdam, Elsevier, p. 129–188.

Balazs, D., 1971, Relief types of karst areas: Symp. Karst Morphogenesis, I.G.U., Hungary.

Beck, B. F., 1977, On the occurrence and origin of karren on granodiorite in Puerto Rico: Proc. 7th Int. Speleo. Congr., p. 28–31.

Birot, P., 1954, Problèmes de morphologie karstique: Ann. Géogr., v. 63, p. 160–192.

Bocker, T., 1969, Karstic water research in Hungary: I.A.S.H. Bull., v. 14, p. 4–12.

Bögli, A., 1964, Corrosion par mélange des eaux: Int. J. Speleo., v. 1, n. 1–2, p. 61–70.

Brook, G. A., 1976, Geomorphology of the North Karst, South Nahanni River region, N.W.T., Canada: unpublished Ph.D. thesis, McMaster University. 627 p.

——, and Ford, D. C., 1976, The Nahanni North Karst: a question mark on the validity of the morphoclimatic concept of karst development: Proc. 6th Int. Speleo. Congr. (1973) II, p. 43–57.

Castellani, V., and Dragoni, W., 1977, Surface karst landforms in the Moroccan Hamada of Guir: Proc. 7th Int. Speleo. Cong., p. 98–101.

Chabert, C., and Maingonat, G., 1977, Grottes et gouffres de l'Yonne: Paris, 320 p.

Chilingar, G. V., Bissell, H. J., and Fairbridge, R. W., eds., 1967, Carbonate rocks: Amsterdam, Elsevier.

Corbel, J., 1959, Erosion en terrain calcaire: vitesse d'érosion et morphologie: Ann. Geogr., v. 68, p. 97–120.

Curl, R. L., 1966, Scallops and flutes: Trans. Cave Res. Gp., G.B., v. 7, n. 2, p. 121–160.

——, 1968, Solution kinetics of calcite: Proc. 4th Int. Speleo. Congr. III, p. 61–66.

Cvijic, J., 1924, Types morphologiques de terrains calcaires: Glasnik Geograph. Drustva, v. 10, p. 1–7.

Drake, J. J., and Ford, D. C., 1974, Hydrochemistry of the Athabasca and North Saskatchewan Rivers in the Rocky Mountains of Canada: Water Resources Res., v. 10, n. 6, p. 1192–1198.

Dunham, R. J., 1962, Classification of carbonate rocks according to depositional texture: Amer. Assoc. Pethol. Geol. Mem. 1, p. 108–121.

Folk, R. L., 1959, Practical petrographic classification of limestones: Bull. Amer. Assoc. Petrol. Geol., v. 43, p. 1–38.

Ford, D. C., 1971a, Characteristics of limestone solution in the southern Rocky Mountains and Selkirk Mountains, Alberta and British Columbia: Can. J. Earth Sci., v. 8, p. 585–609.

——, 1971b, Alpine karst in the Mt. Castleguard area: Arctic Alpine Res., v. 3, p. 239-252.

——, and Ewers, R. O., 1978, The development of limestone cave systems in the dimensions of length and depth: Can. J. Earth Sci. b. 15, n. 11, p. 1738–1798.

Glew, J. W., 1975, Simulation of rillenkarren: unpublished M.Sc. thesis, McMaster University, 97 p.

Goodchild, M. F., and Ford, D. C., 1971, Analysis of scallop patterns by simulation under controlled conditions: J. Geol., v. 79, p. 52–62.

Grund, A., 1914, Der geographische Zyklus in Karst: Z. Ges. Erdkunde, p. 621–640.

Gunn, J., 1977, A model of the drainage system of a polygonal karst depression in the Waitomo area, North Island, New Zealand: Proc. 7th Int. Speleo. Congr., p. 225–229.

Harmon, R. S., and Drake, J. J., 1973, Hydrochemical environments of carbonate terrains: Water Resources Res., v. 9, n. 4, p. 949–957.

Horton, R. E., 1945, Erosional development of streams and their drainage basins; hydrophysical approach to quantitative morphology, Geol. Soc. Amer. Bull., v. 56, p. 275–370.

Jakucs, L., 1977, Genetic types of the Hungarian karst: Karzt és Barlang, Spec. Issue, p. 3–18.

Lehmann, H., 1954, Der tropische Kegelkarst der verschiedenen Klimazonen: Erdkunde, v. 8, p. 130–139.

Maruszcsak, H., 1966, Phénomènes karstiques dans les roches du cretace superieur entre la Vistule et le Bug: Przeglad Geograficzny, v. 38.

Maxey, G., 1966, Occurence and movement of groundwater in carbonate rocks of Nevada: Bull. Nat. Speleo. Soc. Amer., v. 28, n. 3, p. 141–157.

Miller, T., in litt., Karst of Caves Branch, Belize: M.Sc. thesis, McMaster University.

Panos, V., and Stelcl, O., 1968, Physiographic and geologic control in development of Cuban mogotes: Z. Geomorphol., v. 12, n. 2, p. 117–163.

Picknett, R. G., Bray, L. G., and Stenner, R. D., 1976, The chemistry of cave waters: *in* Ford, T. D., and Cullingford, C. H. D., eds., The science of speleology: London, Academic Press, p. 213–266.

Pluhar, A., and Ford, D. C., 1970, Dolomitic karren of the Niagara Escarpment, Ontario, Canada: Z. Geomorphol., v. 14, p. 392–410.

Röglic, J., 1960, Das Verhältnis der Flusserosion zum Karstprozess: Z. Geomorphol., v. 4, n. 2, p. 116–128.

Silar, J., 1965, Development of tower karst of China and North Vietnam: Bull. Nat. Speleo. Soc., v. 27, n. 2, p. 35–46.

Smith, D. I., and Atkinson, T. C., 1976, Process, landforms and climate in limestone regions: *in* Derbyshire, E., ed., Geomorphology and climate: New York, Wiley, p. 367–409.

——, Atkinson, T. C., and Drew, D. P., 1976, The hydrology of limestone terrains: *in* Ford, T. D., and Cullingford, C. H. D., eds., The science of speleology: London, Academic Press, p. 179–212.

Sweeting, M. M., 1972, Karst landforms: London, Macmillan, 362 p.

——, and Sweeting, G. S., 1969, Some aspects of the carboniferous limestone in relation to its landforms: Mediterranée, v. 7, p. 201–209.

Szczerban, E., and Urbani, F., 1974, Carsos de Venezuela. Parte 4: Formas cársicas in Areniscas Precambrias del Territorio Federal Amazonas y Estado Bolivar: Soc. Venezolana Espel. Bol, v. 5, n. 1, p. 27–54.

Thomas, T. M., 1954, Solution subsidence outliers of millstone grit on the carboniferous limestone of the North Crop of South Wales: Geol. Mag., v. 90, p. 73–82.

Thrailkill, J., 1968, Chemical and hydrologic factors in the excavation of limestone caves: Bull. Geol. Soc. Amer., v. 79, p. 19–46.

Verstappen, H. T. L., 1964, Karst morphology of the Star Mountains (Central New Guinea) and its relation to lithology and climate: Z. Geomorphol., v. 8, n. 1, p. 40–49.

White, W. B., and Longyear, J., 1962, Some limitations on speleogenetic speculation imposed by the hydraulics of groundwater flow in limestones: Nittany Grotto Newsletter, v. 10, n. 9, p. 155–167.

Wigley, T. M. L., Drake, J. J., Quinlan, J. F., and Ford, D. C., 1973, Geomorphology and geochemistry of a gypsum karst near Canal Flats, British Columbia: Can. J. Earth Sci., v. 10, n. 2, p. 113–129.

17

BARRIER ISLAND MIGRATION

Bruce P. Hayden, Robert Dolan, and Phyllis Ross

INTRODUCTION

The Atlantic coast of North America is one of the world's most dynamic sedimentary environments. Extratropical and tropical storms generate waves and surge that frequently exceed the capacity of the beaches to accommodate these energies; this, in turn, leads to major alterations of the subaqueous and subaerial parts of the shore zone. During the past several decades there has been a trend toward net coastal recession (erosion) along the Atlantic coast. This has been variously attributed to a recent rise in sea level (Hicks and Crosby, 1974), a reduction in new fluvial sediments (Wolman, 1971), human alterations of coastal geomorphology (Dolan, 1971), and secular changes in storm frequencies and magnitudes (Hayden, 1975).

Coastal erosion and deposition are a function of three interrelated factors: (1) amount and kind of sediment within a coastal segment, (2) magnitude of the sediment flux, and (3) stability of sea level. All of these factors are also directly related to the geologic origin of sedimentary coasts. It is now well established that sea level has oscillated several times during the past half-million years. During the warmer, interglacial periods, continental ice melted, and the shorelines of the world's ocean basins advanced inland across the continental shelves. During the cooler glacial periods, water was withdrawn from the seas and stored in the form of glacial ice, and the shorelines moved seaward across the continental shelves. This process involved great quantities of seawater, enough to move the shoreline across more than 100 km of the coastal plain and continental shelf.

The last phase of glaciation, the Wisconsinan, began about 100,000 years ago and ended 11,000 years ago. As the Wisconsinan came to an end, sea level was about 130 m lower than it is today, and the shoreline of the mid-Atlantic region was 80 to 160 km eastward (seaward) of the present coast (Kraft et al., 1976). With the change from glacial to interglacial, the sea started to rise and continued to rise for 8000 years. It reached to within a few meters of the present level about 4000 or 5000 years ago.

As the level of the sea rose and the shoreline moved across the continental shelf, large masses of sand were moved along with the migrating shorezone in the form of beach deposits. Additional surplus sediment that had been deposited as a series of deltas in the vicinity of the Atlantic coast river systems was reworked by wave action and moved along the shore. Once sea level became fairly stable 4000 or so years ago, waves, currents, and winds, working together on the sand surpluses, formed the beaches and barrier islands stretching from New York to Florida, and along the margins of the Gulf of Mexico. As long as the beach-energy system contained a surplus of sediment, the coastal beaches built seaward until equilibrium was established. Equilibrium in this case is a function of a balance among energy, sea level, and the amount of material in the transport system.

Based upon all lines of evidence, this equilibrium was reached about 3000 to 4000 years ago. At that time the barrier islands were possibly much wider—some, perhaps, as much as 2 km or more. As time passed and the beaches and barrier islands matured, inlets formed, sealed, and reformed in the narrow areas connecting the several regions of fairly complex topography that we see today.

Although sea level has remained fairly stable, a rise of several meters has occurred over the past 2000 years. The slow upward adjustment of sea level has resulted in the progressive recession of shorelines and enlargements of bays and sounds. Over the past 100 years, the rise in sea level has been very rapid, totaling slightly more than 30 cm (Hicks and Crosby, 1974).

MIGRATION OF BARRIER ISLANDS

The mechanics of barrier island formation and migration has been a subject of debate in the geoscience literature for many years (Schwartz, 1973). Nevertheless, there is indisputable evidence that most of the mid-Atlantic barriers are migrating westward. Peat and tree stumps are found on open ocean beaches and island recession is easily measurable from historical maps and aerial photos.

After the Atlantic coast barrier islands were formed during the Holocene stillstand, sea level started to rise again and the seaward-facing shorelines began to retreat. The rate of recession varied with the rate of sea-level rise, the regional sediment budget, slope of the inshore zone, tide range, and wave and storm climate.

Some of the eroded material may have been lost in offshore sinks, such as Diamond Shoals in North Carolina, but most of it remained within the barrier island sediment budget through spit growth, inlet filling, dune building, and storm-surge overwash.

Overwash and inlet formation and migration are the most important processes in barrier island migration. During severe storms, the beach zone and seaward dunes are overtopped by high surge and waves. As this sediment-charged mass of water spills over the shore zone and flows downslope toward the interior of the island, a layer of sediment is removed from one part of the island (the beach), and added to another (the lagoon side); this process tends to conserve the island mass.

While the migration process is, in general, accepted in its broad outlines, the spatial and temporal variations of the process have not been quantified. In this report, high-resolution data on shoreline erosion and storm-surge penetration are used to evaluate a barrier island migration model.

Dynamics of Barrier Island Migration

When viewed from the perspective of geomorphic time, the sedimentary masses called barrier islands may be considered fluid-like in their motion. These motions vary from relatively day-to-day shoreline recession and accretion to occasional threshold-exceeding storm-surge overwashes. Along the Atlantic coast the net long-term motion of the barrier islands is westward or toward the mainland.

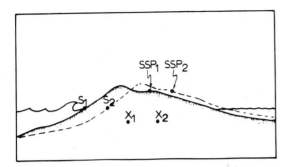

FIGURE 1 *Schematic representation of the hypothetical island center of mass (X), shoreline (S), and storm-surge penetration limit (SSP).*

Figure 1 gives a schematic composite of a migrating barrier island at two points in time, t_1 and t_2. The rate or velocity[1] (v) of island migration is defined by

$$v = \frac{x_1 - x_2}{t_1 - t_2} \tag{1}$$

where x_1 and x_2 are the centers of the island masses at t_1 and t_2, respectively. The momentum transport (kinetic energy) associated with island migration is

$$\text{momentum transport} = MV^2 \tag{2}$$

where M is the mass of the island under the constraint that in the long term

$$\frac{dM}{dt} = O \tag{3}$$

While an estimate of island mass (M) could be obtained from detailed surveys, such measurements are in general not available for extensive reaches of the coast. Therefore, the model developed here assumes a unit mass (m), and the momentum transport is the momentum per unit mass, or, expressed in terms of kinetic energy, the per unit mass kinetic energy of island migration.

The centers of island mass x_1 and x_2 for times t_1 and t_2 are, like island mass, not readily measurable. Consequently, indirect measures based upon the physical processes of migration are substituted. From aerial photography, both the shoreline (S) and the landward limit of storm-surge penetration (SSP) can be defined and mapped (Dolan et al., 1978a, 1978b). The horizontal displacement rate of the shoreline (dS/dt) is here represented by E and the rate of the landward limit of storm-surge penetration [d(SSP)-

---

[1] The vector, velocity, is used here to indicate island migration in the direction of the mainland and at the rate defined in equation 1.

$/dt$] by O. Using measurements from a series of historical aerial photographs, $\Delta t \doteq 10$ years, E and O can be calculated (Dolan et al., 1978a, 1978b).

If it is assumed that in the long term

$$SSP_1 - S_1 = SSP_2 - S_2 \qquad (4)$$

Where SSP_1, SSP_2, S_1, and S_2 are the storm-surge penetrations and shoreline positions at times t_1 and t_2, it follows that

$$E \doteq O \qquad (5)$$

and also that

$$E \doteq O \doteq V \qquad (6)$$

where V is the rate of island migration because in the long term if E is consistently greater than O, the island would become progressively narrower. If E is consistently smaller than O, the island would become progressively wider. Given the equality of E, O, and V in the long term, the equation for momentum transport may be written

$$\text{momentum transport} = mEO \qquad (7)$$

While in the long term the shoreline rate (E) and the storm-surge penetration limit rate (O) must be nearly equal, for smaller time intervals they may differ. Accordingly, selection of either of the two representative rates to stand alone for the island migration rate is difficult to justify. In this paper both E and O are used because they each represent a component of the process of island migration. Using island-wide mean values for the shoreline change and the storm-surge penetration change, the momentum transport of island migration can be calculated. However, neither of these means is very stable, either spatially or temporally, within our measurement period of four decades. Accordingly, the statistical properties of the two rates must be considered.

The values of E and O for any point in time may be represented by

$$E = \bar{E} + E' \qquad (8)$$

and

$$O = \bar{O} + O' \qquad (9)$$

where $\bar{E}$ and $\bar{O}$ are long-term means and E' and O' are the departures from the mean at that point in time. These relationships describe the shoreline and storm-surge penetration rates of change in terms of mean values and the variations about the mean. Major storms, such as the Ash Wednesday Storm of March 1962 would cause departures from the mean. This statistical formulation therefore includes threshold-exceeding events.

In terms of the means and departures from the means of both the shoreline and storm-surge penetration rates of change, the momentum transport of island migration at a given point in time may be specified in statistical terms as

$$mEO = m(\bar{E} + E')(\bar{O} + O') = m(\bar{E}\bar{O} + \bar{E}O' + E'\bar{O} + E'O') \qquad (10)$$

The time-average momentum transport is given by

$$m\overline{EO} = m\overline{\bar{E}\bar{O}} + m\overline{\bar{E}O'} + m\overline{E'\bar{O}} + m\overline{E'O'} \qquad (11)$$

Assuming a Gaussian distribution, it follows that the average of the departure from the mean ($\bar{E}'$ and $\bar{O}'$) is zero. The mean transport is therefore given by

$$m\overline{EO} = m\overline{\bar{E}\bar{O}} + m\overline{E'O'} \qquad (12)$$

where $\overline{mEO}$ is the steady momentum transport and $m\overline{E'O'}$ is the eddy momentum transport. Steady momentum transport is equivalent to the long-term climatic mean transport while the eddy momentum transport component depends on the correlation between E and O as they fluctuate about their mean values with greatest weight given to extreme values. Extreme values are most frequently associated with severe storms which cause shoreline erosion and extensive storm-surge penetrations. Very calm conditions, such as the 1974 and 1975 period, when accretion exceeded erosion along much of the mid-Atlantic coast, are also part of the extreme departures from the mean.

Using the model developed thus far, we may characterize the barrier island system as existing in one or a combination of the following states:

Steady state:	$\overline{EO} = 0; \overline{E'O'} = 0$
Trend state:	$\overline{EO} \neq 0; \overline{E'O'} = 0$
Eddy state:	$\overline{EO} = 0; \overline{E'O'} \neq 0$
Extant state:	$\overline{EO} \neq 0; \overline{E'O'} \neq 0$

$$(13)$$

Since large values of E' and O' arise predominantly when the capacity of beaches to accommodate waves and surges are exceeded, the eddy term $(\overline{E'O'})$ in the model is a measure of the importance of the threshold-exceeding events and resulting barrier island migration.

STUDY SITE

The Atlantic coast of the United States can be divided at Cape Hatteras into two major embayments (Fig. 2). The sites selected for this study, Assateague Island, Virginia, and Core Banks, North Carolina, occupy similar positions within the two coastal arcs. In each case there is an inlet defining the northern terminus. While the similarities of the two islands are significant, their dissimilarities were the reason for their selection: (1) Cape Hatteras extends more than 20 km into the Atlantic Ocean, and is fully exposed to coastal storms. This shelters Core Banks to the south. In addition, storms of the latitude of Assateague Island are more intense than at Core Banks. Consequently, Assateague is a higher-energy environment. (2) Over the last several decades, Assateague has been modified by the construction of barrier dunes, while Core Banks is largely unmodified.

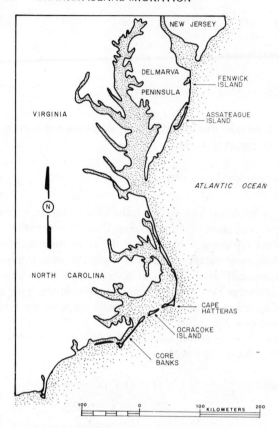

FIGURE 2 *Mid-Atlantic coast of the United States, indicating the locations of Assateague Island and Core Banks.*

(3) The northern 6 km of Assateague has undergone rapid landward migration since a hurricane in the 1930s formed an inlet at Ocean City, Maryland. This provides an excellent within-island difference for comparisons of migration rates. Core Banks is uniform by comparison. Both the similarities and differences between Assateague and Core Banks make these islands ideal for a quantification of within- and between-island migration and island dynamics.

Data Collection

The analysis of spatial and temporal variations of the shoreline and storm-surge pene-tration line requires repetitive sampling in space and time. Information of this type can be obtained from (1) ground surveys, (2) maps and charts, and (3) aerial photographs. Based upon our research, we concluded that metric aerial photography is the only feasible data source for high-resolution regional analyses. For this reason we have developed an Orthogonal Grid Address System (OGAS) to provide a uniform data base for both intra-and interisland comparisons (Dolan et al., 1978a, 1978b).

The OGAS method provides for the rapid and systematic acquisition of shoreline information from historical aerial photographs at 100-m intervals along the coast (Dolan et al., 1977). Comparison of the data derived from a sequence of years permits the definition of statistical properties of shoreline change and storm-surge penetration.

In brief, standard 1:5000-scale base maps of the study region are prepared. These maps are produced by photo enlargement of 7½-minute-series U.S. Geological Survey topographic maps (1:24,000), which provides an area 3500 m by 2100 m. The frame of each base map is oriented with the long axis parallel to the coastline and positioned over the active part of the coast. One long edge, lying entirely over the ocean, serves as the base line from which all measurements are made.

The historical aerial photographs are then enlarged to the exact scale of the base map through the use of a reflecting projector. On a transparent overlay the shoreline and storm penetration line are traced from the projection. These tracings, prepared from 1:5000-scale projections of the sequence of historical photographs, constitute the raw cartographic data base from which subsequent measurements are made. The shoreline is defined as the high-water line, and the storm-surge penetration line is defined as the line separating the active nonvegetated sand areas from the areas of continuous stands of grass and shrub.

On each tracing a transparent grid is overlaid—the grid is rectilinear with 100-m spacings. Any coastal location is thus specified by base map number and coordinates of the grid. The shore and storm-penetration line positions are then measured to the nearest 5 m with respect to the base-map base line. These data are then punched on IBM cards for subsequent analyses (Dolan et al., 1977).

The photographic coverage of Assateague and Core Banks is summarized in Table 1 The five sets of photos for Assateague and six sets of photos for Core Banks are equivalent in that both include the most severe event of the last half-century for each island in the time span covered.

TABLE 1 *Dates of Photography*

Assateague Island	Core Banks
6-2-38	10-21-40
5-3-49	3-29-55
3-14-55	10-10-58
10-5-59	5-3-62
4-21-61	4-14-69
12-3-62	6-20-75
6-4-74	

MODEL APPLICATION

Shoreline Erosion

The rate of shoreline erosion, or velocity (E), at 100-m intervals along Assateague Island and Core Banks is given in Figure 3 and is summarized for each of the islands in Table 2. Assateague Island was subdivided into the northern 60 transects that are highly influenced by the Ocean City jetties and the remaining southern 503 transects. The mean

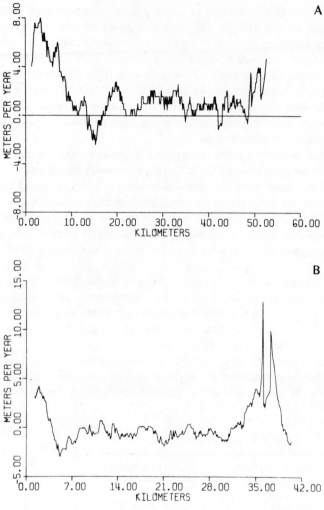

FIGURE 3 *Mean rates of shoreline erosion along (A) Assateague Island and (B) Core Banks.*

rate of shoreline erosion *(E)*, for Assateague is an order of magnitude greater than for Core Banks. The major variation within the island is found in the northern 6 km of Assateague, which are eroding at a rate five times greater than the rate for the remainder of the island. This rapid erosion can be traced to the construction of jetties on Ocean City Inlet, which stop the southward drift of sand that would normally moderate the erosion rate (Fig. 4).

The difference in the mean erosion rate of Assateague and Core Banks is probably due to the lower incident wave and surge energies at Core Banks; however, the standard deviations of the spatial means of the along-the-coast erosion rates are essentially the same for Core Banks, 2.0 m/yr, and Assateague Island, 2.9 m/yr (Table 2). The same

FIGURE 4 *North end of Assateague Island (foreground) and Ocean City, Maryland (distance).*

TABLE 2 *Summary of Spatial Means of Shoreline Change (Velocity) ($\bar{E}$)[a] And Standard Deviations ($\bar{E}'$)[b]*

	$\bar{E}$ (m/yr)	$\bar{E}'$ (m/yr)
Core Banks	0.2 ± 2.0	4.9 ± 3.0
Assateague Island	2.3 ± 2.9	6.6 ± 5.0
North end	8.3 ± 2.5	12.2 ± 8.6
Remainder	1.6 ± 2.0	5.7 ± 3.6

[a]The mean values of E at each transect, based upon measurements from several sets of aerial photographs, were averaged and the standard deviations were calculated.
[b]The standard deviations of E at each transect were averaged and their standard deviations were calculated.

standard deviations for the two units of Assateague are also of the same order (Table 2). This suggests that there is a characteristic along-the-coast variation of shoreline erosion

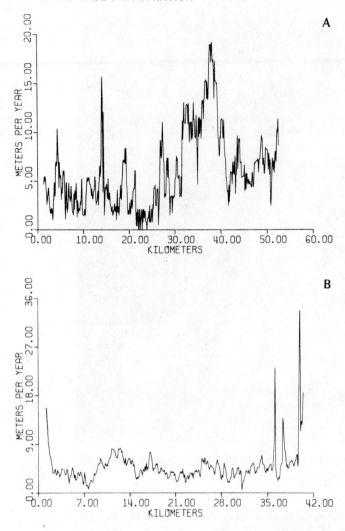

FIGURE 5 *Standard deviations of shoreline erosion rates along (A) Assateague Island and (B) Core Banks.*

rates and that this variation is independent of the magnitude of the mean rate of island erosion. Further evidence of this independence is shown in Figure 3. Along Core Banks there is a net accretion while erosion dominates along Assateague. Yet, the spatial variation of rates as indicated by the standard deviations from the mean shows a similar level of variation (Table 2). We interpret this to mean that there is an along-the-coast differentiation of incident wave energies probably due to bottom topography and to the resulting similar wave shoaling and refraction at Core Banks and Assateague.

The temporal variations of erosion rates at each transect were also calculated. The shoreline erosion rate standard deviations (E') at 100-m intervals are plotted in Figure 5 and the island-wise means $(\overline{E'})$ are summarized in Table 2. It is clear that the temporal variation in erosion rates at the various transects exceeds the along-the-coast variation by a factor of 2.4 for Core Banks and 2.3 for Assateague. Such ratios would be expected if the year-to-year variations in storm-wave conditions greatly exceed the along-the-coast differentiation of wave energies due to shoaling and refraction. In general, temporal erosion rate variations $(\overline{E'})$ are larger for Assateague than Core Banks, and are probably caused by larger and more frequent winter storms. Of the four units summarized in Table 2, the north end of Assateague had the greatest temporal variation of mean erosion rates $(\overline{E'})$. The value for this reach, 12.2 ± 8.6, is nearly twice the level recorded for the other reaches summarized. This is attributed to the interruption of the sediment stream as it

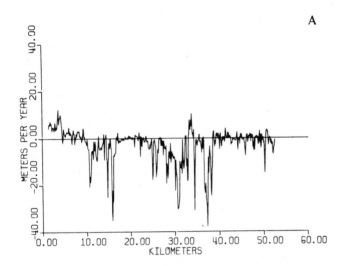

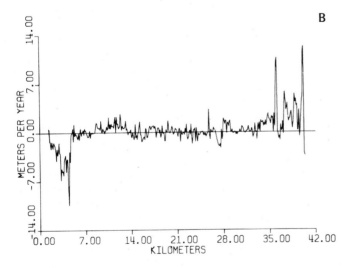

FIGURE 6 *Mean rates of movement of the limit of storm-surge penetration along (A) Assateague Island and (B) Core Banks.*

crosses Ocean City Inlet. Given the absence of an along-shore drift, and perhaps a deficit in the near offshore as well, it is reasonable to view the temporal variations in erosion rates as resulting from slow beach recovery during post-storm periods.

Storm-Surge Penetration

The rates of change of the landward limit of storm-surge penetration (O) for both islands are given in Figure 6, and are summarized in Table 3. Data for the two units of Assateague are also summarized in Table 3. The along-the-coast averages of the storm-surge penetration limit rates for Core Banks with Assateague Island differ by slightly less than an order of magnitude (Table 3). In addition, the direction of movement is, in general, landward on Core Banks and seaward on Assateague. That is, the limit of storm-surge penetration and the resultant overwash sand deposit has been gradually (0.3 m/yr) moving landward on Core Banks, while on Assateague the landward distance of overwash deposits has declined over the last several decades. In more graphic terms, overwash sediments on Core Banks are being spread across the island while on Assateague they have been piling up ever closer to the sea, and shifting the island center of mass seaward. The shoreline is generally accreting on Core Banks (Fig. 3), causing a flatter island profile. Conversely, Assateague is generally eroding, causing a less flat profile with the topographic maximum of hydraulic sediments being moved seaward. A general flattening of an island profile can be interpreted as a response to increased wave heights. Evidence appears in the literature that such increases have occurred (Hayden, 1975; Resio and Hayden, 1975).

TABLE 3 *Summary of Spatial Means of Storm-Surge Penetration Limit Change ($\bar{O}$)[a] And Standard Deviations ($\bar{O}'$)[b]*

	$\bar{O}$ (m/yr)	$\bar{O}'$ (m/yr)
Core Banks	0.3 ± 2.0	7.6 ± 5.3
Assateague Island	−1.9 ± 7.8	16.5 ± 15.1
North end	6.2 ± 8.5	22.5 ± 4.6
Remainder	−2.5 ± 6.7	15.3 ± 14.4

[a]The mean values of E at each transect based upon measurements from several sets of aerial photographs were averaged and the standard deviations were calculated.
[b]The standard deviations of E at each transect were averaged and their standard deviations were calculated.

On Assateague Island the limit of storm-surge penetration has moved seaward over the last several decades (Fig. 6). This can be linked to the construction of barrier dunes along much of Assateague, and the associated stabilization of sand deposits and vegetation (Schroeder et al., 1976). In contrast, the northern 6 km of Assateague is eroding rapidly (Table 3) and now has a profile much like that of Core Banks; however, one difference is apparent. Along the northern 6 km of Assateague, the shoreline is moving landward more rapidly than the storm-surge penetration limit (8.3 versus 6.2 m/yr), and the island is becoming narrower. This reduction in island width is related to the net deficit of sediments.

The along-the-coast variation in the mean rate of movement of the storm-surge penetration limit is given in Table 3 as the σ of $\bar{O}$. The mean temporal variation is given in Table 3 as the average σ at each transect $(\overline{O'})$, which is also the average value of the points plotted in Figure 7. In general, the temporal variance exceeds the spatial variance of storm-surge penetration rates by a factor of two. Both the temporal and spatial variances for Core Banks are much lower than those of Assateague (Table 3 and Fig. 7). These lower variances are attributed to either a lower frequency of storms or to a smaller magnitude of storm surge.

One of the basic assumptions in our island migration model is the long-term equality of shoreline and storm-surge penetration rates. A comparison of the island-wide averages

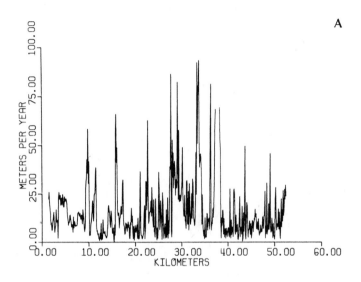

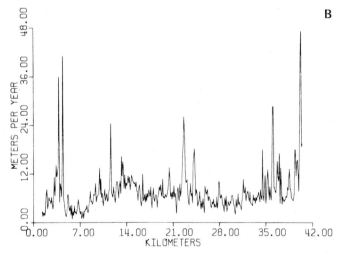

FIGURE 7 *Standard deviations of the rates of change of the limit of storm-surge penetrations along (A) Assateague Island and (B) Core Banks.*

in Table 2 and Table 3 supports this assumption. At Core Banks, an island essentially unmodified by coastal engineering measures, the mean velocities of the shoreline and the storm-surge penetration line are essentially equal (0.22 ± 2.00 and 0.26 ± 2.04 m/yr).

The Overwash Zone

The distance between the shoreline (S) and the inland limit of storm surge (SSP) is defined as the overwash zone. These overwash zone widths are given at 100-m intervals in Figure 8 and are summarized in Table 4 for both islands. The average width along Core Banks is 225 ± 45 m, while the average variation of the width over the last four

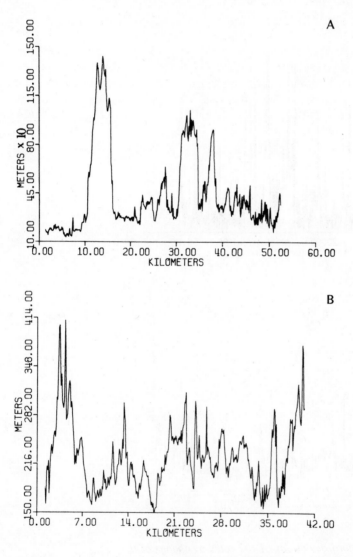

FIGURE 8 *Width of the zone of overwash penetration along (A) Assateague Island and (B) Core Banks.*

decades was only 44 ± 22m. This modest temporal difference supports the assumption implicit in equation 4 of our barrier island migration model.

TABLE 4 *Variations in the Width of the Overwash Penetration Zone*

	Along the coast	Over time
Core Banks	225.2 ± 44.8	43.7 ± 22.4
Assateague Island	429.6 ± 285.7	72.8 ± 60.4
North end	292.1 ± 76.0	106.3 ± 47.6
Remainder	437.7 ± 294.9	66.9 ± 59.1

For Assateague Island, both the along-the-coast and temporal variations in the width of the overwash zone exceeded those recorded for Core Banks. Within Assateague Island the narrowest overwash zone was found along the rapidly changing northern 6 km, while along the remainder of the island there were major variations in the width of the zone. Between 10 and 16 km to the south of the Ocean City Inlet, the width of the overwash zone exceeds 1 km.

The greatest widths of the overwash zone on Assateague are found in areas that have been sand-fenced to encourage barrier dune development. In contrast, the most dynamic areas of Core Banks, with the lowest island profile, have the narrowest zones vulnerable to overwash. Apparently, barrier dunes do not offer the protection to catastrophic storm surges normally attributed to them. The cause of the along-the-coast variations in the width of the overwash zone (Fig. 8) appears unrelated to the heights of the dune deposits along the coast, but rather to the height of the hydraulic sand prism of the island and probably to along-the-coast variations in the orientation of the shoreline (Dolan et al., 1977).

BARRIER ISLAND MIGRATION AS A TRANSPORT PROCESS

The magnitudes of the steady and eddy momentum transport for each 100 m along the coast for Core Banks and Assateague are given in Figures 9 and 10 and summarized on Table 5. It is clear from Table 5 that the eddy momentum transport term dominates barrier island dynamics. The eddy contribution for individual transects can be large when there are (1) large changes in the position of the shoreline, (2) large changes in the position of the limit of storm-surge penetration, or (3) large changes in both the shoreline and the storm-surge penetration limit. While the eddy transport is calculated according to

$$mE'O' = m\frac{1}{n} \sum_{i=1}^{n} E_i'O_i' \tag{14}$$

the large size of the sample used permits an evaluation of equation 14, which is approximately specified by

$$mE'O' \cong m\frac{1}{n} \sum_{i=1}^{n} E_i' \sum_{i=1}^{n} O_i' = m E'O' \tag{15}$$

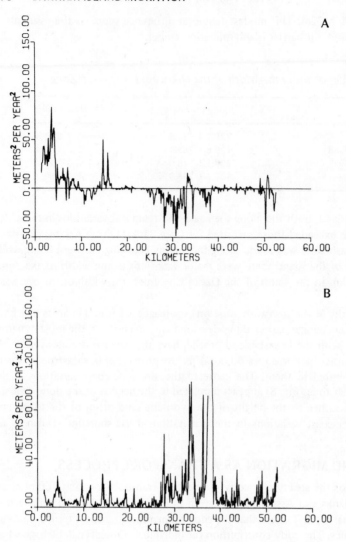

FIGURE 9 *Steady (A) and eddy (B) momentum transports along Assateague Island.*

While the mean E' and O' over time must be near zero, the average of E' and O' along the island is not constrained to be near zero. For a comparison of the small differences using equations 14 and 15, see Table 6.

Accordingly, information in Tables 2 and 3 is used here for interpreting the causes of varying magnitudes of the eddy momentum transport term. With the exception of the rapidly eroding northern end of Assateague, the magnitude of $\overline{E}'$ is of the same order as for Core Banks. We interpret this to result from equally frequent storm excitation of the beach at both Assateague and Core Banks. In contrast, $\overline{O}'$ is twice as large at Assateague as at Core Banks, probably the result of more intense storms at the latitude of Assateague

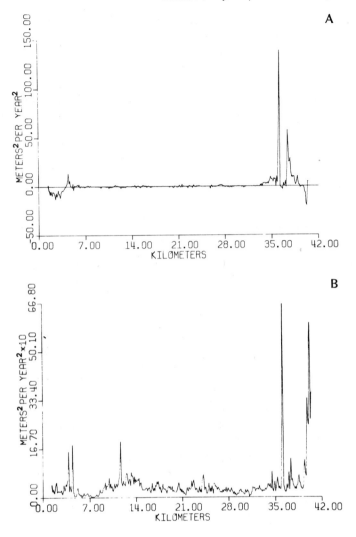

FIGURE 10 *Steady (A) and eddy (B) momentum transports along Core Banks.*

as compared to Core Banks. Thus the higher application of eddy momentum transport at Assateague (Figs. 9b and 10b) is the result of greater wave and storm surge energies. The high eddy momentum transport levels for the northern reach of Assateague (Table 5) is due to the great magnitudes of changes in the shoreline and the storm-surge penetration limit. Examination of Tables 2 and 3 reveals that both $\bar{E}'$ and $\bar{O}'$ are very high. The higher value for the northern end of Assateague compared to the remainder of the island cannot be attributed to either higher waves or more frequent storms. The difference is probably due to the trapping of sediments on the north side of the Ocean City Inlet. Thus sediment loss from northern Assateague, which moves southward with the

TABLE 5   *Average Values of Total, Steady, and Eddy Momentum Transport (g m²/yr²)*

	Momentum transport		
	Steady: $m\overline{EO}$	Eddy: $m\overline{E'\overline{O}'}$	Total: $m\overline{EO}$
Core Banks	0.1	41.8	41.9
Assateague	−4.4	125.6	121.2
North end	51.4	334.6	286.0
Remainder	−4.0	92.3	88.3

TABLE 6   *Comparison of Two Estimates of Eddy Momentum Transfer*

	Eddy momentum transport (g m²/yr²)	
	$m\overline{E}'\overline{O}'$	$m/n \sum_{i=1}^{n} E_i'O_i'$
Core Banks	37.1	41.8
Assateague	102.3	125.6
North end	274.5	334.6
Remainder	88.3	92.3

longshore drift, is not replaced by drift from the north. The negative sediment budget for this reach is responsible for the high eddy momentum transport component.

Steady momentum transport for the two islands varies by more than two orders of magnitude (0.1 g m²/yr² to 51.4 g m²/yr²). Inspection of Tables 2 and 3 reveals that both $\overline{E}$ and $\overline{O}$ for each island are approximately equal. Shoreline erosion and storm-surge penetration are equally important processes in steady-state aspects of barrier island migration.

Dynamic States

In the development of the barrier island migration model, it was suggested that the islands could be classified according to the dominance of one of four dynamical states. Based on the transect by transect evaluation of $m\overline{E}\overline{O}$ and $m\overline{E'}\overline{O'}$, the proportions of Assateague Island and Core Banks in each of the four dynamical states is assessed in Table 7. Assateague and Core Banks differ dramatically in the relative dominance of dynamic states. Core Banks is largely in a steady state (83% of 403 transects), whereas Assateague is dominated by the eddy state and a combination of trend and eddy states (extant state). This difference is equivalent to the difference between a system in equilibrium and one in disequilibrium. Clearly, the northern end of Assateague is in disequilibrium because of the interruption of the longshore drift, but most of Assateague is also in disequilibrium. We believe the stabilization of the barrier dunes is related to the departure from equilibrium.

TABLE 7 *Percentage of Assateague Island and Core Banks in Steady, Trend, Eddy, and Extant States[a]*

		n = 563 Assateague	n = 403 Core Banks
Steady state	($\bar{E}\bar{O}$ = O ± SE; $\overline{E'O'}$ = O ± SE)	12	82.7
Trend state	($\bar{E}\bar{O}$ ≠ O ± SE; $\overline{E'O'}$ = O ± SE)	1	4.7
Eddy state	($\bar{E}\bar{O}$ = O ± SE; $\overline{E'O'}$ ≠ O ± SE)	40	7.4
Extant state	($\bar{E}\bar{O}$ ≠ O ± SE; $\overline{E'O'}$ ≠ O ± SE)	47	5.2

[a]The definition of O here is ± the standard error of the estimate about true O.

ISLAND MIGRATION RATES

Barrier island migration rates (V) can be defined by

$$V = \frac{\bar{E} + \bar{O}}{2} \tag{16}$$

The average island migration rates are summarized in Table 8. Both Assateague and Core Banks have an average rate of 0.2 m/yr. While the migration rate is relatively uniform along Core Banks, there is a substantial difference along Assateague. The northern 6 km are migrating landward at 7.3 m/yr while the southern 50 km are migrating seaward at 0.5 m/yr.

The seaward migration is a temporary state. The southern section of the island is eroding at a rate of 0.6 m/yr, but overwashed sediments are not penetrating landward as far now as during earlier decades. Therefore, the center of mass of the island is moving seaward even though the shoreline is moving landward. At some time in the future, the high dune masses toward the seaward side of the island will be eroded away and breached, leading to an increase in island migration probably to a level consistent with shoreline erosion rates of about 1.6 m/yr. Assateague is dominated by the eddy state $\overline{E'O'} \gg \bar{E}\bar{O}$ and will undergo considerable changes before steady-state behavior dominates.

TABLE 8 *Summary of Island Migration Rates (m/yr)*

	$V = (\bar{E} + \bar{O})/2$
Core Banks	0.2
Assateague	0.2
North end	7.2
Remainder	−0.4

DISCUSSION

The migration rates of Core Banks and Assateague Island are the same (0.2 m/yr).

If one assumes a shoreface slope of 1:60 (Bruun, 1962), and a sea-level rise of 0.3 m/100 yr (Hicks and Crosby, 1974; Susman, 1975), the expected migration rate is similar (0.18 m/yr). The 0.2-m/yr migration rate of Assateague is a composite value which includes a 7.6-m/yr rate for the northern 6 km and a -0.4-m/yr rate for the remainder of the island. The 7.6-m/yr rate is more than two orders of magnitude greater than observed for Core Banks. Since neither sea-level rise nor the shoreface slope can be greatly different, the excess rate above 0.2 m/yr must result from the deficient sand budget. The negative migration rate for the remainder of Assateague results from a negative rate of change of the storm-surge penetration limit. The rate of shoreline recession for Assateague (1.6 m/yr) is, like the north end, much larger than expected, and probably also reflects the net sediment deficiency.

While these mean rates of change are useful as general estimates, it is clear that they are not uniform along the island (see Figs. 3 and 6). These variations have been persistent over decades, and give rise to large-scale variations in shoreline configuration, referred to by Dolan and Ferm (1968) as sandwaves, false capes, and secondary capes (Fig. 9). Earlier we reported correlations in excess of $r = +0.9$ between shoreline orientation and shoreline erosion (E) on Assateague Island (Dolan et al., 1977) and more recently a correlation of -0.9 for Core Banks. Since the σ of along-the-coast migration rates are proportional to migration rates (see Tables 2 and 3), it follows that departures from long straight beaches are indicative of rapid island migration and as such are highly diagnostic of barrier island dynamics.

While the island migration can be adequately characterized by the mean shoreline and storm-surge penetration rates, both the process of shoreline erosion and storm-surge penetrations are largely the result of episodic storms. During such events the capacity (threshold) of the beach to accommodate the elevated water levels and high waves is exceeded. First the active part of the beach is reworked with the subaerial beach sediments moved offshore, resulting in a flattened profile and a wider active zone. If waves and surge increase further, they may top the topographic maximum (a threshold exceeding event), resulting in a storm-surge penetration (overwash).

The importance of threshold-exceeding processes in barrier island dynamics is best characterized by the eddy momentum transport of island migration (Table 5). More than 98% of the extreme event wave and surge energies are involved in eddy state in contrast to steady-state processes.

CONCLUSIONS

The barrier islands of the U.S. mid-Atlantic coast are currently migrating westward at a rate of approximately 0.2 m/yr. While variations of this rate within the islands is large (~± 2 m/yr), the island-wide migration rates of the two islands studied are the same. The migration rates are about as expected for the current rate of sea-level rise (0.3 m/100 yr). In view of the higher wave energies at Assateague Island compared to Core Banks, we conclude that equilibrium island morphologies exist which reflect adjustments to prevailing wave energies, sediment size, and sediment supply. In low-wave-energy environments, such as Core Banks, finer sediments in a lower profile with a narrow active sand zone dominate. In the higher-energy environment of Assateague Island, the width of the active sand zone is much larger (two times), island topographic elevation is higher, and sand sizes are larger. While the character of these two islands differs they are responding to sea-level rise with nearly identical migration rates.

Along much of Assateague Island, barrier dunes have been built to reduce storm-surge penetration. While the effect of this is to retard island migration, shoreline erosion rates along this stabilized reach of coast are high (2.3 m/yr). The landward movement of the shoreline and the seaward movement of the limit of storm-surge penetration are on a "collision course." Catastrophic breaches of the dune system are inevitable.

Along the northern 6 km of Assateague Island longshore sediment supply is limited by the jetties of the Ocean City Inlet. This section of the island has undergone major changes, which include (1) a reduction in island width, (2) an accelerated migration rate (7.3 m/yr), (3) a narrow active sand zone, and (4) a reduction of beach grain size. In short, the absence of a resupply of beach sands from the longshore drift has resulted in an adjustment toward characteristics like that of Core Banks but with a rapid migration rate consistent with the higher wave energies.

Based upon our studies, we conclude that coastal engineering measures that result in a disruption or alteration of the island's sediment budget result in major morphological changes. In order to further verify these findings, we have initiated studies on an additional 400 km of mid-Atlantic coast barrier islands, which includes both heavily engineered and natural coastal reaches.

ACKNOWLEDGMENTS

Support for this research was provided by the National Aeronautics and Space Administration, Contract NAS5-20999; the National Park Service, Contract CX0001-4-0096; and the Office of Naval Research, Geography Programs, Contract N00014-75-C-0480.

REFERENCES CITED

Bruun, P., 1962, Sea level rise as a cause of shore erosion: J. Waterways and Harbors Div., Proc. Amer. Soc. Civil Engineers, v. 88, p. 117–130.

Dolan, R., 1971, Coastal landforms: crescentic and rhythmic: Geol. Soc. America Bull., v. 82, p. 177–180.

——, and Ferm, J., 1968, Crescentic landforms along the Atlantic coast of the United States: Science, v. 159, p. 627–629.

——, Hayden, B., Heywood, J., and Vincent, C., 1977, Shoreline forms and shoreline dynamics: Science, v. 197, p. 49–51.

——, Hayden, B., and Heywood, J., 1978a, Analysis of coastal erosion and storm-surge hazards: Coastal Eng., v. 2, p. 41–53.

——, 1978b, A new photogrammetric method for determining shoreline erosion: Coastal Eng., v. 2, p. 21–39.

Hayden, B. P., 1975, Storm wave climates at Cape Hatteras, North Carolina, recent secular variations: Science, v. 190, p. 981–983.

Hicks, S. D., and Crosby, J. E., 1974, Trends and variability of yearly mean sea level 1893–1972: National Oceanic and Atmospheric Administration NOS 13, 14 p.

Kraft, J. C., Allen, E. A., Belknap, D. S., John, C. J., and Maurmeyer, E. M., 1976, Delaware's changing shoreline: Delaware Coastal Management Program, 319 p.

Resio, D. T., and Hayden, B. P., 1975, Recent secular variations in mid-Atlantic winter extratropical storm climates: J. Appl. Meteorol., v. 14, n. 7, p. 1223–1224.

Schroeder, P. M., Dolan, R., and Hayden, B., 1976, Vegetation changes associated with barrier-dune construction on the Outer Banks of North Carolina: Environmental Management, v. 1, n. 2., p. 105–114.

Schwartz, M. L., ed., 1973, Barrier islands: Benchmark papers in geology, vol. 9, Stroudsburg, Pa., Dowden, Hutchinson Ross.

Susman, K., 1975, Post Miocene subsurface stratigraphy of Shackleford Banks, Carteret County, North Carolina: M.S. thesis, Department of Geology, Duke University, 85 p.

Wolman, G. M., 1971, The nation's rivers: Science, v. 174, p. 905–918.

V

THRESHOLDS AND MAN

18

BLUEBIRD CANYON LANDSLIDE, LAGUNA BEACH—A GEOMORPHIC THRESHOLD EVENT

F. Beach Leighton

INTRODUCTION

This threshold-exceeding event occurred on October 2, 1978, in a residential section of Laguna Beach, California, resulting in an estimated $15 million in damages, but no loss of life. The landslide wrecked many residences (Fig. 1), and many more bank accounts. Of the 69 residences involved in existing or potential sliding, 21 were entirely destroyed, and many more threatened, pending subsurface investigation and implementation of corrective measures.

This landslide event came as a surprise to the community, as none of the reported forewarnings had been connected with possible landsliding. Following the landslide, the stricken area of 23 residences (two of which remained essentially intact) became unin-

FIGURE 1 *General aerial view of Bluebird Canyon landslide of October 2, 1978.*

habitable, and over 13 residences on the periphery became unsuitable for occupancy because of potential landslide retrogression. As a result, the area was immediately declared a "disaster area" by federal and state proclamations.

GENERAL SETTING OF LANDSLIDE AREA

The Bluebird Canyon Landslide occurred in a hillside residential district on a dip slope of the ridge between Rimrock Canyon and Bluebird Canyon (Fig. 2). Development history of the area as ascertained from the examination of aerial photographs, shows it was developed initially in the late 1940s. Many communities, including the City of Laguna Beach, did not require the filing of grading plans as a part of the standard engineering approval process in 1947 when the principal tract map was filed for Tract 1252 of 53 lots averaging 552 m². Grading of roads was initiated shortly thereafter. Lots were developed with a minimum of grading. Most of the 24 residences built in the original tract were constructed on cut lots or "transition lots" that included maximum fill thicknesses of 1.2 to 3 m. Thirty-two residences had been constructed by 1957 and 17 additional residences by 1967. Only one residence was constructed in Tract 1252 since adoption of grading and building codes by the City. All residences within the general slide area were connected to a public sewer system at the time of landsliding, although it is reported that scattered earlier residences had temporarily utilized on-site sewage disposal systems in the late 1950s.

FIGURE 2   *Close-up view of Bluebird Canyon Landslide of October 2, 1978, looking eastward from head scarp at Meadowlark Lane to four overhanging homes (Lot 32 behind mailbox, Lot 39 to immediate left of large tree).*

The preliminary parameters of the landslide were as follows:

Time of occurrence: 5:00–6:00 A.M., October 2, 1978
Duration of movement: 40 to 60 minutes
Dimensions of landslide of October 2:
  Length—145± m
  Width—135± m
  Maximum height of headscarp—10± m
  Area involved—3.6 acres (approximately one-third of Tract 1252)
  Maximum thickness—21 to 24 m
  Volume—223,100 m^3
Amounts of movement:
  Horizontal displacement—15 to 18 m
  Vertical displacement—6 to 7.5 m
Rate of movement: 12± m/hr, decreasing to approximately 5 cm/day
Movement direction: $S25°E$

EYEWITNESSES

No serious injuries were sustained even though the landslide occurred while most residents were still sleeping or awakening between 5 and 6 A.M. This can be attributed to the slow shifting of ground, over a period of 40 to 60 minutes, the ranch-type residences, and the fortuitous position of slopes and fissures with respect to most residences. The boundary scarp left six homes overhanging, but the residents were able to escape through the unsuspended portions, even though the slide destroyed three roads in four places and severed utility lines.

Rumbling sounds, snapping of power lines, falling of power poles, hissing of leaking gas, and sparking of electric wires awakened most residents. Some residents thought it was a tropical rainstorm. Other residents described the sounds of people walking on their roof, breaking glass, gushing and flowing water from broken lines, pounding sledgehammers, and ripping of nails out of wood.

Walls cracked, hanging fixtures swayed, and homes began to creak, groan, and splinter. Utilities became inoperative. Rescue crews led by police with loudspeakers and firemen warned the residents to evacuate quickly without dressing and bringing any belongings. Cracks became fissures slowly, and jigsaw puzzles of broken ground developed in the yards. Driveways, floors, and roofs tilted crazily, and many cars, including one police car, were stranded by the formation of scarps and bulging earth. Some senior citizens were pulled, carried, or led out of the disaster area.

Precursors of movement prior to actual landsliding are fragmentary, but significant. These included cracking of residences and their attachments, "stretching" of the ground, as evidenced by tiny cracks and enlargement of these cracks, slight rattling of windows, broken pipes and other utilities, flashes and interruptions on TV sets, and other shorts in electrical systems. The strange behavior of pets and rapid exits of "dumb" animals were also reported. Locksmiths, plumbers, carpenters, and other repairmen were reportedly busy at irregular intervals in the area before the landslide.

SCOPE OF STUDY TO DATE

The geotechnical investigation to date has included the following steps:

1. Background research (collection, compilation, and review of available data base), including maps and reports, both regional and specific and stereoscopic pairs of aerial photographs taken in 1931, 1938, 1947, 1957, 1960, 1967, 1970, 1977, and 1978.

2. Field investigation, including geologic mapping on base maps and aerial photos of landslide features and surficial soil and bedrock conditions within and around the slide area.

3. Monitoring of 10 tiltmeters, areas of cracking and movement stakes.

4. Documentation of slide conditions by ground and aerial photography.

More detailed investigations are under way, including assessment of a future Phase 2 subsurface exploration program, continued evaluation of peripheral slide hazard and occupancy hazard, consideration of ingress and egress, and erosion and stability control measures on both an interim and permanent basis. This second phase of the investigation should answer more completely questions related to the origin of the slide and future alternatives for stabilization.

PREVIOUS PERTINENT STUDIES

A general geotechnical study of the City of Laguna Beach was part of the overall master plan for the City under a 701 Urban Planning Assisting Study (Leighton, 1969a). It was supplemented by two later studies for the City, the Seismic Safety Element (Leighton and Associates, 1975), and an engineering geologic report of storm damage (Leighton, 1969b). These geotechnical studies were mostly field and office compilations without significant original mapping beyond the earlier agency work (Vedder et al., 1957). Geoenvironmental maps of Orange County (Morton et al., 1973) and engineering geologic maps of the Laguna Beach Quadrangle (Tan and Edgington, 1976) contributed to the knowledge of the area upon publication by the California Division of Mines and Geology. Site-specific reports were limited to essentially cosmetic soil considerations and very little geology in depth.

General geologic conditions and the chief terrain problems of major canyon areas are summarized by Figure 3. This figure was included in the General Planning Study of 1969 to illustrate the dip slope problem within northeast-trending ridges and valleys.

A reduced replica of the Preliminary Geologic Stability Map contained in the general planning report is shown as Figure 1 with the Bluebird Canyon Slide added. It identifies the Bluebird Canyon dip slope as a MAJOR BEDROCK AREA SUBJECT TO POTEN-TIAL INSTABILITY.

Parcel A within Figure 1 is of special note. This parcel of about 10 acres (4 ha) is enlarged in Figure 4 to show the much greater precision and detail demanded by special-purpose mapping involved in the engineering geology of site development. Emphasis was placed in the geotechnical planning report on the future need to secure similar site-specific data prior to further development of the dip slope areas.

To repeat, the Bluebird Canyon Landslide Area had been developed in the 1940s and 1950s, prior to the general planning investigation and establishment of modern building and grading codes.

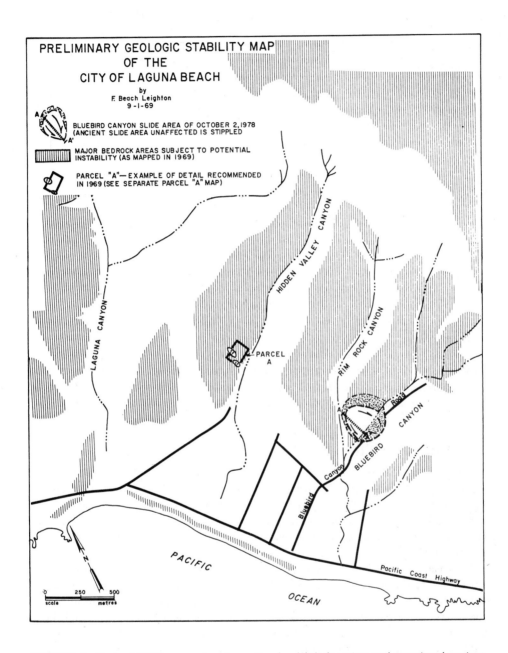

FIGURE 3 *Part of 1969 map, showing active landslide location and associated ancient slides; major bedrock areas subject to potential instability (stippled) were identified in 1969 for the City of Laguna Beach.*

391

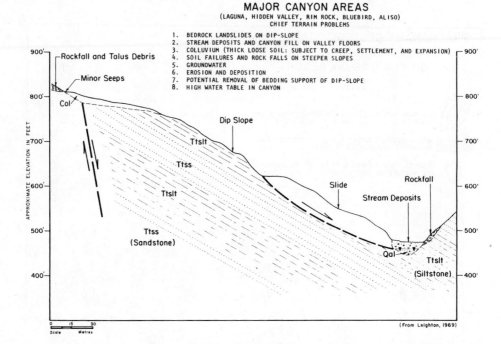

FIGURE 4 *Schematic cross section of chief terrace problems in canyon areas. (Section is from the report to the City of Laguna Beach, 1969. Principal problems for major canyon areas such as Bluebird are enumerated.)*

GENERAL LANDSLIDE FEATURES

Geologic field mapping and surface measurements have defined the boundaries of the October 2, 1978, landslide, permitting an estimation of its three-dimensional geometry (Figs. 5 and 6). The landslide occurred within the Topanga Formation (Middle Miocene) of interbedded siltstone and sandstone. The clayey siltstone interbeds, in turn, contain highly plastic and sheared clay seams roughly 2 to 8 cm thick; where continuous, these form potential surfaces of failure.

It appears that the initial slide mass moved essentially as a unit rather than failing in a markedly retrogressive manner with development of subsidiary graben zones. Tall trees in the central part of the landslide are still essentially vertical, confirming that little rotational movement occurred. Graben-like features occupy not only the slide head, but portions of the eastern and western boundaries. At least a portion of the headward limit of the landslide appears to coincide with the scarp of a preexisting slide.

Graben features occupy not only the slide head but also portions along the eastern and western boundaries. Secondary scarps and fissures formed within the slide block in response to differential movement, but zones of intense shattering are restricted to the head and foot of the slide.

The toe of the slide appears to be located very close to the former channel bottom of Bluebird Canyon (Fig. 6). The slide mounded at its toe where it crossed Oriole Drive,

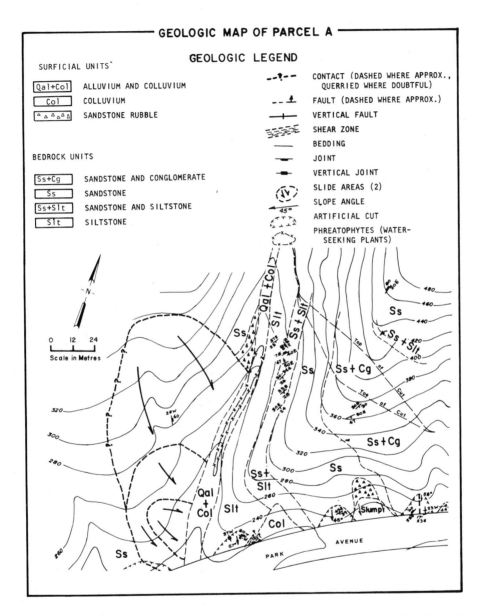

FIGURE 5 *Geotechnical map, showing the detail recommended to the City of Laguna Beach in 1969 for undeveloped lands in the City.*

393

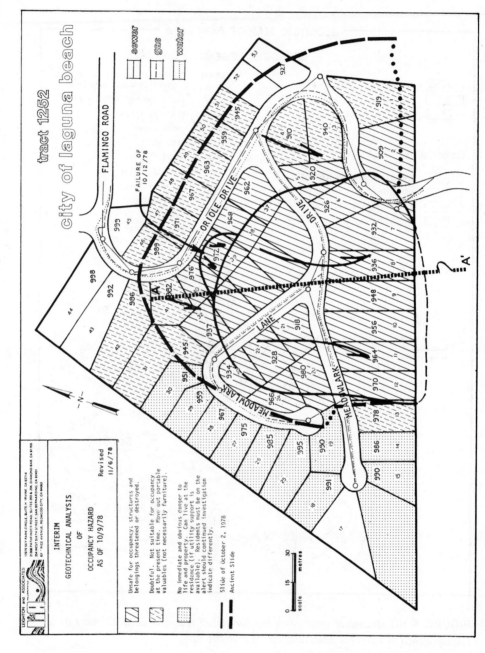

FIGURE 6 *Approximate active and ancient slide boundaries, and degree of occupancy hazard as of November 1978.*

394

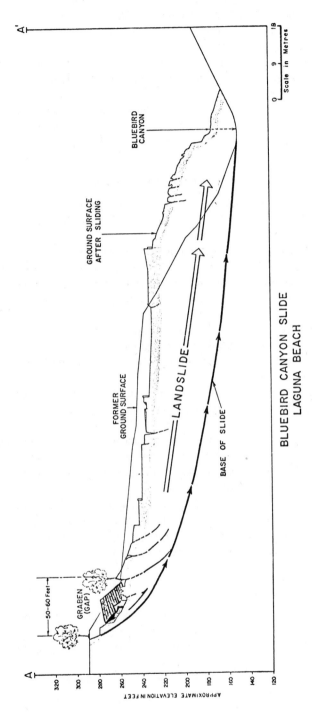

BLUEBIRD CANYON SLIDE
LAGUNA BEACH

FIGURE 7 *Preliminary cross section of landslide, based on surface mapping.*

395

indicating that slide movement was minimized at this location by the 40-ft (12-m)-thick fill prism that extends across Bluebird Canyon and acted as a temporary buttress. The tree shown in cross section A-A' that has shifted 15 to 18 m is the same large tree shown in the graben area of Figure 1.

The basal rupture surface, where it is exposed, is a distinct shear zone of plastic clay whose thickness averages 0.75 to 4 cm. It appears to be smooth, continuous, and parallel to bedding.

The landslide appears to lie within a synclinal structure. Subsurface exploration is expected to document a flatter bedding dip near the foot of the slide mass and slideward dips on the flanks of the slide.

EVIDENCE OF OLDER LANDSLIDE EPISODES

The preliminary investigation has resulted in three major types of evidence of older landslide episodes: (1) aerial photographic studies, (2) identification of older landslide materials, and (3) mapping of old surfaces of rupture.

Eight sets of vertical aerial photographs taken prior to the October 2, 1978, failure were studied during this investigation, as well as photographs taken immediately following the landslide. Stereoscopic overlap of these photographs aided in identifying arcuate scarps, old graben zones, and uneven landslide topography not readily explained by other geologic processes.

Shattered rubble contains veins of gypsum at the margin of the active slide. Older colluvium probably represents infilling of old grabens and fissures. It is believed that a considerable time must have elapsed between the active slide and the earlier episode(s) of sliding.

Fissure fillings and ancient failure surfaces were identified on the surface of the slide in the headwall. Multiple failure surfaces may indicate more than one period of prehistoric failure.

The approximate boundaries of the suspected ancient landslide(s) have been indicated on Figures 2 and 5. It is believed that the earliest episode of landsliding occurred prior to the present erosion cycle over 11,000 years ago when base level was much lower. This landslide may have forced Bluebird channel to the south, cutting and oversteepening the south bank and promoting slumping of what is essentially an anti-dip slope. The 1978 landslide involved a part of the ancient landslide on the dip slope of Bluebird Canyon. In all probability this slide had also moved within the Holocene, as older colluvial graben-like material is exposed in the headscarp of the 1978 landslide.

FIELD-MONITORING PROGRAM

Ten tiltmeter stations were installed around the margin of the landslide under the direction of Bruce Clark. They were located at varying distances from the original head scarp and installed to monitor small ground movements that might indicate additional failures prior to actual cracking and movement. These instruments proved to be the most positive means of establishing what ground outside the slide area would show potential movement. The movement of October 12, 1978 (Fig. 5), was anticipated as a result of increased tilting in that area.

Cracks in structures were painted red in order that deformation of the cracks might be inspected and recorded. Increased strain measurements recorded by tiltmeters were followed by visible cracking that was measurable, particularly before the subsidiary failure of October 12, 1978.

Sets of stakes were also established to measure additional movement of the major landslide block. Movement of this block has not been more than 1 to 2 cm since the October 2 event, but further movement is anticipated with the next heavy rains.

WATER REGIMEN

Cyclic wet years and wet periods are common in southern California. They occur almost every decade. Whereas the average annual rainfall for the City of Laguna Beach is approximately 33 cm, 53 cm fell in 1969, and over 70 cm in 1978. The wet season in southern California is concentrated between November 15 and April 15, and the groundwater year extends between October 1 and September 30.

The 1969 storms were particularly damaging because they combined high total rainfall, long duration, and high daily maxima that deluged wet ground already soaked. The 1978 wet season produced much less storm damage because the rainfall was less intense, with storms of shorter duration and lower daily maxima.

As a result, much more rainfall than normal in 1978 percolated into the subsurface, and a much lower percentage than normal was lost as runoff. Rainfall that infiltrated near the crest of the landslide area during the spring of 1978 has been (or still is) percolating toward the Bluebird Canyon aquifer within permeable beds contained in the dip slope. This is shown by the fact that running water continues to flow in the canyon 6 months after the rains. Other amounts, along with the irrigation water of the summer season, are probably stored beneath perched or normal water tables within the ridge.

No high-water-table, mounding, or artesian groundwater conditions are evident. Dip slope conditions do favor the downdip migration of surface water on the south-facing slopes of Bluebird Canyon, but the clayey siltstone acts more as a barrier than as a conductor. Sparse and small-scale structural controls for groundwater circulation may exist, such as local unconformities and pockets and lenses of sandy material, but no effective structural barrier such as faults are known on the subject slope.

Although a special search was conducted for any natural or man-induced groundwater seepages in the vicinity of the landslide, none was noted. A small amount of wet and muddy soil was present at the slide toe on Oriole Drive after landslide occurrence, probably as a result of water flow from ruptured utility lines uphill under Meadowlark Drive. Because the soil around the pipe was not saturated, it was concluded that the pipe was intact until the landslide occurred. No areas of wet or saturated subsoil were observed around the perimeter of the landslide, along the head scarp, or in the graben area.

ORIGIN OF THE LANDSLIDE

Principal Geologic Conditions Favoring Landsliding

Three principal factors created the potential for landsliding on this south-facing slope of Bluebird Canyon:

1. *Rock weakness.* Clayey siltstone and clay seams are the units of low strength in the

Topanga Formation. An understanding of the spatial distribution of these "weak links" is essential to evaluating slope stability, particularly on the south-facing dip slopes in this area. These weak links commonly have angles of internal friction of 6 to 10° or less and cohesion values of 75 lb/ft² or less.

2. *Adverse geologic structure*. The Bluebird Canyon Landslide occurred in a rock section that contained bedding surfaces that are smooth, continuous, inclined downslope, and unsupported at the lower end. The landslide broke away from a preexisting fracture that probably represented a former head scarp of an earlier landslide. The slide mass probably moved downslope on a bedding surface coated by a plastic clay seam. The plastic clay behaved somewhat like jelly in a sandwich, being squeezed and grooved by the upper slide block as it overrode the seam.

A synclinal warp aided the downslope sliding by forming a spoon-shaped geometrical structure. This adverse structure combined with water-filled materials on the watertight plastic clay is believed to have created the major instability.

3. *Channel entrenchment and erosion of foot of ancient landslide*. Comparison of aerial photographs taken every decade since the 1930s indicate gullying of the Bluebird channel and sloughing and erosion of the foot of the slide. A comparison of topographic maps prepared in 1960 and 1978 corroborates this conclusion and indicates at least 2.4 m of downcutting immediately downstream of the active slide foot, which has buried the former channel.

Possible Initiating Causes of Movement Discounted

1. Heavy rainfall immediately before and during landslide development did not occur. Approximately 5.6 cm of rain fell on September 5 and 6, the only significant rainfall since April 1978.

2. No earthquakes were recorded on dates before and during sliding. Thus seismic activity and active faulting are discounted.

3. No substandard methods of building construction, grading operations, lot alteration, or inferior foundations are believed to contribute genetically to the sliding. Structural engineers and soil engineers have aided in this determination.

4. Overirrigation and subsurface percolation of septic tank effluent are discounted, because evidences of undue erosion and deposition are not present, available water records do not reflect local overirrigation, and the tract is served by public sewers.

5. Broken pipelines are discounted, because the earth around pipes is not saturated and no unusual water losses have been documented.

Summary of Factual Causal Aspects

1. The Bluebird Canyon landslide is a rock block slide that occurred on October 2, 1978, on a dip slope.

2. The active landslide is part of a larger prehistoric slide area.

3. The heavy rains of 1977–78 introduced more water into the slope than at anytime during this century (with the possible exception of 1941–42). Introduction of this water was facilitated by the open structure of the ancient slide materials.

4. Bluebird Canyon was being entrenched by gullying at the foot of the dip slope that slid on October 2, 1978. The foot of that slope was also being actively eroded.

Summary of Inferred Causal Aspects

1. A perched water table exists in at least the graben area of the actual slide.

2. The basal rupture surface of the active slide lies at least partly within the pre-existing slide(s).

3. The combination of (a) rock weakness, (b) canyonward-dipping bedding, and (c) canyon entrenchment and erosion at the foot of the dip slope created the basic conditions for landslide failure.

4. During the period following the heavy rains of January–April 1978, parts of the landslide area were probably little more than of marginal stability.

5. The principal initiating causes of movement are believed to be a combination of (a) groundwater infiltration from the 1978 rains and (b) erosion and piecemeal mass movements at the foot of the slope that slid.

THRESHOLD ASPECTS

Deterioration of an ancient landslide to the threshold of reactivation commonly proceeds as miniscule cumulative change. The slide may lie dormant for years, movement being confined to unrecognizable creep and to gradual consolidation and settlement that may be recognizable. Gradually, the shear strength of the old slide mass is overcome, the potential slide surface extends itself (in this case as a modification of an old slide surface), and finally, reactivation occurs as a slow or rapid movement. The slide is active.

The slowness of culmination of the Bluebird Canyon Landslide should not obscure the fact that a hazardous condition existed prior to October 2, 1978. The ancient slide had all the earmarks of a slide subject to renewed movement. There was no evidence that this slide was extinct or dead as a result of internal solidity, a favorable position as the result of earlier sliding or deposits in front of it that effectively block it.

Once the configuration of the active landslide is known by subsurface investigation, the threshold values of failure can be calculated. In recent years, considerable advancement and refinement have been made in slope stability analysis by soil engineers working with high-speed computers. Based on the geometry of the slide before and after sliding, it is possible to back-calculate with relative ease the strength parameters, cohesion, and angle of internal friction of the rupture zone at the time of initial failure. These steps are as follows:

1. Based on an assumed cohesion of zero, a trial-and-error process is used to estimate the angle of internal friction that yields a safety factor of unity.

2. With an assumed angle of internal friction of zero, again a trial-and-error procedure is utilized to calculate the cohesion value that yields a factor of safety of unity.

3. A plot is made of the locus of all pairs of cohesion and angle-of-internal-friction values that produce a safety factor close to unity.

4. Threshold values are achieved by analysis of this plot as well as laboratory test results.

ACKNOWLEDGMENTS

I am indebted to members of our landslide team, Larry Cann, Iraj Poormand, Bruce Clark, Jim Knowlton and Jim Fisher, for their help on the project and their critical reading of the original manuscript; and to Dorothy Bergman for doing double duty with her drafting skills.

REFERENCES CITED

Leighton, F. B., 1969a, Final geologic report on the general plan study for the City of Laguna Beach, 36 p.

——, 1969b, Engineering geologic report of storm damage, City of Laguna Beach, 23 p.

Leighton and Associates, 1975, Background data and geotechnical information for the seismic safety and safety element: presented to the City of Laguna Beach, 33 p.

Morton, P. K., and others, 1973, Geo-environmental maps of Orange County, California: Preliminary Report 15, California Division of Mines and Geology.

Tan, S. S., and Edgington, W. J., 1976, Geology and engineering geologic aspects of the Laguna Beach Quadrangle, Orange County, California: California Div. Mines Geol. Spec. Rept. 127.

Vedder, J. G., and others, 1957, Geologic maps of the San Joaquin Hills–San Juan Capistrano area, Orange County, California: U.S. Geol. Survey Oil and Gas Invest. Map OM 193.

19

REGIONAL LANDSLIDE—SUSCEPTIBILITY ASSESSMENT FOR WILDLAND MANAGEMENT: A MATRIX APPROACH

Jerome V. DeGraff and H. Charles Romesburg

INTRODUCTION

Land-use planning and land management in many areas requires large area or regional landslide-susceptibility (threshold) evaluation. This is especially true of federally administered wildlands in the western United States. Pressure from energy development and population growth is increasing the need for slope-stability information in management decision making. In wildlands management, landslides are traditionally a factor relating to practices such as timber harvesting (Gray, 1970; Bailey, 1971; Swanson and Dyrness, 1975) and watershed improvement. Landslides are an increasingly important factor as the demand for roads, structures, recreational facilities, and mining grows on federally administered wildlands (Fig. 1).

Landslide studies usually can be categorized as a site investigation, tract evaluation,

FIGURE 1 *Representative rockslide along U-153, a Utah state highway constructed across part of the Fishlake National Forest. This 18-year-old section provides easy access to a ski resort and other recreation sites. Landsliding occurs annually along a 2.7-mi (4.3-km) highway section.*

or regional assessment (Leighton, 1976). Site investigations of particular landslide mechanisms provide insight into specific geologic conditions or physical and engineering parameters (Bolt et al., 1975). Studies of particular circumstances and threshold conditions are useful for land management decisions and can be facilitated by use of such information as topographic and materials data (Fig. 2). Tract evaluations provide the specific data needed for such decisions (Cooke and Doornkamp, 1974; Leighton, 1976). Unfortunately, the time and money involved for drilling, testing, and detailed studies precludes the wide application of this method. The areas involved in wildlands management are too large to use tract evaluation as a general evaluation technique. A typical management unit may range between 200 to 600 mi^2 (518 to 1554 km^2). Tract evaluations are restricted to use on specific problems or projects. Regional assessment is needed to provide landslide-susceptibility information for general planning purposes. Regional assessment for environmental planning is undergoing extensive development (see Coates, 1977, Part 5). Some efforts appear to require more data for meeting wildlands management needs than time and budgetary considerations will allow. Most of the current studies concentrate on assessing landslide susceptibility in urban and suburban areas or areas that will soon be urbanized (Brabb et al., 1972; Nilsen and Brabb, 1973, 1975; Briggs et al., 1975; Lessing et al., 1976). The needs of wildland areas involve different constraints requiring modification of established methods.

A MATRIX ASSESSMENT APPROACH

The matrix-assessment approach to evaluating landslide susceptibility is a quantita-

FIGURE 2 *View of massive downslope movement within Sheep Creek drainage, Fishlake National Forest. Landsliding was reactivated by test project efforts to increase water yield through vegetative manipulation.*

tive method for establishing an index of instability over an area. It lacks the ability to predict landslide susceptibility in terms of probability or confidence intervals. However, the matrix-assessment approach does allow relative potential to be evaluated over large areas using a few key measurable factors. When this approach is used, the landslide susceptibility classes are defined by discrete combinations of measurable attributes. Thus the determination of terrain thresholds for displacement is a vital ingredient.

The conceptual basis for matrix assessment is the intuitive process by which a geologist establishes relative landslide susceptibility in an area. Rather than being a new idea, matrix assessment is an articulation of this intuitive process. It is common to find the results of this process illustrated while the process itself is undescribed or sketchily defined (Van Horn, 1972; Williams, 1972; Radbruch-Hall, 1976).

Significant differences in interpretation for the same area can arise from variations in the educational perspective and experience background of different workers. Matrix assessment seeks to avoid this possibility by articulating the process, relating the factors to a few basic measurable attributes, and following an objective numerical procedure.

There are several assumptions and basic decisions behind the matrix assessment approach to landslide-susceptibility evaluation. Landslides can be viewed as a threshold phenomenon. Landsliding occurs when the sum of the forces resisting downslope movement are exceeded by the sum of those creating it (Cooke and Doornkamp, 1974). The value of the point at which the resisting forces are exceeded is the threshold value. Because each set of forces represents the sum of all forces acting in a particular direction, the summation values of opposing forces and the related threshold value varies among points in an area. How close the difference between these opposing forces is to the threshold value is the relative susceptibility. Although it is difficult to establish the threshold value and the difference between the opposing force values in absolute terms, the relative proximity of these two values can be assessed. Matrix assessment established this relative difference by comparing the conditions involved in existing landslides to the conditions existing throughout the area of interest. For this reason and statistical considerations, matrix assessment uses all disturbed and undisturbed terrain.

Sharpe (1938) listed an extensive group of conditions favoring landsliding. This list was divided into basic and initiating conditions. Initiating conditions (Sharpe, 1938) include human activities, earth tremors, floods, and other highly unpredictable factors. Sharpe (1938) categorized basic conditions favoring landsliding as lithologic, stratigraphic, structural, topographic, and organic. Organic conditions included climatic factors. Matrix assessment used three geologic attributes: bedrock, slope, and aspect (slope orientation). These attributes are assumed to include the five basic conditions outlined by Sharpe (1938). Bedrock is used by formation or mappable member. Particular units with greater susceptibility are differentiated. This greater susceptibility may result from physical and chemical factors, including permeability, fractures, and cementation. Greater susceptibility may be a function of landslide-prone regolith derived from a particular bedrock (Nilsen and Brabb, 1975). Sharpe defined lithologic and stratigraphic conditions, respectively, as conditions resulting from inherent weakness in a bedrock unit, and circumstances involving weaker bedrock interlayered with stronger bedrock. Bedrock, used as a matrix attribute, is assumed to incorporate lithologic and stratigraphic conditions into the landslide-susceptibility assessment. Slope identifies inclinations of the land surface conductive to landsliding. It is expressed in percent and grouped in 5% or 10% classes. Slope inclination is equivalent to the topographic basic condition (Sharpe,

1938). Aspect is the compass direction a slope faces. It is expressed as 8 or 16 compass direction classes (N, NE, SW, etc.) defined by degrees of azimuth. Aspect is used to include any significant slope orientation that might enhance landsliding by interaction with climatic or structural variables. By using aspect as a matrix attribute, Sharpe's (1938) structural and organic basic conditons are incorporated into the assessment. Structural condition refers to bedding planes, fault planes, and similar factors which enhance landsliding (Bailey, 1971; Eisbacher, 1971; DeGraff, 1978). The organic condition includes climatic factors such as rainfall distribution or slope drying characteristics that influence landsliding (Beaty, 1956, 1972). Bedrock, slope, and aspect are used in conjunction with the distribution of landslides within the region being evaluated. Matrix assessment evaluates landslide susceptibility, in the context of the five basic condition categories, by using bedrock, slope, and aspect as measurable attributes. Omission of initiating conditions from this matrix implies that the basic conditions control stability. This assumes that the slopes most susceptible to initiating conditions are ones with unfavorable basic conditions. Although this is not always true, difficulty in evaluating future initiating conditions precludes their explicit use.

METHODOLOGY

The area to be investigated must be defined and boundaries established. In wildlands management, this defined area may be a typical administrative unit for the land management agency. All landslides are then identified and inventoried within the area (Fig.

FIGURE 3 *Part of landslide scarp area at the head of the Left Fork of Rilda Canyon, Manti-LaSal National Forest. This is one of the landslides inventoried for matrix assessment. (DeGraff, 1977.)*

3). This is best accomplished through a combined program of aerial photo interpretation and field work.

From the inventory, a matrix of bedrock, slope, and aspect is assembled. Total landslide acreage values are placed in each appropriate cell (Fig. 4A). Within this landslide matrix, each set of bedrock, slope, and aspect conditions is described by the amount of landslide terrain with the same particular combination of conditions. For example, there are 98 ha of landslides on bedrock C, slope M, and aspect Z.

A similar matrix of all bedrock, slopes, and aspects is developed for the management unit (Fig. 4B). In the management unit, 141 ha is described by the same attribute combination bedrock C, slope M, and aspect Z used in the example of landslide acreage. All the combinations found for landslide locations will be part of the management unit matrix. The total acreage for the management unit will be included within this matrix.

The landslide-susceptibility matrix will have a cell corresponding to every cell in the management unit matrix (Fig. 4C). Each cell in the management unit matrix is divided into the corresponding cell in the landslide matrix. A value of 0.70 is obtained for the combination of bedrock C, slope M, and aspect Z when the landslide acreage (98 ha) is divided by the management area acreage (141 ha). This value and all others computed for the landslide-susceptibility matrix represent the proportion of area in the management

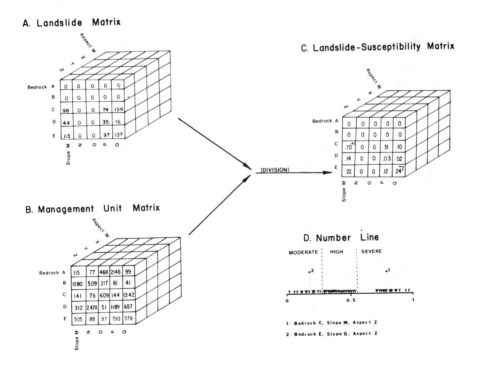

FIGURE 4 *Schematic representation of the matrix assessment method for establishing regional landslide susceptibility.*

unit where landslides have occurred. Where a cell in the management unit matrix has no corresponding cell in the landslide matrix, a value of zero is placed in the appropriate cell of the landslide-susceptibility matrix. The completed landslide-susceptibility matrix will contain zero values at all cells, with bedrock, slope, and aspect combinations that have no associated landslide terrain. Values greater than zero and less than 1 will be found in all cells where the combination of bedrock, slope, and aspect corresponds to some inventoried landslide acreage.

All combinations of bedrock, slope, and aspect that have zero values in the landslide-susceptibility matrix constitute the low-susceptibility class. All other values in the matrix are placed figuratively on a number line between zero and 1 (Fig. 4D). These values are grouped to produce three basic clusters. Grouping is achieved by a nonhierarchical clustering method which begins with an initial partition and attempted interative improvements (Anderberg, 1973, p. 44). An initial grouping is achieved by creating three equal divisions of the range of proportion values. This one-dimensional grouping takes the form of minimizing the sum of squared deviations about the three group means (W function). The cells in the group near zero constitute the moderate susceptibility class. Severe susceptibility is composed of the cells grouped near the value of 1. High susceptibility is defined by the cells grouped between these two end groups. In each class, the combinations of bedrock, slope, and aspect for each cell defines the susceptibility conditions. In other words, bedrock E, slope Q, and aspect Z of cell 2 and the combinations describing all other cells in this group identifies areas with moderate landslide susceptibility. Bedrock C, slope M, and aspect Z describes cell 1, an example defining one of the condition sets for high landslide susceptibility.

Because each class is defined by a set of discrete combinations of bedrock, slope, and aspect, the relative susceptibility of a point can be readily determined. The combination of bedrock, slope, and aspect for the point is compared to the identical combination in the landslide-susceptibility matrix (Fig. 5). This identifies the susceptibility class for that point. In the figure, the point description is identical to cell 2 in the earlier example. Because this cell defines the moderate susceptibility class, the point is identified as having a moderate susceptibility for landsliding. Similarly, a landslide-susceptibility map can be produced for the management unit (Fig. 6). Every point in the unit is described by its particular combination of bedrock, slope, and aspect. Each point is then compared to the corresponding cell in the landslide-susceptibility matrix. This identifies the susceptibility class for each point in the same manner as for an individual point. After all points are assigned to classes, boundaries are drawn around contiguous points with identical susceptibility ratings. This produces a map of severe, high, moderate, and low landslide susceptibility for the management unit.

Mathematically, almost any reasonable number of categories could be objectively defined on the number line. Selection of four susceptibility categories is based on the utilization made of matrix assessment results. It is a matter of perception on the part of the user. Clearly, two categories are the minimum to enable a comparison. With more than six to eight categories, it becomes too difficult to keep track of their relative differences. It is necessary to label each category with terms that convey the relative landslide susceptibility to the nonspecialist user. Social psychology has shown that this type of labeling should be characterized by opposing polar terms, and show the direction and distance between these polar terms (Osgood et al., 1975). Therefore, four categories are defined between the polar terms of severe and low, with high and moderate as inter-

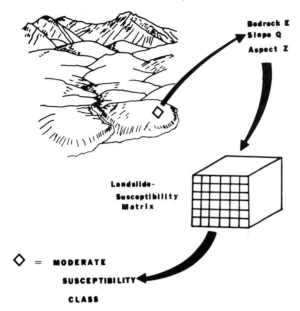

FIGURE 5 *Diagrammatic illustration of landslide-susceptibility identification of a selected point.*

mediate labels showing direction and distance between these polar terms. This may seem a minor point; however, landslide-susceptibility assessments are made for planners, politicians, managers, and other nongeologists. If the information available in the assessment is not perceived by these users, it is of little consequence whether the assessment system or results are valid.

Matrix acreage determination and manipulation can be done by hand and by using overlays on a map. The resulting landslide-susceptibility class map may be produced in the same way. Obviously, computer manipulations are advantageous. Programs capable or nearly capable of these operations exist in some management agencies. Similar programs have been developed for research applied to other land-use problems (Tilmann et al., 1975). Computer manipulation has the additional advantage of being able to extract slope and aspect data from digital terrain tapes.

APPLICATION

Matrix assessment for landslide susceptibility is an outgrowth of the ECOSYM project. ECOSYM is a natural resources classification and data storage system developed by Utah State University for the Surface Environment and Mining Task Force (SEAM) of the USDA Forest Service and the Office of Biological Services, USDI Fish and Wildlife Service. Matrix assessment is currently being applied to a 400-mi^2 (1036-km^2) planning unit of the Manti-LaSal National Forest (DeGraff, 1977) and a 514-mi^2 (1331-km^2) planning unit of the Fishlake National Forest in central Utah.

A completed assessment for part of the Wasatch-Cache National Forest in northern Utah illustrates the application of this method. Landslides within a 215-mi^2 (557-km^2)

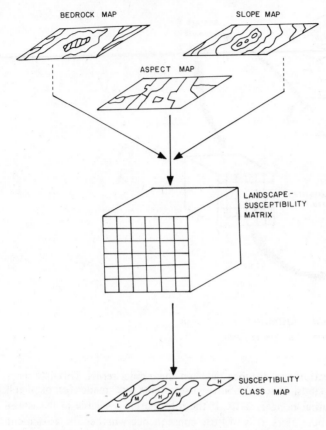

FIGURE 6 *Schematic diagram of the process followed to map landslide susceptibility over an area in which an assessment has been completed.*

area of the Bear River Range were mapped and inventoried (Fig. 7). Nineteen landslides of varying size and diverse setting were identified (Table 1).

Following the methodology outlined for matrix assessment, a landslide-susceptibility map was produced. Figure 8 shows a part of that map and matrix assessment elements used. Final partitioning was 0.01 - 0.29, 0.29+ - 0.58, and 0.58+ - 0.88, with $W = 0.0723$.

DISCUSSION

To provide consistent results in regional landslide-susceptibility evaluations, it is necessary to employ some numerical method. The nature of landslide phenomena presents some troublesome points when applying numerical techniques.

To derive valid evaluations employing numerical methods, attributes must be measured on both landslides and undisturbed terrain. Evaluation using only attributes measured on landslides is like evaluating the incidence of measles in children by only exam-

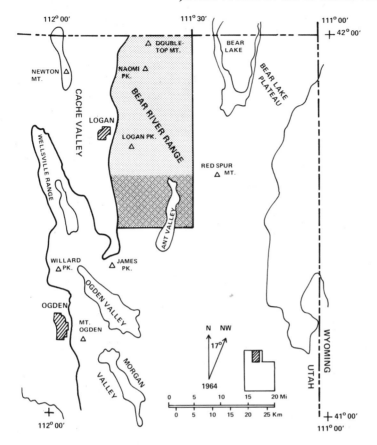

FIGURE 7 *Index map showing the location of the landslide-susceptibility example on the Wasatch-Cache National Forest in north-central Utah. Shaded area is the Bear River Range, the darkest section being the assessment area. Major valley and mountain areas are delineated along the 5000-ft (1524-m) contour.*

ining children with measles. A comparison of landslide attributes with the same attributes measured on undisturbed terrain normalizes the data. This reveals whether a specific circumstance is actually more landslide-prone or just so widespread that it comprises an excessive amount of the sample.

Some standard quantitative techniques require the assumption of random distribution for the phenomena sampled. The assumption that identified landslides in an area are a random sample of all past or future natural landslides is open to question. This problem can be avoided by using only a randomly selected part of the total landslide sample or by using all points, with or without landslides, in an area. Unfortunately, there are frequently too few landslides to allow use of a randomly selected subset. If all points are used, the landslide sample size often becomes diminishingly small compared to the

TABLE 1 *Characteristics of Some Landsides in the Bear River Range (DeGraff, 1976)*

Landslide	Size (ha)	Slope (%)	Aspect (°)	Bedrock (formation)
1	5	34	269	Brigham
2	2	31	256	Brigham
3	5	32	240	Brigham
4	2	16	4	Wasatch
5	2	27	351	Brigham
6	49	19	298	Bloomington
7	18	21	224	Langston
8	2	53	313	St. Charles
9	2	58	10	Brigham
10	2	21	271	Wasatch
11	5	34	288	Wasatch
12	49	19	310	Wasatch
13	8	32	281	Wasatch
14	39	27	276	Wasatch
15	8	11	110	Wasatch
16	5	27	136	Brazer (Great Blue?)
17	18	36	231	Brazer (Great Blue?)
18	38	16	344	Salt Lake
19	26	17	325	Salt Lake

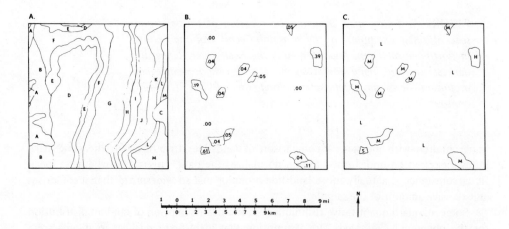

FIGURE 8 *Part of the Wasatch-Cache National Forest assessment map. (A) is the bedrock map with different bedrock formations represented by different letters. (B) shows the proportions obtained in the landslide-susceptibility matrix. (C) is the landslide-susceptibility map.*

410

undisturbed-area sample size.

Recent work to evaluate regional landslide susceptibility have employed several strategies for incorporating numerical methods.

Leighton (1976) assesses landslide stability as part of evaluation procedures in a land-use planning program. Although this process employs the more detailed tract level of evaluation rather than regional level, it does involve the use of matrices. Leighton refers to the use of matrices in defining landslide stability ratings. These accompany maps to graphically illustrate the relative stability rating of an individual slide or slope area. These matrices are expressions of the interpreted stability rather than a means for assessing landslide susceptibility. In this case, it is a numerical representation of intuitive judgment. Leighton uses matrices in the numerical sense in defining land-use categories that include landslide stability ratings as part of the equation.

A recent study in southern Italy used bedrock as part of the evaluation procedure for landslide assessment (Carrara and Merenda, 1976). Bedrock units were normalized by comparing landslide area to outcropping area. In other words, the number of acres of landslide involving a certain bedrock unit was divided by the number of acres exposed in the study area. This value was converted to percentage form. Because further analysis is in progress, it was not detailed how normalized bedrock information would be combined with other data to define landslide susceptibility.

Nilsen and Brabb (1977) have been conducting landslide evaluation studies in the San Francisco Bay region for many years. Their program uses bedrock, slope, and landslide distribution to assess susceptibility. Like Carrara and Merenda (1976), susceptibility of bedrock units is established by determining the percent of each exposed bedrock unit that is involved in landslides. The resulting percentages, ranging from 0 to 70% are grouped into six arbitrary categories. These bedrock units are adjusted to a higher or lower category based on their coincidence with susceptible slope categories. Slope susceptibility is first established in the same manner as bedrock units. Therefore, the final six categories are defined as certain associations of bedrock and slope resulting from this stepwise evaluation.

Lessing and Erwin (1977) also used bedrock, slope, and landslide distribution to evaluate urban area landslide susceptibility in West Virginia. Slopes, rock types, and soil types conducive to landsliding were defined by study of recent and old landslides. Investigators identified slide-prone areas by the presence of these conditions and interpretation of past landsliding. The absence of these same conditions defines areas of relatively stable ground (Lessing and Erwin, 1977). By mapping the area as either slide-prone or relatively stable ground, two landslide-susceptibility classes are established.

Matrix assessment uses bedrock, slope, and landslide distribution like Nilsen and Brabb (1977), and Lessing and Erwin (1977). The additional factor of topographic aspect is included with these factors. Matrix assessment actually computes landslide susceptibilities. If it were only a graphical representation of intuitive judgment, different evaluators might arrive at different conclusions for the same area. An objective, numerical evaluation process attempts to avoid differing conclusions being drawn from the same data. To obtain an accurate view of the susceptibility of each factor, the landslide data must be normalized, as is done in matrix assessment and by Carrara and Merenda (1976) and Nilsen and Brabb (1977). Matrix assessment described here differs from the approach of Nilsen and Brabb (1977) in adding aspect as a factor. It also evaluates all factors simultaneously rather than in a stepwise manner. This incorporates interactions and variations among factors that might be missed in a stepwise procedure. Another difference is in the establishing of classes or categories. Nilsen and Brabb (1977) divide the range of values at arbitrary points. This produces numerically defined categories, but introduces a subjective element into the procedure. Matrix assessment creates four numerically defined categories based on an objective mathematical procedure. The procedure described by Lessing and Erwin (1977) is a statistically valid nominal scale model producing two classes. Matrix assessment differs mainly in offering four rather than two susceptibility classes and is more firmly based on a numerical evaluation procedure.

CONCLUSIONS

The matrix assessment approach is designed to satisfy the need for large area or regional landslide-susceptibility information for wildlands management purposes. It has the capability of defining discrete susceptibility classes (thresholds) for identifying points and mapping areas. Matrix assessment reduces the subjective judgments in the evaluation process. Because it employs a few key measurable attributes, it is efficiently and economically applied over large wildlands areas. Computer compilation of acreage values used in the matrices greatly enhances the time and cost effectiveness of using this approach. It must be remembered that attribute combinations defining susceptibility classes in one area cannot be indiscriminately applied to areas that may seem similar. Each area requires a separate matrix assessment.

As a general planning tool, matrix assessment provides preliminary identification of potentially unstable areas. Detailed tract or site evaluations for specific projects will incorporate additional significant factors, define areas of higher susceptibility unrecognized due to the resolution of preliminary assessment data, and identify possible remedial measures.

ACKNOWLEDGMENTS

Robert W. Fleming, Donald R. Coates, and Peter Lessing reviewed an earlier version of this paper. Their constructive criticism is gratefully acknowledged.

REFERENCES CITED

Anderberg, M. R., 1973, Cluster analysis for applications: New York, Academic Press, 359 p.

Bailey, R. G., 1971, Landslide hazards related to land use planning in Teton National Forest, northwest Wyoming: USDA Forest Service, Intermountain Region, 131 p.

Beaty, C. B., 1956, Landslides and slope exposure: J. Geol., v. 64, p. 70–74.

———, 1972, Geographical distribution of post-glacial slumping in southern Alberta: Can. Geotech. J., v. 9, p. 219–224.

Bolt, B. A., Horn, W. L., Macdonald, G. A., and Scott, R. F., 1975, Geological hazards: New York, Springer-Verlag, 328 p.

Brabb, E. E., Pampeyan, E. H., and Bonilla, M. G., 1972, Landslide susceptibility in San Mateo County, California: U.S. Geol. Survey Misc. Field Studies, Map MF-360.

Briggs, R. P., Pomeroy, J. S., and Davies, W. E., 1975, Landsliding in Allegheny County, Pennsylvania: U.S. Geol. Survey Circ. 728, 18 p.

Carrara, A., and Merenda, L., 1976, Landslide inventory in northern Calabria, southern Italy: Geol. Soc. Amer. Bull., v. 87, p. 1153–1162.

Coates, D. R., ed., 1977, Landslides: Geol. Soc. Amer. Rev. Eng. Geol. III, 278 p.

Cooke, R. U., and Doornkamp, J. C., 1974, Geomorphology in environmental management: Oxford, Clarendon Press, 413 p.

DeGraff, J. V., 1976, Quaternary geomorphic features of the Bear River Range, north-central Utah: unpublished M.S. thesis, Utah State University, Logan, Utah, 215 p.

——, 1977, A procedure for developing an empirical mass failure rule, Appendix 33: *in* Henderson, J. A., Davis, L. S., and Ryberg, E. M., ECOSYM—An ecosystem classification and data storage system for natural resources management: Dept. Forestry and Outdoor Recreation, Utah State Univ., Logan, Utah, 49 p.

——, 1978, Regional landslide evaluation: two Utah examples: Environ. Geol., v. 2, p. 203–214.

Eisbacher, G. H., 1971, Natural slope failure, northeastern Skeena Mountains: Can. Geotech. J., v. 8, p. 384–390.

Gray, D. H., 1970, Effects of forest clearcutting on the stability of natural slopes: Assoc. Eng. Geol. Bull., v. 7, p. 5–67.

Leighton, F. B., 1976, Urban landslides: targets for land use planning in California: *in* Coates, D. R., ed., Urban geomorphology: Geol. Soc. Amer. Spec. Paper 174, p. 37–60.

Lessing, P., and Erwin, R. B., 1977, Landslides in West Virginia: *in* Coates, D. R., ed., Landslides: Geol. Soc. Amer. Rev. Eng. Geol. III, p. 245–254.

——, Kulander, B. R., Wilson, B. D., Dean, S. L., and Woodring, S. M., 1976, West Virginia landslides and slide-prone areas: West Virginia Geol. Econ. Survey Envir. Geol. Bull. 15, 64 p.

Nilsen, T. H., and Brabb, E. E., 1973, Current slope-stability studies by the U.S. Geological Survey in the San Francisco Bay region, California: Landslides, v. 1, p. 2–10.

——, and Brabb, E. E., 1975, Landslides: *in* Borcherdt, R. D., ed., Studies for seismic zonation of the San Francisco Bay region: U.S. Geol. Survey Prof. Paper 941-A, p. 75–78.

——, and Brabb, E. E., 1977, Slope-stability studies in the San Francisco Bay region, California: *in* Coates, D. R., ed., Landslides: Geol. Soc. Amer. Rev. Eng. Geol. III, p. 235–243.

Osgood, C. E., Suci, G. J., and Tannenbaum, P. H., 1957, The measurement of meaning: Urbana, Ill., Univ. Illinois Press, 346 p.

Radbruch-Hall, D. H., 1976, Maps showing areal slope stability in a part of the northern Coast Ranges, California: U.S. Geol. Survey Misc. Invest. Series, Map I-982.

Sharpe, C. F. S., 1938, Landslides and related phenomena: New York, Columbia Univ. Press, 137 p.

Swanson, F. J., and Dyrness, C. T., 1975, Impact of clearcutting and road construction on soil erosion by landslides in the western Cascade Range, Oregon: Geology, v. 3, p. 494–496.

Tilmann, S. E., Upchurch, S. B., and Ryder, G., 1975, Land use site reconnaissance by computer-assisted derivative mapping: Geol. Soc. Amer. Bull., v. 86, p. 23–34.

Van Horn, R., 1972, Relative slope stability map of the Sugar House quadrangle, Salt Lake County, Utah: U.S. Geol. Survey Misc. Invest. Series, Map I-766-E.

Williams, P. L., 1972, Map showing landslides and areas of potential landsliding in the Salina quadrangle, Utah: U.S. Geol. Survey Misc. Invest. Series, Map I-591-L.

20

HUMAN CAUSES OF ACCELERATED WIND EROSION IN CALIFORNIA'S DESERTS

Howard G. Wilshire

> No more can small beginnings, the incre-
> mentals of time, and the web of the
> world be ignored.
>
> Garrett Hardin
> *Exploring New Ethics for Survival*

INTRODUCTION

Human influence over natural landforms is so pervasive that it is commonly accepted as part of the natural scene. Our abilities to modify processes that are near natural thresholds of change and to simply overwhelm others that are in equilibrium have grown with our numbers and with our technological prowess. Generally, the ability to influence geomorphic processes, either to promote or retard an impending event, is viewed as beneficial to land-use planning. On the other hand, unplanned and poorly planned human impacts are daily being felt by delicately adjusted systems in ways that overwhelm the natural balances. The systemic adjustments to such human impacts are bringing about widespread changes in the biological productivity of the land through destruction of the soil mantle, and concomitant deterioration of environmental quality. Nowhere are these changes more openly displayed than in arid lands (United Nations, 1977; Eckholm and Brown, 1977; Dregne, 1978), where wind joins water as a principal agent of erosion and deposition.

The roles of wind and water erosion are closely intertwined, but the magnitude of wind erosion has long been underestimated. As recently as the late 1960s, figures were published indicating that water transports 50 times as much sediment as wind. The view that water is so much more important than wind in transporting sediment has been challenged of late, perhaps because of the combination of three factors: (1) many more people are engaged in quantitative studies of wind erosion than before; (2) global views of dust storms have been made possible by satellite photography; and (3) arid lands may be absorbing ever-increasing human loads as natural systems in more favorable climatic settings become saturated.

Observations of modern dust storms have led to the surprising conclusion that, within the continental masses, wind is transporting an amount of sediment more nearly comparable to that transported by water than was previously believed. For example, Pimental et al. (1976) estimate conservatively that of 5 billion tons of soil lost annually by erosion in the United States, fully 1 billion tons are removed by wind. In some parts of the world windborne sediment far exceeds that carried by water (Carlson and Prospero, 1972;

415

Prospero and Nees, 1977). Individual wind storms of short duration may travel 2000 km and more and carry several million tons of sediment (Yaalon and Ganor, 1980). Phoenix, Arizona, is the reluctant recipient of several dust storms annually that bring perhaps 150,000 tons of particulate matter to the city each year (Péwé et al., 1980). These awesome storms are reminiscent of the 1930s Dust Bowl storms. The conclusions of Eckholm (1976), Pimental et al. (1976), and Brink et al. (1977), among others, that we are losing the battle to protect the soil mantle were heavily underscored by last winter's return to massive dust storms in the Great Plains (Carter, 1977). A single storm that occurred on February 23–25, 1977, was estimated to have carried several million tons of soil from eastern New Mexico (McCauley et al., 1978). This storm's trajectory carried it across northern Florida to the mid-Atlantic, where the dust pall was 1 to 5 km high on February 25. In December 1977, the southern San Joaquin Valley was struck by a massive windstorm of a magnitude never before experienced by valley residents (Nakata et al., 1980). Grazing and agricultural lands in an area of 2000 km^2 were deeply eroded by this storm, and laid bare to subsequent rain storms. The severity of erosion in those storms is traceable to human activities that altered the stability of the surface making the land more vulnerable to erosion.

The effects of these storms on the quality of life are staggering. In the Dust Bowl days, attention was, properly, concentrated on the deleterious effects of dust storms on the productivity of the land from which they originated. Today other concerns are added to the basic one: hazard to life and health, not only to man but also to wildlife (Lave and Seskin, 1970; Reinking et al., 1975; Eckholm, 1976; Fennelly, 1976; Leathers, 1980; Hyers and Marcus 1980); hazard to property (Senate Fact Finding Committee on Transportation and Public Utilities, 1963); the symptomatic evidence dust storms provide of the spread of the desert condition as well as their contribution to it (Dregne, 1976; Paylore, 1976); and the effects of dust upon the climate (Fennelly, 1976; Jackson et al., 1973; Prospero and Nees, 1977).

CAUSES OF DUST STORMS[1]

The geologic record bears testimony to the existence of dust storms long before man became a significant agent of erosion, and natural erosion remains a major contributor to dust storms today. However, perhaps 85% or more of natural desert surfaces are wind-stable. These areas include bare rock outcrop (Fig. 1), surfaces mantled by desert pavement, and surfaces securely anchored by vegetation (Fig. 2). Even sand dunes, which mantle only comparatively small parts of desert surfaces, do not yield dust storms, because they contain little material small enough (<20 microns) to be transported long distances.

The natural sequence of events that leads to desert dust storms includes the collection by running water of the small amount of fine particles that are constantly in process of formation by weathering on the wind-stable surfaces, and concentration of them in main stream channels or at the bottoms of closed basins. As the fine-grained sediments deposited in these channels and basins dry, they become susceptible to wind erosion,

---

[1]Only the surface sources of dust will be considered, not the complex problems of atmospheric dynamics that dictate whether available wind-transportable material is actually picked up or not.

FIGURE 1 *Bare rock exposures (granite) west of Lucerne Valley, western Mojave Desert.*

and such places are the main sources of dust and sand storms (Yaalon and Ganor, 1980).

Man has brought to the desert the ability to modify the natural surfaces on a massive scale. By altering the stability of typical wind-stable surfaces two things happen: (1) there is an immediate expansion of the surface area capable of generating dust; (2) the amount of wind-erodible material available for water transport to the sites of natural dust generation is increased many fold.

FIGURE 2 *Desert surface stabilized by shrubs and intervening grass, near Red Mountain, western Mojave Desert.*

MAN-INDUCED EROSION IN THE CALIFORNIA DESERTS

Considering the radical effects of increased erosion on environmental quality, it is appropriate to examine what is happening in the California deserts that is increasing the rates of water and wind erosion.

Figure 3 is a photograph of the western Mojave Desert taken by the LANDSAT/-ERTS-1 satellite from an altitude of 850 km on January 1, 1973. The photograph shows six separate dust storms produced by a Santa Ana wind. According to our studies (Nakata et al., 1976) each of these storms is largely a product of human disturbance of formerly stable desert soil. The local causes of these dust storms will be used to describe the general problem that each exemplifies.

Roads and Utilities Corridors

Dust plume numbers 1 and 5 (Fig. 3) are remarkable for their nearly linear sources at high angles to the wind direction. These were located as accurately as photograph resolution allowed and then examined in the field. What we found were roadways cutting through surfaces that were generally stable with respect to wind erosion. The roads are in the position and orientation to have yielded the dust plumes, and they expose substantial quantities of wind-erodible material.

In the desert as elsewhere in California (California Division of Conservation, 1971), roads and utility corridors probably rank as the most important surface modifier. There are approximately 50,000 km of roadways (R. J. Badaracco, oral communication, 1976), and an additional 8000 km of utility corridors in the California deserts (Hastey, 1977).

Figure 4 illustrates the gamut of related problems. At the far left (N) of the photograph is a telephone cable corridor more than 12 m wide. Next south is a petroleum pipeline corridor about 14 m wide (Fig. 5). South of the pipeline is a corridor of one bull-dozer-blade-width along the telephone line. South of that lies an access road, paved for a short distance, with steep, cut banks, and stripped shoulders. On the right (S) side of the photograph (Fig. 4) is the north easement of Interstate Highway 15, in which deep drainage channels were cut and surface areas as wide at 37 m were stripped of vegetation and rock crusts.

These corridors cross an area of alluvial fan material that is stabilized by a mature desert pavement and by a creosote bush community. The desert pavement, like the plants, intercept rainfall and dissipate its kinetic energy, thereby slowing the rate of erosion. Each of the construction zones, by breaching the natural rock and plant barriers to erosion, has greatly accelerated water erosion (Fig. 5). The finer-grained materials beneath the desert pavement have been exposed to erosion and now flood the natural drainages, where they are susceptible to wind erosion. While concentrating the corridors in one place has reduced the aesthetic impact, the fundamental problem of erosion caused by the constructions has not been addressed.

Elsewhere road shoulder maintenance and road construction methods lead directly to accelerated erosion (Meyer et al., 1975). The effects commonly extend far beyond the immediate cause. For example, digging of borrow pits to raise the level of roadbeds along Interstate 15 and other desert highways created channels lower than the local stream beds. This has initiated gullying of the alluvial fan surface that has reached distances up the fan many times greater than the width of the borrow pit as the entire alluvial fan surface attempts to regrade to a level commensurate with the pit depth.

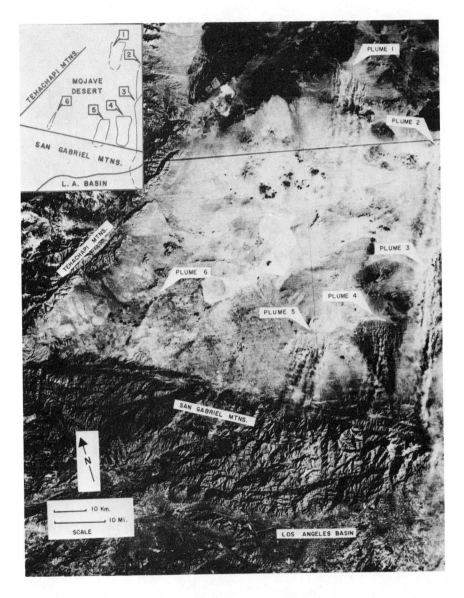

FIGURE 3 *Parts of two NASA LANDSAT/ERTS-1 images (1162-18013 and 1162-18011), showing six dust plumes produced by a Santa Ana wind, January 1, 1973. Distribution of dust plumes outlined (inset).*

FIGURE 4 *View east toward Afton Canyon exit, Interstate 15, showing concentration of utility and roadway corridors. From left to right: cable corridor, pipeline corridor, telephone line access road, secondary road, freeway easement zone.*

FIGURE 5 *View west along pipeline corridor in Figure 4.*

Thus this apparently small modification has far-reaching effects on the total system. In steeper terrain, the common practice of dumping runoff from roadway surfaces at the nearest convenient point has led to incision of previously adjusted drainages.

The construction of roadways leads to an immediate acceleration of erosion that diminishes in time, as shown in Figure 6. The general shape of this curve, not the absolute units on the abcissa and ordinate of an individual experiment, is important because it illustrates the change in rate of erosion with time for many other types of surface disturbances, such as agricultural, urban, flood control, and off-road vehicle disturbances. It should be emphasized that even though the rate of erosion has greatly diminished after 2 years for the experiment illustrated in Figure 6, the last measured rate is still 50 times that of adjacent unmodified surfaces (Megahan and Kidd, 1972), and further disturbance, such as grading the roadway, shoulders, and cuts, pushes the erosion rate back up the curve.

The study results illustrated in Figure 6 indicate that land managers should question the need for new roads, especially those that will remain unpaved, the need for continual regrading of shoulders, the degree and timing of temporary agricultural and urbanization disturbances, and the need for and effectiveness of flood control constructions. Also, the concept of "rest and rotation" of off-road vehicle use areas should be scrutinized in the light of erosion rates. This concept assures that use areas will remain in the regime of highest erosion rates by constantly shifting the impact to new ground.

Agriculture

Dust plume numbers 2 and 6 (Fig. 3) are related directly or indirectly to agricultural uses of the land. Plume 2, which was nearly dissipated when the photograph was taken,

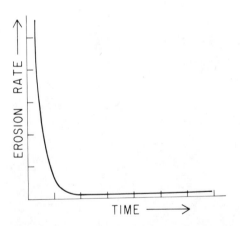

FIGURE 6 *Curve illustrating the change in rate of erosion of disturbed land with time. The experiment from which this curve was taken measured erosion rates from an un-improved road over a period of 6 years. (Megahan, 1974.)*

was derived from Harper Dry Lake. Agricultural land use on the west side of the dry lake (upwind for prevailing winds) has stripped a mature plant cover and replaced it with seasonally bare surfaces that expose highly wind-erodible soil. A number of fields have been abandoned and the windbreaks are deteriorating with the result that long-term dust and sand sources to the dry lake surface are assured. As the disturbed land lies both uphill and upwind of the dry lake, sediment will be supplied to the dry lake surface by water and wind erosion. There it is vulnerable to erosion by the strong northerly winds that produced the dust plume shown in Figure 3.

Although no definite source for the isolated dust plume number 6 (Fig. 3) could be established, the dominant cause of surface instability in its source area is agriculture. Evidence of wind erosion of these lands is omnipresent.

Wind erosion (Fig. 7) of the desert's marginally arable lands has the doubly negative effect of stripping topsoil as well as selectively removing organic and nutrient components, leaving a residue that is substantially less productive than the original soil (Finnell, 1951; Eck et al., 1965; Gillette et al., 1974; Lyles, 1975; Fryrear, 1980).

Flood Control and Urbanization

Dust plume number 3 (Fig. 3), which extended 75 km across Cajon Pass into San

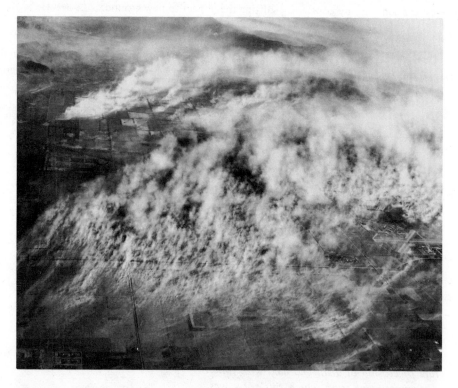

FIGURE 7 *Dust blown from cultivated fields during a Santa Ana wind, near Ontario, California. Surface wind velocity 108 km/hr. (Photograph by John S. Shelton.)*

Bernardino, has two basically different causes: (1) river channel modifications, and (2) urban development.

Levees built in the Mojave River bed channel the wind as well as water, thereby increasing the carrying capacity of both. The levees also increase the surface area exposed in the river channel, and traffic and other activities continually break the fragile crusts that develop on undisturbed sand surfaces, making them more vulnerable to wind erosion.

Flood control constructions are widespread in the California desert. Many of them appear to have created conditions favorable to increased water and wind erosion. Recent construction of a flood control project near Desert Hot Spring involved widening of the active channel (to about 70 m), and stripping of plants across zones 45 m wide on both sides of the channel and from within the channel. This stripping serves no useful flood control purpose and adds substantially to an already chronic wind erosion problem. Another example is shown in Figure 8, where constructions for flood control not only greatly increase the rate of erosion, but by stripping vegetation and constricting the channels will likely add to the problem of flooding that they are intended to solve, and will ultimately add to the problem of wind erosion.

The second contributor to dust plume number 3 was the Silver Lakes urban development (Fig. 9), which bared about 650 ha (Nakata et al., 1976). This development continues to be a significant source of windblown sediment, illustrated by the sand drifts in the areas cut for housing lots and on the lee sides of fences.

The erosional problems resulting from unused and little-used remote urban develop-

FIGURE 8 *Flood control project along Kelbaker Rd., southeast of Baker, California.*

FIGURE 9   *Part of the Silver Lakes subdivision, west of Helendale, California.*

ments in the desert are major problems in California as elsewhere in the United States (Campbell, 1972). Figure 10 shows a portion of the California City development. The first ground was broken in this development in 1958, and in 1964 490 km^2 was incorporated in the city, still sparsely inhabited. The entire central area of the project was stripped of vegetation, and extensive peripheral areas were exposed to erosion through construction of an intricate network of roads that are likely never to be used for their intended purpose. Severe sand storms originating within these destabilized areas are commonplace. Figure 11 shows a dust storm along the west shore of Salton Sea, and its cause: urban development that has stripped the vegetative cover from the land.

Off-Road Vehicular Recreation

Dust plume number 4 (Fig. 3), derived from El Mirage Dry Lake, had substantial contributions from off-road vehicular destabilization of the surface. Large areas around the periphery of the dry lake have been exposed to wind erosion by direct vehicular destruction of vegetative cover (Figs. 12 and 13). Windblown sand on the upwind side of the Shadow Mountains has been similarly destabilized by vehicles and contributed to the dust storm in Figure 3. Natural plant and desert pavement erosion barriers over large areas on the upwind side of the dry lake have been destroyed or disturbed by vehicle use assuring a continuing supply of wind-erodible material to the dry lake surface.

According to the Bureau of Land Management (unpublished data), significant off-road vehicle use is now taking place or is authorized within 2 to 3 million acres of California desert lands (see Sheridan, 1979). This type of land use has developed historically with a minimum of direct controls and site-specific planning. Random, non-competitive use and activities of spectators at competitive events are not supervised.

FIGURE 10 *Central part of California City, western Mojave Desert. Part of the extensive peripheral road system of the development is visible at the top of the photograph.*

Areas of permitted use have not been analyzed in terms of the capacity of the soil to sustain the use, or in terms of the erosional consequences. None of the common-practice techniques of erosion mitigation have been employed to conserve soil or to protect neighboring public and private property from the adverse effects of erosion. No systematic monitoring of the direct and indirect physical impacts of vehicle use has been undertaken, and no program of rehabilitation of closed areas has been initiated. The consequences of a decade of off-road vehicular recreation in the desert are vividly displayed by the land (Geological Society of America, 1977).

Although the Barstow to Las Vegas cross-country motorcycle race course has not been used since 1974, the areas of surface disturbance have begun to erode. Water erosion will enlarge some of the directly disturbed zones and will release fine-grained materials, formerly securely anchored, to the natural drainages where they will become susceptible to wind erosion. This accelerated erosion will probably be with us for centuries.

Figure 14 shows a vehicle trail on a steep wall of Jawbone Canyon, a desert-facing canyon in the southern Sierra Nevada Mountains. Plant cover has been stripped and the soil churned into a loose, unstructured aggregate. Figure 15 is a view down the same trail after a rain. The rain formed a thick slurry of the loose soil, which flowed out over the canyon floor much like a lava flow, burying plants and trapping burrowing animals.

FIGURE 11 *Dust being blown out over the Salton Sea from an urban development. (Photo by John S. Shelton.)*

Both the stripped canyon walls and new deposits on the canyon floors are vulnerable to wind erosion. There are 20 km of canyon walls in Jawbone Canyon, and more in the nearby Dove Springs Canyon, that are in similar or worse condition. Many stretches of canyon wall show complete loss of the soil mantle. These areas were mapped for the El Paso Framework Management Plan as watershed areas severely susceptible to erosion, which were to be protected by limiting uses that will disturb vegetation and the soil surface (Bureau of Land Management, 1976a). Approximately 45% of the areas so mapped for the El Paso and Red Mountain planning units (Bureau of Land Management, 1976b), and more than 50% of the area so mapped for the Yuha Desert planning unit (Bureau of Land Management, 1975) are areas of sanctioned off-road vehicle use.

In areas of high relief and sandy or gravelly soils, the combined effects of direct mechanical erosion (mass wasting) by the vehicles and water erosion of the destabilized surface results in notching of the slopes. In heavily used areas such as Ballinger Canyon (Stull et al., 1978; Fig. 16), Hungry Valley (Sheridan, 1978), Jawbone Canyon, Dove Springs Canyon, and Red Rock Canyon, vehicle notches more than 2 m deep have been cut in less than 10 years of use. These vehicle notches are geomorphically unstable, and natural erosional and depositional processes tend to eliminate them by a general reduction of the surrounding ground surface; this is accomplished at the expense of the soil mantle between the notches. Some favorably located vehicle notches have captured sufficient catchment area that they have become permanent parts of the drainage system.

FIGURE 12 *South end of the Shadow Mountains at the east side of El Mirage Dry Lake. Extensive vehicular destabilization of eolian sand contributed to dust plume number 4, Figure 3.*

FIGURE 13 *Surface destabilization by off-road vehicular recreation on the upwind side of El Mirage Dry Lake.*

FIGURE 14 *Jeeps on steep canyon wall in Jawbone Canyon, southern Sierra Nevada Mountains. Photo taken December 1975.*

FIGURE 15 *View down same trail shown in Figure 14, photographed September 1976. Note burial of valley floor vegetation by sand eroded from the trail.*

Other Causes of Surface Destabilization

Other human activities that are contributing significantly to accelerated erosion in the California deserts, but which are not represented by dust storms in Figure 3, include mining and grazing.

Mining operations in dry lake areas of high natural wind-erosion hazard greatly exacerbate the problem of dust yield. For example, Reinking et al. (1975) have shown that an area of brine evaporators on Owens Dry Lake is the largest dust contributor per unit area to dust storms derived from Owens Dry Lake. Other dry lake surfaces that are disturbed by mining operations include Searles, Koehn, Bristol, Cadiz, and Danby. Open pit mining for iron ore in the Eagle Mts., Kingston Mts., Afton Canyon, and near the Providence Mts., for borates near Boron, California, and for clay in the Castle Mts. yields extensive poorly stabilized spoils that are vulnerable to both water and wind erosion. The products of abandoned and active underground and alluvial mining operations, and of widespread borrow pits dug for highway and railway construction, remain centers of erosion (Fig. 18). The demand for ornamental stone in urban centers in and around the desert has now resulted even in stripping of varnished desert pavement in the Little Maria Mts. Still more bizzare, but akin to mining, are the contributions to erosion of "earth art" (*Newsweek*, November 18, 1974), the practice of using natural desert landscapes as the medium for sculpting designs with a bulldozer.

FIGURE 16 *Ballinger Canyon, eastern Cuyama Valley. Notches to more than 2 m deep cut by motorcycles over a period of 8 years.*

FIGURE 17 *Rand open area, northwestern Mojave Desert. Nearly completely denuded pit area used for camping and staging of off-road vehicle events, and vehicle trails radiating from the pit area.*

FIGURE 18   *Atolla mining area near Red Mountain, California.*

The effects of grazing on erosion in the California desert have not been assessed. The impacts of a century of sheep and cattle grazing are, however, widespread and conspicuous. Bedding and watering areas are denuded and closely resemble off-road vehicle "pit" areas; these areas also have substantial changes in physical properties of soils like those measured in off-road vehicle areas (Edmong, 1962; Ranzi and Smith, 1972; Webb and Stielstra, 1978). Thus on flat terrain, areas denuded by grazing may create wind erosion fetches, whereas on steeper terrain the physical modifications of soil by grazing are likely to increase runoff (Sharp et al., 1964).

SUMMARY

Human causes of accelerated erosion in California's deserts include mechanical disturbances of the surface by road building, construction of utilities corridors, agriculture, urban expansion, construction of flood control devices, off-road vehicular recreation, and mining. Although most of the land involved is under federal jurisdiction (Bureau of Land Management), management practices necessary to mitigate the adverse environmental impacts of these human disturbances have not kept pace with the increasing rate of disturbance. Inspection of land-use habits in the California deserts suggests that the responsible agencies have not made use of the technical understanding of natural desert systems available to them to control the steady expansion of surface disturbance. With expanded disturbance come accelerated erosion and environmental deterioration.

ACKNOWLEDGMENTS

I am indebted to Elizabeth Schwarzman and Jane E. Nielson Pike, U.S. Geological Survey, for their technical review of the manuscript, and to Genny Smith for her editorial assistance. Special thanks are due Robert Berry for his skillful piloting on our aerial reconnaissance, and to Charles Bell and Lyle Gaston for the ready availability of their knowledge of the desert.

REFERENCES CITED

Brink, R. A., Densmore, J. W., and Hill, G. A., 1977, Soil deterioration and the growing world demand for food: Science, v. 197, p. 625–630.

Bureau of Land Management, 1975, California desert plan program, Yuha Desert management framework plan: U.S. Dept. Interior, 45 p.

——, 1976a, California desert plan program, El Paso management framework plan: U.S. Dept. Interior, 40 p.

——, 1976b, California desert plan program, Red Mountain management framework plan: U.S. Dept. Interior, 28 p.

California Division of Conservation, 1971, Problems of the soil mantle and vegetative cover of the State of California: Resources Agency, State of California, 110 p.

Campbell, C. E., 1972, Some environmental effects of rural subdividing in an arid area: a case study in Arizona: J. Geogr., March, p. 147–154.

Carlson, T. N., and Prospero, J. M., 1972, The large-scale movement of Saharan air outbreaks over the northern equatorial Atlantic: J. Appl. Meteorol., v. 11, p. 283–297.

Carter, L. J., 1977, Soil erosion: the problem persists despite the billions spent on it: Science, v. 196, p. 409–411.

Dregne, H. E., 1976, Desertification: symptom of a crisis, *in* Paylore, P., and Haney, R. A., Jr., eds., Desertification: process, problems, perspectives: Tucson, Univ. Arizona, p. 11–24.

——, 1978, Desertification: man's abuse of the land: J. Soil Water Conserv., v. 33, p. 11–14.

Eck, H. V., Hauser, V. L., and Ford, R. H., 1965, Fertilizer needs for restoring productivity of Pullman silty clay loam after various degrees of soil removal: Soil Sci. Soc. Amer. Proc., v. 27, p. 209–213.

Eckholm, E., 1976, Losing ground: Environment, v. 18, p. 6–11.

——, and Brown, L. R., 1977, Spreading deserts—the hand of man: Worldwatch Paper 13, 40 p.

Edmond, D. B., 1962, Effects of treading pasture in summer under different soil moisture levels: New Zealand J. Agric. Res., v. 5, p. 389–395.

Fennelly, P. F., 1976, The origin and influence of airborne particulates: Amer. Scientist, v. 64, p. 46–56.

Finnell, H. H., 1951, Depletion of high plains wheatlands: U.S. Dept. Agriculture Circ. 871.

Fryrear, D. W., 1980, Long-term effect of erosion and cropping on soil productivity: *in* Péwé, T. L., ed., Desert dust: origin, characteristics, and effects on man: Geol. Soc. Amer. Memoir.

Geological Society of America, 1977, Impacts and management of off-road vehicles: Committee on Environment and Public Policy, 8 p.

Gillette, D. A., Blifford, I. H., and Fryrear, D. W., 1974, The influence of wind velocity on the size distributions of aerosols generated by the wind erosion of soils: J. Geophys. Res., v. 79, p. 4068–4089.

Hastey, E., 1977, Planning for land management on the California desert: *in* Conf. on alternative strategies for desert development and management: v. 1, Sacramento, Calif., May 31–10 June.

Hyers, A. D., and Marcus, M. G., 1980, Land use and desert dust hazards in central Arizona: *in* Péwé, T. L., ed., Desert dust: origin, characteristics, and effects on man: Geol. Soc. Amer. Memoir.

Jackson, M. L., Gillette, D. A., Danielsen, E. F., Blifford, I. H., Bryson, R. A., and Syers, J. K., 1973, Global dustfall during the Quaternary as related to environments: Soil Sci., v. 116, p. 135–145.

Lave, L. B., and Seskin, E. P., 1970, Air pollution and human health: Science, v. 169, p. 723.

Leathers, C. R., 1980, Plant components of desert dust in Arizona: *in* Péwé, T. L., ed., Desert dust: origin, characteristics, and effects on man: Geol. Soc. Amer. Memoir.

Lyles, L., 1975, Possible effects of wind erosion on soil productivity: J. Soil Water Conserv., v. 30, p. 279–283.

McCauley, J. F., Breed, C. S., and Grolier, M. J., 1978, The great dust storm of February 23–25, 1977 [abst.] : Geol. Soc. Amer., Abstracts with Programs, v. 10, p. 116.

Megahan, W. F., 1974, Erosion over time on severely disturbed granitic soils: a model: USDA Forest Service Res. Paper INT-156, Intermountain Forest and Range Experiment Sta., 14 p.

——, and Kidd, W. J., 1972, Effect of logging roads on sediment production rates in the Idaho batholith: USDA Forest Service Res. Paper Int-123, Intermountain Forest and Range Research Experiment Sta., 14 p.

Meyer, G. J., Schoeneberger, P. J., and Huddleston, J. H., 1975, Sediment yields from roadsides: An application of the universal soil loss equation: J. Soil Water Conserv., v. 30, p. 289–291.

Nakata, J. K., Wilshire, H. G., and Barnes, G. E., 1980, Origin of Mojave Desert dust plumes photographed from space: J. Geol., v. 4, p. 644–648.

Paylore, P., 1976, Desertification: What, where, why, who: *in* Paylore, P., and Haney, R. A., Jr., eds., Desertification: process, problems perspectives: Tucson, Univ. Arizona, p. 1–10.

Péwé, T. L., Péwé, E. A., Péwé, R. H., Journaux, A., and Slatt, R., 1980, Desert dust: characteristics and rates of deposition in central Arizona: *in* Péwé, T. L., ed., Desert dust: origin, characteristics, and effects on man: Geol. Soc. Amer. Memoir.

Pimental, D., Terhune, E. C., Dyson-Hudson, R., Rochereau, S., Samis, R., Smith, E. A., Denman, D., Reifschneider, D., and Shepard, M., 1976, Land degradation: effects on food and energy resources: Science, v. 194, p. 149–155.

Prospero, J. M., and Nees, R. T., 1977, Dust concentration in the atmosphere of the equatorial north Atlantic: possible relationship to the Sahelian drought: Science, v. 196, p. 1196–1198.

Ranzi, F., and Smith, R. M., 1972, Infiltration rates: three soils with three grazing levels in northern Colorado: J. Range Mgmt., v. 26, p. 126–129.

Reinking, R. F., Mathews, L. A., and St.-Amand, P., 1975, Dust storms due to the desiccation of Owens Lake: Int. Conf. Environmental Sensing and Assessment, Sept. 14–19, 1975, Las Vegas, Nev., IEEE Publishers, p. 1–9.

Senate Fact Finding Committee on Transportation and Public Utilities, 1963, Transcript of proceedings, hearing on sand blasting and blowing on State highways, Nov. 20, Palm Springs, Calif.: State of California: Sacramento, Calif., 59 p.

Sharp, A. L., Bond, J. J., Neuberger, J. W., Kulman, A. R., and Lewis, J. K., 1964, Runoff as affected by intensity of grazing on rangeland: J. Soil Water Conserv., v. 19, p. 103–106.

Sheridan, D., 1978, Dirt motorbikes and dune buggies threaten deserts: Smithsonian, v. 9, p. 66–75.

——, 1979, Off-road vehicles on public land: Council on Environmental Quality, Washington, D.C., 84 p.

Stull, R., Shipley, S., Hovanitz, E., Thompson, S., and Hovanitz, K., 1978, Effects of off-road vehicles in Ballinger Canyon, California: Geology, v. 7, p. 19–21.

United Nations, 1977, Desertification: its causes and consequences: Oxford, Pergamon Press, 448 p.

Webb, R. H., and Stielstra, S. S., 1978, Some effects of sheep grazing on Mojave Desert vegetation and soils: Bureau of Land Management, Final Rept., Contract CA-060-CT8-000041, 29 p.

Wilshire, H. G., Nakata, J. K., and Hallett, B., 1980, San Joaquin Valley's Christmas windstorm, 1977: U.S. Geol. Survey Circ., in press.

Yaalon, D. H., and Ganor, E., 1980, Origin and nature of desert dust: *in* Péwé, T. L., ed., Desert dust: origin, characteristics, and effects on man: Geol. Soc. Amer. Memoir.

21

NON-POINT-SOURCE POLLUTANTS WITHIN THE GREAT LAKES—A SIGNIFICANT INTERNATIONAL EFFORT

Richard R. Parizek

INTRODUCTION

Environmental management programs designed to restore and preserve Great Lakes water quality require identification and control of pollutants derived from both point and nonpoint sources. Judging from the magnitude of diffuse source pollution problems identified for the Great Lakes by the Pollution from Land Use Activities Reference Group (PLUARG, 1978), the diversity of land and material uses within the watershed, and the diversity of physical characteristics of tributary drainage basins, it is concluded that diffuse sources of pollution of streams, lakes, and shallow marine environments must be widespread throughout the world.

The control and management of these upland sources of pollutants will require the talents of individuals trained and experienced in various disciplines, not previously routinely called upon, to solve point-source pollution problems. Geologists, in particular those with training in hydrogeology, hydrology, aqueous geochemistry, and geomorphology, together with agronomists, civil and sanitary engineers, and others, must become involved. Nonpoint pollution problems are tied to the landscape and to surficial processes. These have complex physical, geochemical, and biochemical interactions and time dimensions that must be understood when identifying the nature and sources of pollutants, and when selecting and designing remedial measures to control diffuse source pollutants. Some will argue that it will be easier to meet water quality objectives set for lakes, streams, and estuaries by attacking the problem at easy-to-identify point sources and upgrading treatment at these locations. It should be remembered, however, that liquids and sludges are inevitable by-products of secondary or higher levels of waste treatment. These wastes must be returned to the land by land application systems, buried in landfills, injected into deep wells, or incinerated. In all cases of land application, pollutants must be completely attenuated by biochemical processes or chemically combined to produce stable or rather inert substances that will not contaminate or pollute surface or subsurface waters. If this conversion of wastes is not completely achieved, pollutants may be remobilized in leachates, where they are free to migrate within surface, soil-water, or groundwater flow systems between regions of recharge to their ultimate discharge locations, or they may be carried with soil particles eroded from these waste "disposal sites".

The original wastes or their daughter products of weathering will have an upland storage-time-constant, or retardation factor, that may be measured in hours to thousands of years, depending upon the physical–chemical–biochemical characteristics of the land areas upon which they are applied. We have a choice of how to maximize waste retention times through site selection and by waste management practices that isolate pollutants within predetermined portions of watersheds, within lakes, or soil-water and ground-water flow systems. The residence times of leachates entrained within groundwater flow systems may also be measured in days to thousands of years, depending upon the magnitude of the flow system selected and initial boundary conditions governing these flow systems, which may be local, intermediate, to regional in scale.

No uniform "threshold" as such separates these individual pathways of subsurface water flow, but rather they are the consequence of the hydrogeological setting: water table configuration, magnitude and regions of recharge and discharge, presence of hydraulic boundaries, sequences of beds, and their hydrologic properties. Two drops of water that enter the water table side by side may have vastly different fates within these flow systems which are dictated by their entry points within these predetermined flow channels. Assuming that pollutants entrained with these water droplets are conserved, their pathways and history of travel within their respective flow channels can be adequately predicted given sufficient knowledge of the controlling parameters and existing theory.

These subsurface systems can also be unsteady in that boundary conditions and initial conditions can be altered by a new man-imposed threshold or by seemingly slow changes in climate or geological processes. Environmental changes that respond so slowly that they are gaged against geological time scales are being factored into high-level radioactive waste storage site selection studies in North America and elsewhere. However, time dimensions measured in hundreds to tens of thousands of years have been largely ignored in most other previous waste disposal and management programs, even though wastes may linger in the environment in one form or another long after the designed engineering life of the facility that generated the wastes. The premature breakout of pollutants from waste buried sites to land surface should not be regarded as being controlled by threshold conditions if errors in judgment were made in selecting the waste disposal site. These wastes, for example, may have been applied to a groundwater discharge area where their leachates were flushed back to the land surface through a thin soil cover rather than being attenuated within a deeper flow system. Such a short circuit of pollutants through the waste-containing or -attenuating porous media was dictated by the groundwater flow system, which was not understood or taken into consideration by the designer of the disposal operation. A *threshold can* be involved, however, when the terrain at the burial site is altered. A change in topography that influences the configuration of the water table, amounts of recharge, and so on, can cause a groundwater recharge area to shift to a discharge area. The combination of conditions that causes this shift and the crossover of this threshold are entirely predictable to hydrogeologists.

Chemical Thresholds on the other hand, can govern the presence and concentration of dissolved substances migrating within surface or subsurface waters. A host of physical, geochemical, and biochemical processes can act on waste substances applied to the land that will account for their "storage" on mineral grain surfaces, precipitation from solutions, and so on (Table 1). Changes in Eh or pH conditions, for example, can account for the later remobilization of previously tied up substances. It is entirely reasonable therefore, in waste management programs to be required to investigate the probable changes in environmental conditions that might occur at a waste management site, that, in turn, would release pollutants at an accelerated or difficult-to-control rate. A shift in the acidity of precipitation is one such mechanism that could account for resolubilization of previously stable pollutants applied to the soil. Similarly, a drop in the pH of groundwater influence by the oxidation of pyrite exposed by surface or underground coal mining can result in the leaching of trace elements from soil and rock bathed by these mine waters.

Trace elements—lead, mercury, arsenic, and so on—may be attached to sediment particles and be transported from upland areas to ultimate points of deposition and long-term (geological) storage within the bottoms of lakes. Their concentration in lake water may be in trace or not detectable amounts because of their low solubility. However, it appears that some elements can be remobilized by methylation processes. Jensen and Jernolov (1967) discovered that microorganisms in lake sediments are able to convert inorganic mercury in sediments into a very potent human nerve poison, methyl mercury. This appears to be a common process in aquatic environments, and recent studies by Wong, et al. (1975), and Chau et al. (1976) indicate that there is a possibility that lead, selenium, and arsenic may also undergo methylation, hence have the potential of being released from their storage sites. For this reason, lead contained in lake bottom sediments is regarded as an in-place contaminant that may become remobilized under changing environmental conditions, hence cannot be regarded as in "dead storage" in the bottoms of lakes where they should no longer be of concern to man. Such pollutants must be controlled at their source: removal of lead from gasoline; proper disposal of municipal and industrial sewage effluents and sludges on the land, where they can be retained.

The entire problem of the biological availability of metals in uplands, streams, lakes, and estuaries under various environmental conditions (i.e., pH, Eh, etc.) must be better understood as well as the metal concentrations and types associated with various size fractions of organic and inorganic sediments and their chemical and or minerological composition. The kinetics of metal complexes must be better defined, as these kinetics will shed light on metal speciation and metal behavior within aqueuos systems.

Wastes not converted to acceptable by-products during incineration pose another example of a time-dependent storage problem. Dilution and dispersion processes were adequate to reduce pollutant concentrations to acceptable levels in the past, but it is now clear that the atmospheric pathways for pollutant transport account for a significant part of the total diffuse source pollution load that falls on uplands and waterways. This fallout is indiscriminate and free to be transported by surface and subsurface waters and sediment, depending upon whether it falls on the *hydrologically active* parts of watersheds or less active portions. The size and location of hydrologically active areas depends on a number of geomorphic and hydrological factors, to be outlined later. These are separated in time and place by thresholds that account for the long-term storage or more rapid migration of pollutants and sediments within watersheds, streams, and receiving lakes.

The entire matter of rates of soil erosion, drawing largely upon sediment delivery ratios, and entrapment efficiencies of lakes and marshes are involved in this process of pollutant transport with sediments. This is further complicated by the physical, geochemical and biochemical processes that operate on pollutants during their history of stepwise transport and temporary storage within upland soils, colluvium, alluvium, and so on, and their ultimate transport to more permanent storage sites. The sediment delivery and entrained pollutant delivery ratios need not be equal. In fact, a fractionation process

TABLE 1 *Probable Effect of Various Processes on the Mobility of Constituents in Subsurface Waters Contaminated by Waste Disposal (D. Langmuir, 1972)*

Physical processes

    Dilution—Favors reduction in solute concentrations. Less effective in unsaturated materials where caused by mixing with preexistent or infiltrating moisture in amounts generally below field capacity. Most effective in saturated materials of high porosity and permeability at high groundwater flow rates.

    Dispersion—Causes moisture movement at right angles to primary flow direction. Favors reduction in solute concentrations. Moisture flow mostly vertical in unsaturated soil and rock, and mostly horizontal in saturated soil and rock.

    Filtration—Favors reduction in amounts of substances associated with colloidal or larger-size particles. Most effective in clay-rich materials. Least effective in gravels or fractured or cavernous rock.

    Gas movement—Where it can occur, favors aerobic breakdown of organic substances, and increased rates of decomposition. Constituents mobile under oxidized conditions will then predominate. Restriction of gas movement by impermeable, unsaturated materials or by saturated materials can produce an anaerobic state, and reduced rates of organic decay. This will mobilize substances soluble under anaerobic conditions.

Geochemical processes

    Complexation and ionic strength—Complexes and ion pairs most often form by combination of ions, including one or more multivalent ions and increase in amount with increased amounts of ions involved. Ionic strength is a measure of the total ionic species in a water. Both ionic strength and complexation increase the total amounts of species otherwise limited by processes such as oxidation, precipitation, or adsorption.

    Acid—base reactions—Most constituents increase in solubility and thus in mobility with decreasing pH. In organic-rich waters, the lower pHs (4–6) are associated with high values of carbonic acid and often also of organic acids. These will be most abundant in moisture-saturated soils and rock.

    Oxidation-reduction—Many elements can exist in more than one oxidation state. Conditions will often be oxidized or only partially reduced in unsaturated soils, but will generally be reduced under saturated conditions when organic matter is present. Mobility depends on the element and pH involved: chromium is most mobile under oxidizing conditions, whereas iron and manganese are most mobile under reduced conditions.

    Precipitation—solution—The abundance of anions such as carbonate, phosphate, silicate, hydroxide, or sulfide may lead to precipitation especially of multivalent cations as insoluble compounds. The abundance of major multivalent cations such as calcium and magnesium may cause precipitation of some trace element oxyanions, including molybdate and vanadate. Dilution, or a change in oxygen content where precipitation has involved oxidation or reduction, may return such constituents to solution.

    Adsorption—desorption—Ion exchange can temporarily withhold cations and to a less extent anions, on the surfaces of clays or other colloidal-size materials. Amounts of adsorbed metal cations will increase with increasing pH. Molecular species may be weakly retained on colloidal-size materials by physical adsorption. Adsorbed species will tend to return to solution when more dilute moisture comes in contact with teh colloidal material.

Biochemical processes

    Decay and respiration—Microorganisms can break down insoluble fats, carbohydrates, and proteins, and in so doing release their constituents as solutes or particulates to subsurface waters.

    Cell synthesis—N, C, S, P, and some minor elements are required for growth of organisms, and can thus be retarded in their movement away from a waste disposal site.

438

will be at work that can either increase or decrease the pollutant delivery ratio when compared with the sediment delivery ratio.

Thresholds are involved in these fractionation processes as well as in the entire sediment erosion problem.

Land-use practices and sediment production, sediment control, and consequent stream and lake loadings of sediment-related quality parameters are dependent upon time-specific cause-and-effect relationships within watersheds that are highly variable and difficult to quantify either within a deterministic or statistical framework. The International Joint Commission (1976) sponsored a fluvial transport of sediment-associated nutrients and contaminants workshop which identified the nature of these difficulties. The group concluded that the major deficiency, both of existing land use-loading models and of site-specific process-response studies of sediment production by specific land-use practices, is the inability to link upstream or source water quality observations with river mouth loadings observed in lake systems such as the Great Lakes. Although the principles of sediment transport mechanics are well known, the downstream movement and modification of sediment-related water quality variables are poorly understood. Load modification is not only related to physical processes, but also includes biological and chemical interactions which are still poorly understood.

Contemporary data are inadequate to meet the needs of in-stream process research at the national level. These data inadequacies pertain both to poor characterization of variables, such as particle size, inorganic/organic ratios, speciation of key components, and aerosol input, and to lack of definition of spatial and temporal variations associated with in-stream processes such as sediment storage-remobilization mechanisms, time of travel of particle-size classes, short anthropogenic and long-term climatic variations in sediment flux and storage, and other factors.

Thresholds are involved in the erosion and sediment transfer mechanism both in uplands and within streams. However, much still remains to be learned about the variation in rates of these processes between hydrologically active and immediately adjacent parts of watersheds as well as about the mechanisms that can change sediment quality and quantity during stream transport in complex fluvial systems.

Some of the concepts presented here are further supported by the PLUARG (1978) 6-year investigation, whose results were presented to the International Joint Commission in July, 1978.

BACKGROUND

The International Joint Commission (IJC) was organized pursuant to the Boundary Waters Treaty of 1909 between Canada and the United States. The Commission's responsibilities under the treaty fall into two principal categories: one is that of applications and Orders of Approval; the other, References, that is, the undertaking of investigations and studies of specific problems or questions of differences referred to the IJC by the Governments. These differences can arise anywhere along the more than 8000 km (5000 mi) of common border between the two nations.

The Commission consists of three Canadians appointed by the Government of Canada, and three Americans appointed by the president with the advice and consent of the Senate. The Commissioners act as a single, unified body which seeks common solutions in the joint interests of Canada and the United States, not as separate national delegations under instruction from their respective Governments as is the case for most similar bodies in the rest of the world.

References involve the study of questions or matters of difference brought to the Commission by either or both Governments. Under the Treaty, either Government may refer to the Commission questions or differences between them which involve the rights, obligations, or interests of either in relation to the other or to the inhabitants of the other country. The two Governments usually consult and agree on the terms of a Reference and send a joint Reference to the IJC.

Studies requested by the IJC concerning water quality in Lakes Erie and Ontario (i.e., lower Great Lakes), completed and submitted to the Commission in 1969, demonstrated that diffuse land drainage sources of pollutants were significant and extremely variable, and therefore difficult to measure. Subsequent improvements in municipal wastewater treatment facilities for point sources of pollution manadated by the 1972 Water Quality Agreement magnified the relative importance of the land drainage sources of many pollutants, necessitating a clearer definition of the impact of land-use activities, practices, and programs on water quality in the Great Lakes. For this reason, the Governments of Canada and the United States, on signing the 1972 Great Lakes Water Quality Agreement, requested the IJC to investigate pollution of the Great Lakes system from agriculture, forestry, and other land-use activities and to make their recommendations to Governments concerning remedial actions and programs they deem necessary to abate these pollution problems (Great Lakes Water Quality Agreement, 1972).

The PLUARG Reference

In November 1972, the IJC appointed an International Reference Group on Great Lakes Pollution from Land Use Activities (PLUARG), composed of nine Canadian and nine U.S. representatives, to conduct the study under the Great Lakes Water Quality Board (Fig. 1).

The purpose of this study was:

1. To determine and evaluate the causes, extent, and locality of pollution from land-use activities.

2. To gain an understanding of the relative importance of various land uses in terms of their diffuse pollutant loads to the Great Lakes.

3. To determine the most practicable remedial measures for decreasing the diffuse pollutant loads to an acceptable level and the estimated costs of these measures.

Detailed plans for this study were developed early in 1973, and assignments made to both Canadian and U.S. agencies and more than 350 engineers, scientists, and technicians to commence studies on specific tasks and programs formulated by PLUARG (1974, 1976). The detailed study plans were implemented, and approximately $18 million committed and a final report presented in July 1978 that summarized the findings contained in a host of publications (PLUARG, 1978).

The PLUARG study considered diffuse (i.e., nonpoint) sources of pollutants, including surface runoff from all land uses and groundwater inflows from the entire Great Lakes basin. The atmospheric loads were also evaluated to determine their magnitude, relative to the total pollutant load to the Great Lakes. The term *diffuse* and *nonpoint* are used interchangeably here. Pollutants from diffuse sources are those polluting materials conveyed to the Great Lakes by natural runoff to tributaries, ditches, groundwater, storm sewers, or as combined sewer overflows. In comparison, point sources define those sources of pollutants that are "pipeline" in nature, such as municipal sewege treatment plants and industrial wastewater discharges, regardless of whether they were discharged directly to the Great Lakes or to tributaries draining to the lakes.

During the PLUARG study, supporting technical papers and reports of public consultation panels were developed, including reports on (1) pilot watershed studies; (2) tributary and shoreline loadings; (3) the assessment of problems, management programs, and research needs concerning the effects of land-use activities on Great Lakes water quality; and (4) the legislative and institutional frameworks of the Great Lakes basin jurisdictions. An overview of the complexity and interrelationship of these activities is shown in Figure 1.

AREA OF STUDY

All five Great Lakes, their connecting channels, and the entire Great Lakes basin, as well as drainage to the international section of the St. Lawrence River, were considered (Fig. 2). In order to provide background information of characteristic basin properties, an inventory of major and specialized land uses and land-use practices in the Great Lakes basin was conducted. This inventory included information of geology, soils, mineral resources, climate, hydrology, vegetation, wildlife, waste disposal operations, high-density nonsewered residential areas, recreation lands, economic and demographic characteristics, and use of pesticides, commercial fertilizers, agricultural manures, and highway salts. Trends in land-use patterns and practices were assessed and projections of economic and demographic factors to 1980 and 2020 were made. This information base, never before adequately assembled, was necessary to gain a better understanding of the combination of factors that affect pollution from land drainage sources (PLUARG, 1976, 1977; Chapra and Dubson, 1978).

Sixty-one percent of the basin consists of forested/wooded land. Agricultural land, including cropland and pasture, comprise 24% of the basin area. Urban land, including residential, commercial, and industrial areas, comprise about 3% and the remaining 12% consists of recreational lands, wetlands, transportation corridors, waste disposal sites, extractive industries, and idle lands (Fig. 2).

Major jurisdictions involved in the Great Lakes basin are the federal governments of Canada and the United States, the province of Ontario, and the states of Illinois, Indiana, Michigan, Minnesota, New York, Ohio, Pennsylvania, and Wisconsin. In 1975, there were approximately 6,900,000 and 29,6000,000 residents in the Canadian and U.S. portions of the Basin, respectively. A significant part of the GNP of both nations is generated within the watershed of the Great Lakes, which contains the world's largest single supply of fresh water.

TYPES OF POLLUTANTS AND THEIR PATHWAYS

Diffuse source pollutants that cause Great Lakes water quality problems have been identified as part of the PLUARG activities (Table 2). These are derived from land drainage sources and include phosphorus, sediments, some industrial organic compounds,

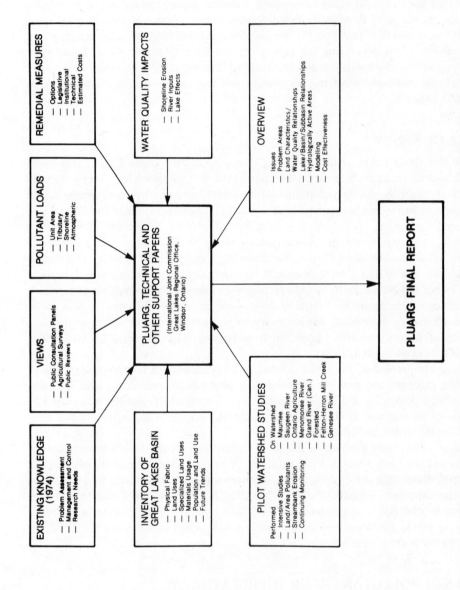

FIGURE 1 *Overview of the Pollution from Land Use Activities Reference Group investigation of Great Lakes pollution from nonpoint sources. (PLUARG, 1978.)*

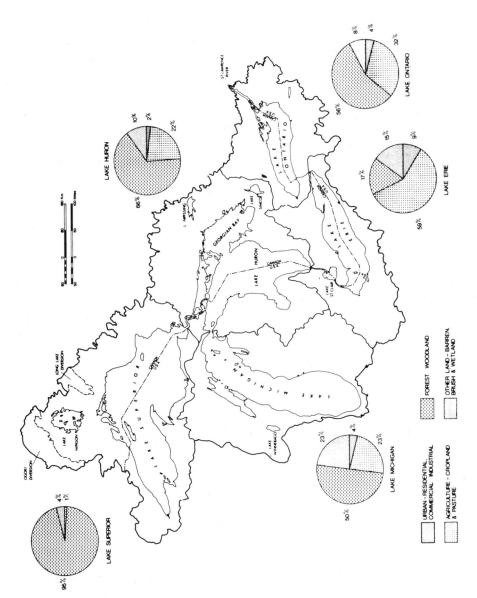

FIGURE 2 *Location of the Great Lakes watershed and major land uses for each lake basin (percent). (PLUARG, 1978.)*

TABLE 2 *Great Lakes Water Quality Pollutants Defined by the Pollution from Land Use Activities Reference Group (PLUARG, 1978)*

POLLUTANT	PROBLEM		SOURCES				REMARKS
	Lakewide	Nearshore or Localized	DIFFUSE			POINT	
			Land Runoff	Atmosphere	In-Lake Sediments		
Phosphorus[1]	Yes	Yes	Yes	Yes	Yes[a]	Yes	a percentage unknown; not considered significant over annual cycle
Sediment[b,1]	No	Yes	Yes[c]	Negligible	Under some Conditions	Negligible	b may contribute to problems other than water quality (e.g., harbor dredging) c including streambank erosion
Bacteria of Public Health Concern	No	Yes	Minor[d]	No	No	Yes	d land runoff is a potential, but minor source; combined sewer overflows generally more significant
PCBs[1]	Yes	Yes	Yes	Yes	Yes	Yes	
Pesticides[1] (Past)	Yes[e]	Yes[e]	Yes	Yes	Yes	No	e some residual problems exist from past practices
Industrial Organics[1]	Yes	Yes	Yes	Yes	Yes	Yes	
Mercury[1]	Yes	Yes	Minor	Yes	Yes	Yes	
Lead[1]	Potential[f]	Potential[f]	Yes	Yes	Yes	Yes	f possible methylation to toxic form

II   Parameters for which no Great Lakes water quality problem has been identified, but which may be a problem in inland surface waters or groundwaters

POLLUTANT	Lakewide	Nearshore or Localized	Land Runoff	Atmosphere	In-Lake Sediments	POINT	REMARKS
Nitrogen	No	No[g]	Yes	Yes	Minor	Yes	g some inland groundwater problems
Chloride	No	No[h]	Yes	Negligible	No	Yes	h some local problems exist in nearshore areas due to point sources
Pesticides[1] (Present)	No	No	Yes	No	No	Yes	i new pesticides have been found in the environment; continued monitoring is required
Other Heavy Metals	Potential[f]	Potential[f]	Yes	Yes	Yes	Yes	j see Upper Lakes Reference Group Report[37]
Asbestos[j]	No	Yes	No	?	Yes	Yes	k better detection methods needed
Viruses[k]	◄——— No Data Available ———►					Yes	
Acid Precipitation	No	No[m]	No	Yes	No	No	m a potential problem for smaller, soft water, inland lakes

[1] Sediment per se causes local problems; phosphorus and other sediment-associated contaminants have lakewide dispersion

444

previously used pesticides, and, potentially, some heavy metals (PLUARG, 1978). The table indicates whether or not the pollutants are known to be causing lakewide problems or only nearshore or localized problems such as in bays and shallow water river mouths. It also indicates if the pollutants are derived from diffuse or point sources or both. A distinction is made between lake water and upland water quality problems, problems in inland surface waters, or groundwaters where pollutants are present but have not yet reached the lakes in significant quatities.

Unit area loads were used to estimate the total tributary loadings to the Lakes. These are calculated by dividing total pollutant contributions from a given land area by the size of the land area. Two years of monitoring data were obtained from U.S. and Canadian pilot watersheds to determine a large number of pollutant unit area loads for areas with a single dominant land use. Unit area loads for suspended sediment, phosphorus, and nitrogen from intensive agricultural and urban land uses were approximately of the same order of magnitude, which were 10 to 100 times greater than those of forested and/or idle land. These were at or near background levels. The unit area loads for improved pasture overlapped the upper range of forested and/or idle land categories and the lower range of the cropland land-use category. Unit area loads of lead from general urban lands were about 10 times greater than the upper range of general agricultural and croplands. Phosphorus unit area loads for wastewater spray irrigation approximate the loads from general agriculture, cropland, and urban categories, while nitrogen unit area loads for spray irrigation sites were up to 10 times greater than those from other land uses (PLUARG, 1978).

BASINWIDE DISTRIBUTION OF SOURCE AREAS

Figures 3, 4, and 5 indicate the primary sources of phosphorus from the main contributing land-use activities that yielded the highest phosphorus loads to the lakes. These include general agriculture, livestock operations, and urban development. Figure 3 is based on row crop density, mainly corn, soybeans, tobacco, and vegetables, and the clay content of soils. The agricultural contribution of total phosphorus to streams is highest on the intensively farmed clay soils of northwestern Ohio and southwestern Ontario. Moderate loading areas include southeastern Wisconsin, the Niagara peninsula of Ontario, and the lowlands of New York at the eastern end of Lake Ontario. Coarse-texture soil areas, and areas with knob and kettle or hummocky topography lacking an integrated stream network show as low phosphorus loading regions.

An estimate of diffuse phosphorus loads from livestock operations are shown in Figure 4; PLUARG (1978) extrapolated phosphorus loads based on a livestock contribution of 0.2 kg P/ha/yr per animal unit, which was determined by their study results. The density of livestock operations indicates that phosphorus loads from this source are smaller than from general agriculture and other sources. Both Ontario and southeastern Wisconsin show appreciable livestock source areas.

Diffuse phosphorus and lead loads from urban lands are estimated in Figure 5. Contributions from these sources were estimated by multiplying the percent of urban area in a watershed by a fixed urban unit area load of 2 kg P/ha/yr. Major metropolitan areas of Detroit, Toledo, Cleveland, and the Hamilton–Toronto areas have the highest urban diffuse phosphorus loadings to the lakes.

The PLUARG overview approach of extrapolating pilot watershed data to pollutant

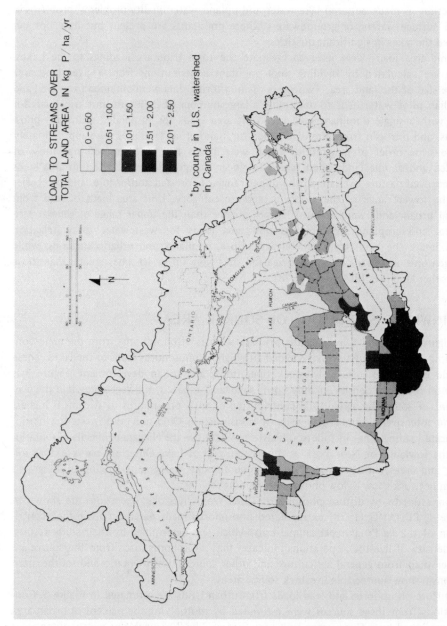

FIGURE 3 Locations of estimated agricultural contributions of total phosphorus to stream loadings within the Great Lakes basin. (PLUARG, 1978.)

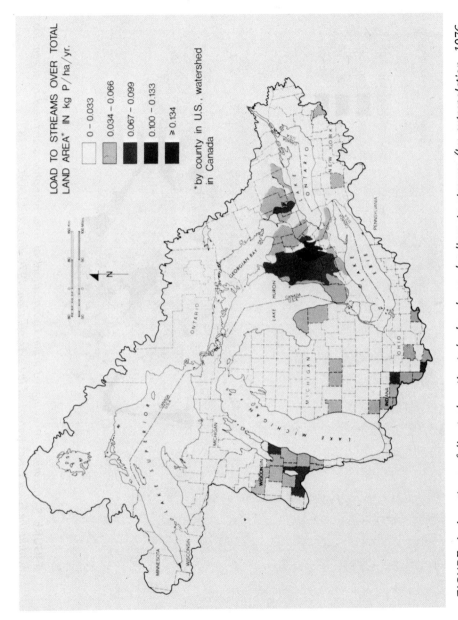

FIGURE 4 *Locations of livestock estimated phosphorus loadings to streams (by extrapolation, 1976 data). (PLUARG, 1978.)*

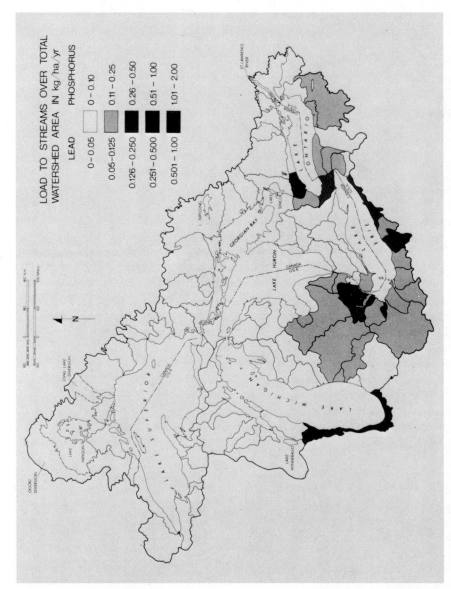

FIGURE 5 *Locations of estimated urban contributions of total phosphorus and lead to stream loadings within the Great Lakes Watershed (by extrapolation, 1976 data). (PLUARG, 1978.)*

loading estimates to the Lakes did not involve deterministic models, although progress was made on this overall effort. Different phosphorus loads other than those predicted by the overview modeling method used can occur because of variations in watershed characteristics, but the study provided the best estimates possible within the time constraints (Hetling et al., 1978; Logan, 1978; Konrad and Chesters, 1978; Bahr, 1978; Hore and Ostry, 1978a, 1978b; Nicolson, 1977; Frank and Ripley, 1977; Agriculture Canada, 1978).

Good agreement was found, however, between predicted phosphorus loads using the unit area method and areas monitored for phosphorus contributions based on land-use intensity and physical characteristics of individual drainage basins. River mouth loading data of major streams also support these conclusions, despite the fact that both point and nonpoint sources of phosphorus are combined as total loads at river mouths.

Nonpoint tributary phosphorus loads expressed as unit area loads are highest for tributaries draining Lake Erie, southern Lake Huron, southern Lake Michigan, and part of Lake Ontario.

The most important factors influencing the magnitude of pollution from land-use activities are the (1) physical, chemical, and hydrological characteristics of the land; (2) land-use intensity; and (3) the intensity and types of materials usage within tributary watersheds. Meteorological conditions also affect annual and seasonal variations in pollutant contributions from land-use activities. For example, 100-year frequency storm in a portion of the Maumee basin of the southwestern portion of Lake Erie in 1975 caused as much as a 100-fold greater sediment yield for 1975 than in 1976. Structural remedial measures (sediment basins, terraces, etc.) must be designed with an assumed flood frequency in mind, but the problem of retaining sediment and entrained pollutants on the land is complicated because fine-grained sediments act as scavangers for pollutants. These are hardest to control using existing sediment control measures. Soil and water conservation measures developed over nearly a 35-year period by the U.S. Department of Agriculture were developed largely to maintain soil fertility, *not* for pollution control.

Diffuse source pollutants within the Great Lakes system use all pathways by which pollutants migrate within the environment. These include the atmosphere where they move in liquid, solid, or gaseous form, surface water systems in liquid or solid forms, and largely in dissolved form in soil-water and groundwater flow systems. Pollutants may be transported within lakes and dispersed by processes of shoreline, beach, and lake bottom erosion. Lakewide currents account for the rather rapid migration of pollutants within lake water and bottom sediments and their transport along connecting channels from upper to lower lakes in the system. The wide distribution of mercury in lake bottom sediments, for example, illustrates this distribution pattern (Fig. 6).

Biological pathways are also important in the dispersion process from uplands to the lake system and within the lakes. Animals, plants, fish, insects, and so on, all selectively accumulate or biomagnify contaminants which are carried by them in life and after death (Table 3).

It is still not clear where all of these pollutants originate within the Great Lakes watershed, because they may be derived from a host of land uses and activities of man since settlement of the region. They are derived in large quantities from metropolitan and rural areas alike, hence cannot be controlled by a single land management program. Their control is further complicated by significant parts of airborne pollutants from outside the watershed from farming regions and metropolitan and industrial complexes

TABLE 3 *Concentrations of Lead in Great Lakes Fish (PLUARG, 1978)*

Lake	Number of fish analyzed	Concentration[a] (mg/kg)
Superior	70	0.012–0.066
Michigan	23	N.D.[b]–0.54
Huron	50	0.04–0.10
St. Clair	34	0.47–0.63[c]
Erie	49	0.04–0.12[c]
Ontario	219	<1.0

[a]The accepted guidline concentration is 10 mg/kg.
[b]Not detected.
[c]Range of mean values.

such as St Louis, Mo., Pittsburgh, Pa., and Albany, N.Y. Attenuation mechanisms, also still poorly understood, may alter pollutants and contaminants during their transit and interim storage between their sources and ultimate sinks. Wastes applied to the land may be acted upon by physical, geochemical, and biochemical processes operating within surface water, soil-water, and groundwater flow systems that can partly or completely cleanse them from the system. Some organic substances become biodegraded during storage and transport; other substances are conserved but may be trapped with sediment in upland lakes, marshes, and related wetlands, or be placed into temporary storage with colluvium or soil moving along slopes, or sediment deposited along floodplains, in deltas, bays, and esturaries or within lake bottoms.

PLUARG (1978) documented the magnitude of the total phosphorus load to the Great Lakes in 1976 and compared these loads from various sources with the recommended target loads that must be achieved to maintain or restore the lakes to a healthy trophic state (Fig. 7). The percent of diffuse sources of phosphorus was found to constitute a significant part of the total loads, which contained important contributions of nonpoint tributary and atmospheric sources. Figure 8 shows that these were highly variable for each lake basin, reflecting variations in the physical framework (soils, geology, etc.) and the intensity and type of land uses involved. It should be clear that no single phosphorus cleanup program could be applied to all areas of the watershed. The prolonged contribution of phosphorus to the lake system from all sources has altered the trophic status of the nearshore waters of these lakes, which has direct impact on all users of the lakes. The trophic state was defined by a composite of several parameters indicative of the algal productivity of water bodies (Jenson and Jernolov, 1967), including total phosphorus concentration, chlorophyll concentration, and Secchi depth (a measure of water clarity). The poorest water quality (eutrophic to mesotrophic/-eutrophic) appears in Lakes Ontario and Erie, especially opposite major metropolitan areas and intensively farmed tributary watersheds containing fine-grained soils and draining into bays. Mesotrophic conditions characterize nearly all of the remainder of their shoreline waters. Lakes Superior, Huron, and Michigan are generally oligotrophic, except in Green Bay, Saginaw Bay, southern Georgian Bay, Lake St. Clair, the Bay of Quinte, and the south shore red clay area of Lake Superior, because of local sediment problems (Fig. 8).

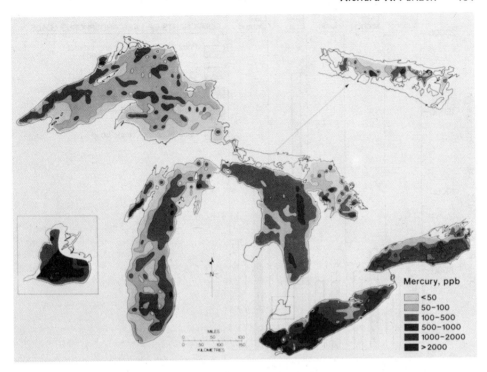

FIGURE 6 *Mercury concentrations in surface sediments of the Great Lakes (ppb-ug/kg). (PLUARG, 1978.)*

REMEDIAL PROGRAMS FOR POLLUTION ABATEMENT

The control of diffuse pollutants entering the Great Lakes will be possible only after those parts of the watershed that contribute to water quality problems have been delineated. Despite the fact that the Great Lakes are an interconnected system, each basin is unique in terms of its limnology, the socioeconomic characteristics of its communities, the type and degree of pollution, the physical characteristics of individual upland tributary basins, hence the kinds of required control measures. Diffuse source pollutants are not derived uniformly from entire watersheds or even subbasins within individual tributaries.

One of PLUARG's environmental management strategies was developed with these points in mind. Criteria were developed for the identification of *potential contributing* or threshold areas and within these, the most *hydrologically active areas*, which are the zones most likely to produce water pollution from a wide range of land-use activities.

A *hydrologically active* area is one that contributes directly to surface runoff even during minor storm and snowmelt events, owing to its steep slopes, low infiltration rate and infiltration capacity, and rather common location within groundwater discharge areas. A shallow water table can quickly promote surface runoff as available storage space is filled by infiltrating water. These areas help promote runoff by rejecting some or all precipitation. Areas of rejected recharge are free to expand and contract on a seasonal

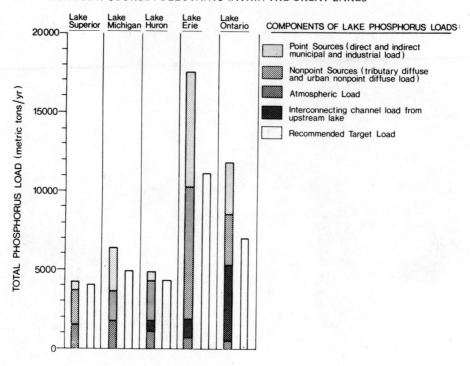

FIGURE 7 *1976 phosphorus loads and recommended target loads for the Great Lakes. (PLUARG, 1978.)*

basis as water is added to or discharged from the groundwater reservoir. Areas with shallow water tables may extend well beyond stream channels, floodplains, and valley walls, depending upon the hydrogeological setting. In glaciated areas containing low relief, poorly permeable glacial drift and/or bedrock and long, gentle slopes leading to upland surface-water divides, poorly drained areas may extend hundreds to thousands of meters beyond valley walls. Emergent lake plains containing poorly permeable lacustrine clays with only minor slopes typically fall in this category. From 10 to 80% of the total land area of some basins tributary to the Lakes fall within poorly drained areas that contribute directly to surface runoff. Pilot watershed data showed, for example, that 15 to 20% of the land surface may contribute up to 90% of the total sediment load from a watershed. Tile lines and drainage canals are required to allow cropping practices in these settings. These regions may still be hydrologically active areas because the residence time between soil-water infiltration and groundwater discharge from tile lines may be but a few days or weeks.

Groundwater discharge areas also may be hydrologically active areas because these areas contribute to diffuse surface seeps and springs, and support shallow water tables.

Nutrients, trace elements, toxic substances, organic chemicals used for pest and weed control, and sediments are all mobilized within these zones and tend to be flushed to nearby streams and lakes. These potential pollutants are kept from entering the shallow soil-water and groundwater zone, where they may be acted upon by biochemical,

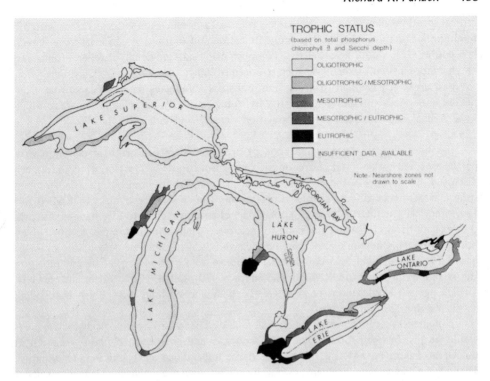

FIGURE 8 *Nearshore trophic conditions of the Great Lakes. (PLUARG, 1978.)*

geochemical, and physical processes that may transform them into harmless by-products of weathering. They may also be assimulated by plants and soil microbes or physically-chemically filtered out and tied up in soils and bedrock.

These same substances applied to groundwater recharge areas, by contrast, become entrained in infiltrating surface water and percolating soil- and groundwaters. Their residence times within shallow soils may be measured in days to weeks, where they may be absorbed by soils, transformed by biochemical processes, precipitated out on soil grain surfaces, or physically retained at or near land surface. Alternatively they may leach into groundwater flow systems, causing upland water pollution problems.

PLUARG (1977) has shown through its field investigations and modeling activities that the land area contributing contaminants to streams from agricultural practices, for example, frequently is <20% of the total land area during periods of average rainfall. During heavy rainfall periods when ground is frozen, this part can increase to 100% of the available land area. These hydrologically active surfaces are potentially major source areas for Great Lakes contaminants. The design and implementation of remedial measures to abate Great Lakes pollution and water quality degradation will have to consider this concept of variable source area together with variations in delivery ratios of sediments and other contaminants moving from the land surface.

It should be apparent that the pollution abatement and control dollars should not be distributed equally over the land area of selected drainage basins but rather applied in critical locations likely to yield the largest returns for the investment. Thus areas with different pollution thresholds need careful delineation and cautious selection of management procedures. This would be easier to accomplish using state, regional, or federal funds, but may be more difficult to achieve when required of individuals.

Vast areas of pavement of urban and industrial complexes serve the same role. Connected impervious surfaces help identify hydrologically active areas in urban areas, which in a general way correlate with population density and land-use intensity (PLUARG, 1978).

The hydrologically active area concept presented here has much in common with partial-area and variable-source-area models of storm runoff production which have been discussed by various authors. This concept has important implications for sediment source areas, hence diffuse sources of pollution. The concept recognizes that when surface runoff is generated from only a small part of the basin close to the channel network, the area of the watershed significant for sediment yield will be limited to this contributing area. This network also must be reviewed as a dynamic element of the drainage basin that can expand and contract in response to variations in catchment moisture status and storm magnitude (Dunne and Black, 1970; Ragan, 1967; Nutter and Hewlett, 1971; Engman, 1974; Dunne et al., 1975; Gregory and Walling, 1968; Blyth and Rodda, 1973; Walling, 1977; and others)

Implementation of remedial measures on lands adjacent to hydrologically active areas would be justified to protect upland surface water and groundwater quality, but not in the Great Lakes. PLUARG (1978) stressed the importance of (1) the long-term nature of the solutions to most problems of pollution from land-use activities; (2) their ramifications through most sectors of society; (3) the involvement of many agencies in the implementation of these solutions; and (4) their public consequences in such policy areas as food production, housing, and public health. Population growth and distribution, continued industrial development, and technological innovation will all have impacts on the loadings of pollutants to the lakes from land-use activities. These factors will affect both the need for nonpoint-source control and the ability to control some of these sources. We recognized that as populations grow and industrial and economic activity continues, given current technology, pollutant inputs from point and nonpoint sources will undoubtedly both continue to grow. Further, the lakes have a finite capacity to accept these imputs, as the past record shows.

Because of the international, national, and local jurisdictional and political boundaries involved, effective environmental management strategies must be developed that are flexible, that are capable of incremental adjustments in response to a changing environment, and that recognize that pollutants are not uniformily derived from the land. "Management of notpoint source pollutants will require a dramatic departure from the traditional approach followed for the control of point sources" (PLUARG, 1978).

The equity problem is revealed by the sediment–phosphorus problem. Intensive agricultural operations have been identified as the major diffuse source contributor of phosphorus. Erosion associated with intensive crop production, especially on fine-textured soils and from rapidly urbanizing areas, however, contribute to lake-wide phosphorus pollution. Areas of high phosphorus loading from intensive agricultural activities were isolated for northwestern Ohio (Maunee River basin) and southwestern Ontario (PLUARG, 1978). The total phosphorus pollution load can be reduced for these lakes by either attacking further point sources which are easier to identify, but which place a disproportionate burden on the urban sector by controlling soil erosion from lands undergoing urbanization, or by attacking the problem on a farm-by-farm basis.

The urban sector has already been assigned a greater obligation for phosphorus removal through the 1972 Water Quality Agreement. More than $3 billion has been committed by governments to the task of upgrading municipal sewage treatment plants within the Great Lakes basin, including effluent phosphorus concentration reductions. To go from a 1 mg/l to 0.5 or 0.3 mg/l phosphorus treatment level for sewage effluent will require a vast outlay of capitol and costly annual treatment. Reducing soil erosion and entrained phosphorus in urbanizing regions does have its appeal because the costs of these land management practices can be assigned to housing costs, costs of industrial goods, and so on. The farm sector has other problems recovering costs for environmental protection measures because it does not control markets nor can it pass on increased operating costs mandated by governments. Here, education, technical assistance, and government subsidies will be required to reduce the outfall of pollutants from agricultural land uses.

To be equitable, phosphorus and other pollutants probably should be reduced by all sectors of society, but this will be costly. Urban nonpoint remedial measures can be incremental and include (1) pollutant source reduction (primary street cleaning); (2) detention of storm water through watershed storage, downstream storage, and treatment of runoff by settling; and (3) all preceding measures, augmented by advanced treatment of storm water runoff (American Public Works Association and Ontario Ministry of the Environment).

The first program may cost $50,000 to $100,000 per metric ton of phosphrous removed, and the second- and third-level programs may have unit costs of $125,000 and $250,000 per metric ton removed. Rural remedial programs can also be staged. Level 1 efforts involve sound soil and nutrient conservation practices which include (1) using soil test results to quide fertilizer application, (2) incorporation of manures into the soil to avoid runoff, (3) avoiding the spread of manures and fertilizers on hydrologically active areas (frozen ground, near streams, on steeply sloping ground), (4) using crop residues to build soil organic matter and a protective mulch, and (5) cross-slope tillage and minimizing tillage for reducing erosion and obtaining optimum yields. Costs for these programs are low, to the advantage of farmers, and could reduce agricultural phosphorus loads by 10% (PLUARG, 1978; Ontario Ministry of Agriculture and Environment, 1978).

Level 2 efforts will require implementation of additional field and structural measures, especially on fine-textured soils. In addition to level 1 sound management practices, improved drainage practices will be required, including the use of expanded buffer strips along drains and natural water courses. Changes in farm cropping practice will be required on some agricultural lands. Unit costs will vary widely from $5000 to $6000 per metric ton reduction in phosphorus loads attributable to strip cropping programs in certain regions to in excess of $100,000 per metric ton reduced for other remedial measures (PLUARG, 1978).

Choices for a combination of both point- and nonpoint-source control programs are available. For Lake Erie a phosphorus target load of 11,000 metric tons/yr is recommended. To achieve this, 2400 metric tons of phosphorus must be removed using 1976 loading data. Expanded, but costly point-source control programs to reach a 0.5 mg/l effluent phosphorus concentration from municipal wastewater treatment plants could account for 1300 metric tons of the 2400-metric-ton total to be reduced to meet target loads. Increased population by the turn of the century would reduce this to about 900 metric tons. The remaining 1000 metric tons of reduction required by that time could be acheived by various combinations of nonpoint programs and by further reductions in point sources from 0.5 mg/l to 0.3 mg/l of phosphorus (Fig. 9).

PLUARG (1978) does not favor across-the-board measures for nonpoint-source pollution control as was required for point-source pollution control. Rather, a comprehensive strategy for management of the Great Lakes ecosystem and a methodology to identify priority management areas to be treated was developed. This strategy recognized the importance of developing management plans that include:

1. A timetable indicating program priorities for the implementation of the PLUARG recommendations.

2. Need to define agencies responsible for the ultimate implementation of programs designed to satisfy the recommendations.

3. Formal arrangments that have been made to ensure inter- and intra-governmental cooperation.

4. The programs through which the recommendations will be implemented by federal, state, and regional levels of government.

5. Sources of funding.

6. Estimated reduction in loadings to be achieved.

7. Estimated costs of these reductions.

8. Provisions for public review.

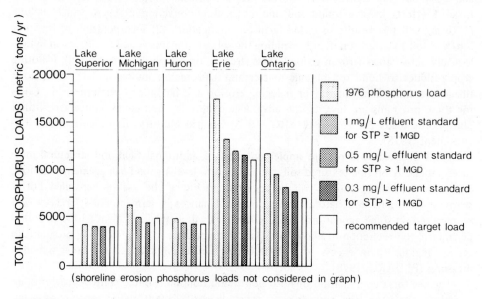

FIGURE 9 *Great Lakes 1976 phosphorus loads under point-source-reduction scenarios. (PLUARG, 1978.)*

Essential elements of the PLUARG management plan include the planning function; development of fiscal arrangements, an expansion of information, education, and technical assistance programs; and strengthening existing regulations. Provisions for the implementation of the management plan stress the need to attack the problems on a regional priority basis; and the need to control phosphorus, sediment, toxic substances, and microorganisms. Programs were called for to develop and implement water quality plans for agricultural and urban areas, to preserve wetlands and prime farmlands, which will help reduce pollutant runoff, and to address local problem areas. The need to review and evaluate the effectiveness of these management plans, and to provide for long-term surveillance and public review, was also emphasized (PLUARG, 1978).

AREAS OF POTENTIAL CONFLICT

The following remarks were taken largely from the author's contribution to the PLUARG annual report, submitted to the Commission in July 1977. It reviews potential areas of conflict and challenges that must be faced by all jurisdictions faced with the problem of waste management and diffuse source pollutant control.

Both the United States and Canada are looking into existing remedial measures that can be adopted to abate or minimize pollution from nonpoint sources. These have been identified as part of the PLUARG effort (PLUARG, 1978). The interest in this effort is great in the United States because many study areas within the Great Lakes watershed and beyond have been designated as 208 planning areas where area-wide management treatment plans are required under Public Law 92-500, Section 208. Managers of 208 plans, subcontractors who are conducting the required studies in designated regions, and others involved in this total effort face many of the same problems encountered by PLUARG researchers: poorly documented field information, inadequate existing literature on the subject, and so on. There is one notable distinction that can be made between the PLUARG mandate and that of other concerned investigators, and that is our required emphasis on the resulting water quality in the Great Lakes. The importance of this distinction cannot be overlooked, as many land uses that may cause only local water quality degradation or pollution problems within upland watersheds may never be shown to reach the open waters of the Lakes. These pollutants could be ignored from a more narrow perspective as being beyond the responsibility of PLUARG and the IJC. Land uses that currently are having only minor or local impacts on upland water quality were regarded as low-priority concerns requiring no further attention in the time span of PLUARG but which may require longer-term monitoring from an IJC time perspective.

It is necessary to understand the variety and nature of nonpoint sources of pollution issues that will arise from the PLUARG and similar other diffuse-source pollution control efforts. Some of these are of a social-political nature; others are of a geochemical nature, in that they involve interaction of natural and man-altered processes that operate within uplands, at or near the land surface, within the shallow soil zone, within groundwater flow systems, and within streams and lakes. These must also be considered from a time perspective, as will be made clear.

People's Perception of Their Environment

It is reasonable to expect that many people will deny that their range of land-use activities will have a water quality impact on the Great Lakes or any other water body, hence that they should be excluded from financial burdens that may be imposed on them by outsiders who wish to reduce pollution problems in lakes and streams. Some upland landowners are not even aware that they live and work within the Great Lakes watershed, let alone have a personal stake or interest in the Lakes. This level of awareness will pose an educational problem for years to come. The public must gain insight into these problems if we are to win their support for costly remedial measure programs.

The impact of land uses on lake water quality is expected to increase as the proximity of this land use to the Lakes decreases. Major metropolitan centers located on or near the shore—Rochester, Buffalo, Toronto, Cleveland, Detroit, etc.—may therefore be required to have a larger burden for pollution abatement programs to bring about further improvement in lake water quality than metropolitan centers more remote from the lakes. Will this mean that nearshore land uses will have to come under more stringent controls and that individuals involved bear a higher cost for pollution abatement programs than those in more remotely located areas in the same tributary or even within the same state or political jurisdiction? Double standards are not popular either with the public or with the governmental body that must administer these programs, and yet a double standard may be required to meet a stated water quality objective or standard set for the lakes (PLUARG, 1977).

Subsidies by federal governments may be required to make up the difference in costs involved for these more stringent remedial measures that may be prescribed for nearshore land uses. On the other hand, the same water quality standards and remedial measures may be desirable to protect areas more remote from the lakes to ensure adequate local water quality within lakes, streams, and groundwater. These local water supplies may appear to have greater immediate value to the nearby community than to the more distant Great Lakes.

Hydrologically Active Areas

The question of the appropriateness of a double standard will become even more of a conflict when applied at the individual landowner level, where a pollution issue cannot be hidden or obscured within the complexity of a larger governmental body. Immediately adjacent neighbors with identical land uses may apply the same total pollutant load to the land but with vastly different contributions to the Lakes, depending upon the hydrogeological and environmental geological setting. How should each individual be treated under the law?

Both may be required to adopt and implement a similar level of remedial measures to meet local water quality requirements prescribed by their respective government. Should the individual, by chance of living and working in a hydrologically active region, be required to adopt additional, more stringent remedial measures than his neighbor to meet Great Lakes Water quality requirements in addition to those of the local jurisdiction? Who will decide what and where the thresholds and physical boundaries, of these hydrologically active areas are located? Furthermore, because hydrologically active areas are likely to display degrees of "activeness" on a probabilistic basis, who will decide what "degree" of threshold or additional remedial measures may be required to contain his potential source of pollutants? The concept of hydrologically active areas is still new enough and will pose problems to water quality managers, regulatory agencies, and others who may wish to be guided by the concept when formulating pollution abatement plans and programs.

Adequacy of Existing Remedial Measures

PLUARG was asked to review the existing remedial measures available to control pollution from land-use activities, not to evaluate their effectiveness by field demonstrations. Structural control measures and legal and institutional measures were all reviewed. However, gaps in knowledge still exist in defining and engineering the best systems for achieving diffuse source pollutant reduction goals.

Atmospheric components added to the Lakes, for example, are ubiquitous, and accurate estimates of sources are likely to be exceedingly difficult to define. This will require follow-up investigations to pinpoint the portion of the total pollution problem derived from individual land uses and activities (i.e., fossil-fuel combustion, farming, manufacturing, etc.).

PLUARG has identified sheet erosion as a major mechanism for shedding up to 90% of the total phosphorus load leaving agricultural areas. Most of this phosphorus transport is tied to the sediment load leaving the land, which increases with decreasing grain size of particles. Much of this sediment may be contributed by farmed row cropping systems, which have the highest erosion potential, especially directly adjacent to streams. As the pressure for farm productivity increases on a worldwide basis, these more-pollutant-yielding regions are bound to be subjected to cropping practices that promote erosion and runoff.

Remedial measures may have to involve the mandated restricted use of these lands, prevention of plowing too close to streams, drainage ditches, and field borders, and the use of sediment control borders (structures and vegetation) that will help trap these sediments. Much will have to be learned about retaining colloid-sized soil particles using available sediment control structures other than larger lakes and reservoirs. PLUARG (1978) has shown through its pilot watershed efforts that these finer-grained sediments also contain many of the pollutants transported by overland flow mechanisms. These pollutants will have to be controlled to reduce upland loadings to streams and lakes. These same uplands receive atmospheric fallout which contributes to the pollutants available.

FORCES AT WORK AND THE TIME PERSPECTIVE

The environmental framework of the Great Lakes is a dynamic system which is driven by natural and man-induced stresses. In the short term, the total system and its many subsystems may appear to be in a state of equilibrium where the rates of inputs are balanced by the rates of outputs, whether this is the hydrologic balance of the Lakes, nutrient balance, sediment balance, or the like. The system is so large that any one activity at the local level appears to have no affect on the Lake system. However, PLUARG and other Reference Groups would not have been created were this indeed the case. The evidence is abundant that water quality degradation has occurred through time from point and nonpoint sources and impaired some uses of the Lakes.

The dynamic nature of the total lake system can be appreciated when viewed in a time perspective long enough so that environmental changes are measurable or obvious. This is most obvious from a geological time perspective, for here one considers the complex events that led to the origin of the Lakes during the Pleistocene Epoch. The many tilted beach ridges, abandoned glacial spillways and channels, emergent lake plains, and drift sheets that surround the Lakes all attest to the complexity of this birth event. As geological time is allowed to run its course, individual lake basins will be subjected to prolonged siltation and natural processes of aging and/or eutrophication that will produce a shrinking lake surface and cause large delta plains, meandering stream channels, and vast wetlands. The downcutting and headward migration of the Niagara River gorge would in time cause a drop in the level of Lake Erie and, in time, reduce levels of all upstream lakes, and produce headward stream erosion. Alternatively, a return of continental glaciers would again modify the Lakes. These are academic considerations from the fleeting time perspective of PLUARG and the IJC but serve to remind us that the benefits of all remedial measures that may be recommended and adopted to abate pollution in the Lakes or adjacent uplands are, at best, ephemeral. The hope is to maximize the duration and impact of these remedial measures.

Two important observations should be made with a longer time perspective than has been used in pollution abatement programs in the past. A time frame for water quality improvement and/or quality maintenance must be defined that is shorter than geological time but longer than the design line of typical engineering structures—25 to 50 years or so.

1. Remedial measures should be selected and implemented that will provide the greatest pollution abatement possible for the money spent when considered in the long term.

2. Remedial measures should be selected and implemented with a clear realization that the basic laws of physics, chemistry, and so on, will not be violated. It may be hard to justify spending vast sums of money implementing one pollution control measure only to find in a short time that a secondary pollution problem has been created which may have an adverse impact equal to or greater than the first at some other location within the environmental system. This will be illustrated to emphasize the importance of considering the time dimension of natural processes and man-induced changes in these processes.

It is important to consider here, therefore, the difference between a concept of "permanent storage and transient storage of pollutants" and the pathways and mechanisms that may be available to aid in their release and transport or the transport of their daughter or granddaughter products of weathering within the environment.

Sewage effluents and/or sludges derived from secondary or tertiary treatment plants are being applied to the land in increasing volumes throughout the Great Lakes watershed. The removal of these pollutants from rivers and streams leading to the Lakes and from direct discharge lines can have either permanent or temporary benefits in the Lakes, depending upon whether these pollutants are placed into permanent or temporary storage and/or whether renovation mechanisms operating at the new treatment or disposal sites are adequate to render these waste materials harmless before they are ultimately recycled to waterways leading to the lakes. Sludges, for example, might be mistakenly applied to hydrologically active areas only to be flushed into nearby streams with little or only partial improvement in quality.

Sludges and effluents may be flushed by intermittent surface runoff events, or entrained within groundwater discharged to the land surface in seeps and springs, or intercepted by shallow tile fields placed to improve the soil drainage. The transit time to streams within this upland surface runoff transportation system may be measured in but hours to months. Alternatively, these waste substances may be applied to the land at rates that exceed natural renovation processes operating within soil and biological systems in upland areas. Pollutants may migrate within soil-water and groundwater systems causing local water quality degradation problems that are merely transferred from the lake system to a more local upland setting. This may represent an acceptable trade-off to protect lake water quality but may be totally unacceptable to residents in the upland area who now must bear the burden or environmental consequences of this management decision.

Hydrological systems may be selected that provide "permanent" upland containment to such waste materials, even where renovation of waste substances is incomplete. Some upland areas within the Great Lakes watershed, for example, are internally drained. That is surface runoff may be confined to wetlands, marshes, ponds, and lakes that are drained by groundwater underflow. Groundwater flow rates may be so slow as to be regarded as imperceptible.

Pollutants can be placed into transient storage for years if not generations from points of recharge within groundwater flow systems to points of ultimate discharge. Many waste disposal practices, such as deep-well injection of liquid wastes, use of "sanitary" landfills to dispose of solid and liquid wastes, evaporation basins intended to evaporate organic and inorganic liquid wastes but that in fact leak, and similar practices appear suitable for the waste management practice adopted. This is largely due to the fact that years may pass before a secondary pollution problem is discovered. Ideally, these pollutants should be placed into permanent storage when they can be absorbed by organic or inorganic particles contained within soil or bedrock. The entire question of sludge application to productive agricultural lands, to marginal and idle lands, or to disturbed lands must be considered. Disturbed lands might include mine spoil deposits, highway right-of-ways (embankments and cuts), areas undergoing subdivision, and so on. The concentration and total volume of more toxic substances, high concentrations of heavy metals and trace elements, for example, that should be applied to lands with various soil types, hydrogeological settings, and agricultural uses will have to be evaluated. The intent should be to alter biochemically these substances to harmless by-products without degrading productive agricultural land, crops, or local water resources, or to store them permanently within soil and rock.

All available transportation mechanisms must again be understood to be assured that these wastes have been placed into permanent storage. Trace elements and heavy metals that may become entrained within sediment can be remobilized by soil erosion processes. Here, their residence time along with soil may be highly variable because sheet erosion, gully development, streambed and channel erosion can all be prolonged processes that are controlled each by its own threshold, or physical mechanisms. The sediment must first be entrained with water in uplands or moved by processes of mass wasting to stream channels, where it again must be transported in stages to the open lakes. Some of this sediment may be placed into temporary storage within uplands in local depressions, as colluvium deposited on the slopes of hills, and in sediments of floodplains, channels, bays, lagoons, marshlands, and deltas near Lakes.

The PLUARG studies have shown that much of the undesirable organic and inorganic waste substances, phosphates, nitrates, trace metals, pesticides, and so on, tend to be entrained with the finer-grained or colloidal-sized soil particles. The rate of transport of these fine-grained soils compared with the total sediment load yielded from the land is still poorly documented. It is well known that the erosion rate from isolated soil test plots exceeds the sediment delivery rate to more distant streams and lakes. The delivery ratio for fine-grained sediments probably far exceed that for the total volume of sediment and coarse-grained fractions eroded from uplands. The trap efficiency of sediment control structures and soil erosion control management practices is less for colloid-size particles than for silt-, sand-, and gravel-size particles.

Little is known about the rate of this grain-size fractionation process and the effectiveness of existing technical remedial measures to contain this finer-grained pollutant-enriched fraction. Temporary storage of these sediments, however, should not be confused with permanent storage. Hummocky areas with interspersed undrained depressions, which resulted from glaciation in the Great Lakes watershed, may provide "permanent" sediment traps or storage sites that can be relied upon to contain contaminated sediments for centuries until they are breached by headward-eroding streams. Hummocky drift regions of the Great Lakes watershed, for example, still lack an integrated stream drainage network despite the 14,000 years or more that have elapsed since the retreat of continental glaciers. Hundreds, if not thousands, of years may still be required for headward stream erosion to interconnect these upland depressions. Sediments enriched with toxic substances that are largely nonbiodegradable may be allowed to erode from adjacent uplands where waste treatment facilities are placed and be transported to these sediment traps or sediment sinks that can act as "closed chemical systems," at least for the distant future.

Man-made lakes, and most sediment control basins, lagoons, terraces, and so on, by contrast, should be regarded only as more nearly temporary storage sites for soils enriched with toxic substances. Rates of erosion and siltation from land areas with a well-developed intergraded drainage network can be appreciable. Sheet and gully erosion of sediment containing trace and heavy metals added to uplands to abate existing pollution problems should be expected to fill these sediment traps in time until their trap efficiency is reduced or eliminated. A design life for siltation basins must be viewed with this point in mind. Will sediment traps contain toxic substances? If so, are these siltation basins to be cleaned to renew storage space? How often will this have to be done, and what is to be done with these new toxic-substance-enriched sediments?

Smaller siltation control structures will be no match for the persistent geological processes of erosion which, in time, will cause them to be breached and their sediments again regurgitated into the local waterways. Even larger lakes and reservoirs have a design life from a siltation viewpoint. These, in time, even from an IJC perspective, can become sites for alluvial plains containing meandering rivers graded to present spillways if new storage space is not provided.

Most existing siltation and sediment control programs in upland areas are designed to slow rates of geological erosion, not to eliminate it. Even the U.S. Department of Agriculture's soil and water conservation programs have aimed to suppress soil erosion enough to maintain soil fertility, but out of necessity have allowed for an acceptable rate of erosion. This "acceptable, or tolerable, rate of erosion" undoubtedly will prove excessive from a water quality protection perspective. The temporary storage of sediment as hillslope colluvium, alluvial fans, and stream alluvium is being investigated in various areas in North America and should provide a valuable insight into the problem of diffuse sources of pollution derived from erosion and sedimentation. Extensive cutting of virgin timber and prolonged agricultural activity since the middle and late 1700s in the Appalachian region, for example, have been responsible for 2 to 3 m or more accumulations of sediment observed at the base of hill slopes and within floodplains. These sediments overlie soil profiles with wood fragments which can be dated by radiocarbon methods, and channel lag gravels containing the artifacts of early white settlers. Rates of erosion were greatly accelerated during the transition from virgin forests to timber cutting, farming, and urbanization. Some streams appear to be entrenching themselves within these sediments and re-eroding the sediments that have been in temporary storage for approximately 200 years.

Accelerated rates of erosion accompany processes of urbanization and industralization. As more rural lands adjacent to urban growth corridors receive wastes of a variety of types and complexity resulting from various land uses, these shallow soils will become enriched in toxic substances and will add to diffuse sources of pollution resulting from accelerated rates of erosion and siltation as urban sprawl increases and intensive farming spreads to more marginal lands. The influx to the lakes of these pollutants should increase as these new urban area sediments are disturbed and removed from sites of temporary storage downvalley from urbanizing areas. Important geomorphic questions still remain. How can these thresholds or accelerated rate of erosion be identified? Will threshold conditions develop within these sediment–fluvial systems that can greatly accelerate the rates of erosion of temporarily stored sediments enriched with a variety of wastes? In the near future, surges of sediment may be delivered to streams and lakes in volumes far in excess of present loading rates.

Atmospheric Loadings

Studies on the atmospheric loadings of pollutants to the Lakes by the Upper Lakes Reference Group and continuing under PLUARG (1978) provide but another example of a substitute pollution problem that results in part from the incineration of municipal and industrial wastes rather than the treatment or disposal of wastes using alternative methods. The atmospheric fallout of airborne pollutants to the water surfaces of the Lakes has been documented. The residence time of atmospheric pollutants may be measured in but a few hours to days, and pollutants tend to be uniquely widespread in their distribution. The atmosphere contains the least storage potential for pollutants and the shortest residence time of the various storage and transportation systems available for pollutant transport studied by PLUARG.

The potential to magnify the importance of airborne pollutants within the lakes should be apparent when one considers the total land area that comprises the Great Lakes watershed available to receive airborne fallout. Particulates and substances in solution that fall on the land are available for remobilization to the Lakes by surface runoff, groundwater flow, and soil erosion.

Atmospheric pollutants are not restricted in their source to the Great Lakes watershed but may be derived from metropolitan and industrial centers remote from the Lakes. A reduction of this source of loadings to the Lakes will have to be accomplished in and beyond the region of jurisdiction of IJC.

Atmospheric pollutants derived from the incineration of wastes is another example of a transfer of pollution problems from one segment of an environmental system to another, which by first analysis may have appeared to be cost effective and cost beneficial.

From a longer perspective, recycling and resource recovery of these same wastes may prove to be the best and ultimate solution to this "disposal problem." In the long run this may even prove to be the least costly from an overall environmental protection and cost viewpoint. Such "ultimate remedial measures," on the other hand, may appear at present to be too costly to local citizens, who will resist their adoption. The regulatory agencies and ultimate decision makers must be cautious not to adopt and recommend to governments expedient remedial measures and pollution abatement programs that appear economically attractive in the short term, but which, in fact, rely upon built-in pollutant delay or transfer mechanisms that in time will have adverse environmental consequences and prove more costly. The credibility of our institutions is at stake for approving past waste management practices which were regarded as the best available practice just a few years earlier.

Rates of Change Within the System

The biological, geochemical, and physical processes operating within the Great Lakes system are not static and constant but have changed through geological time. Changes in rates of erosion and sedimentation have been induced by drops in lake levels accompanying the advance and retreat of continental glaciers. A drop in lake level represents a drop in the ultimate base level of erosion. A southerly tilt in the earth's crust due to glacial rebound has occurred, as evidenced by tilted breach ridges, marine sediments elevated above present sea level, and other observations. A change in the surface slope of the land surrounding the Lakes has not only accelerated local erosion and the incision of streams, but has also accentuated beach erosion.

Climatic fluctuations have also been documented since the last retreat of continental glaciers that have influenced the type and distribution of fauna and flora, precipitation, evapotranspiration, and runoff (Table 4). Man-induced changes in the system are superimposed on these natural processes and influence the effectiveness of remedial measures recommended to abate pollution in the lakes. More pronounced influences of man may be confined to tributaries of the Great Lakes basin, as is the case in urban areas, where storm runoff tends to be accentuated by buildings and paved areas and where rates of erosion and sedimentation are altered within the fluvial system. The impact of urbanization tends to be a maximum for smaller streams and should diminish as the basin area increases.

The transfer of water from one segment of a basin to another as a source of water supply and the disposal of sewage effluent in another part of the basin will influence the stream-flow characteristics of smaller streams within the affected regions but may be of little consequence along lower reaches of rivers. However, these local changes in the stream and watershed regimen may have a significant bearing on selecting remedial measures for pollution abatement at the local level.

Man-Induced Changes in Precipitation

Man-induced changes in local rainfall and runoff are also possible which can increase the importance of local nonpoint sources of pollution. Existing and proposed electrical power generating plants will continue to rely upon evaporative cooling to dissipate waste heat. Large-scale "energy parks" of the future might contain a cluster of coal-fired and nuclear-powered generating plants that can place a significant daily consumptive draft on local and regional water resources. The volume of water available in the Lakes is more than adequate to meet these growing demands for cooling water.

TABLE 4 *Climatic Changes Since the Last Retreat of Continental. Glaciers Related to Vegetation Sequences in Parts of North America During the Last 14,000 Years (Adapted from Ogden, 1965, in Flint, 1971)*

Time (^{14}C BP)	Midwest	New England	Climate compared with present
0-	Oak-hickory-beech-maple	Oak-hemlock-chestnut	Warmer Cooler,
-	Ragweed	Ragweed	moister
2000-	Oak-beech	Oak-chestnut	
-			
4000-	Oak-hickory	Oak-hickory	Warmer, drier
-			
6000-	Oak-beech	Oak-chestnut	
-			
8000-			
	Oak-elm	Oak-hemlock	Warmer, moister
			Cooler,
10,000-	Pine	Pine	drier
-	Spruce-pine	Spruce-fir	Cooler, moister
12,000-	Spruce-fir-birch	Spruce-Parkland	Cooler
-			Cooler
14,000-	Spruce-parkland	Tundra	

Delays and costs of clearing four or five smaller power plants through environmental review procedures might prove greater than the time and money required to win approval for a single 20,000±-megawatt power complex. C. Hosler, Dean of the College of Earth and Mineral Sciences at The Pennsylvania State University, indicates that the waste heat dissipated might approach the 40,000-megawatt range for such large energy parks. This could require a 500-cfs (14-cms) makeup water demand for cooling water dissipated to the atmosphere. A 10 billion gallon/day (437,000-l/sec) or more consumptive water demand could be met at a number of locations along the shores of the lakes. However, as Hosler points out, such amounts of waste heat can produce convective patterns that would have important effects on downwind weather conditions. Precipitation increases of approximately 5 inches (12.7 cm) would be possible 20 to 25 mi (32 to 40 km) downwind, especially during hot summer months. This, together with increased cloud cover, fog, and so on, will supress evapotranspiration, increase infiltration, groundwater recharge, and surface runoff that can accelerate local rates of surface erosion and the leaching of waste substances accumulating on and within the soil in nearby metropolitan and rural areas.

Salts and other dissolved mineral matter contained in cooling water would accumulate in the atomosphere and be returned by precipitation a short time later. Fresh river water might deliver approximately 45 to 136 k (100 to 300 lb) of salt per acre through precipitation fallout. These same salts would find their way into the lakes in any event, but consideration would have to be given to the possibility of accelerating the total waterborne pollution load derived from local watersheds impacted by such energy parks.

Changes in Precipitation Chemistry

A more regional change in precipitation chemistry may have even more important consequences in Great Lakes water quality that may have to be factored into Great Lakes water quality management programs. Harr and Coffey (1975) concluded that acid precipitation represents a major potential hazard to a significant part of the New York State ecology, and it may also inflict substantial economic losses, if the trend continues. They suggest the problem could become of crisis proportions, as has been documented for parts of Scandinavian countries. They conclude that both empirical observation and hypothetical models indicate that current rates of deposition of inorganic acids in precipitation can be sufficient to produce significant environmental effects.

Acid in precipitation can lead to an acidification of lakes and streams, producing pHs of less than 5 and thus damage to aquatic ecosystems, including fish populations (Bolin, 1971; Beamish and Harvey, 1972; Almer et al., 1974). A drop in pH can result in reduced growth, lower survival rates, and/or elimination of fish populations. Such acidity can cause increased leaching of nutrients from foliage (Easton et al., 1973; Wood and Bormann, 1974), disruption of leaf physiology and growth (Wood and Bormann, 1974), and increased leaching of soil cations (Overrein, 1972) which can contribute to lessened plant growth over a wide area, as postulated for Scandinavia (Jonsson and Sundberg, 1972). The main implication of these findings is that the rain and snow of Canada and northeastern United States appears no longer to be characterized by the weak, highly buffered, carbonic acid, but instead is dominated by the stronger, unbuffered, sulfuric and nitric acids, at a greatly reduce pH (Harr and Coffey, 1975). They point out the ecological effects of this change are not yet completely known, but potentially they are manifold and very complex.

Increased acidity of rainfall in Scandinavia has been attributed to industrial activity performed hundreds or thousands of kilometers away, and it has been shown that most of the acidity is explained by formation of H_2SO_4 and NHO_3 (Wood and Bormann, 1974; and Oden, 1968). Data are still scarce for North America and only a few reports are available in the published literature (Likens, 1972; Likens and Bormann, 1975; Lazrus et al., 1974; Klein, 1974). Aside from the potential direct impact of acid precipitation on limnology and aquatic biology and vegetation, forests and soils, human health, and so on, a special concern may develop around the problem of leaching of soils and their nutrients in the presence of acid rains.

Glacial drift within the Great Lakes watershed contains an abundance of calcium and magnesium carbonate eroded from carbonate bedrock, hence is alkaline in character. In the presence of acid rains, alkaline metals should be leached from the shallow soil and replaced by hydrogen ions, and the soils will tend to become more acidic as a result of the action on rainwater. For soils already acid, podzolic soils prevalent in areas of heavy rainfall, the buffering capacity has already been exceeded and there is a concomitant sharp drop in soil pH and fertility (Harr and Coffey, 1975). This can be counteracted by the application of lime and other fertilizers.

Oden has indicated there is acidification of forest soils with considerable leaching of calcium from the surface and subsurface soil horizons. He estimates that basic soils will turn acidic in 125 to 1000 years and that forest soils will lose their buffering capacity completely in 30 to 50 years. Overrien (1972) also studied the rate of leaching of calcium on forest soils as a function of soil type and pH in rainfall. The leaching rate of calcium was greatly increased as the pH of precipitation dropped less than 4. Groundwater was also found to become more acidic as the buffering capacity of the soil, a sandy clay loam, was exceeded.

Major concerns that may face us in the future, if the present trend in acid precipitation continues, is the problem of maintaining trace and heavy elements in an oxidized state within shallow soils and bedrock, and preventing them from becoming remobilized as the soil pH drops. Significant pH changes could remobilize pollutants that had been formerly deposited by atmospheric fallout and the noxious materials in sludges and effluents, solid wastes, dredging spoil, and reclaimed areas near the shore zone. This situation could produce a pollution timebomb with a threshold that would greatly accelerate the rates of loadings to upland water supplies and local streams in the future. This problem is of potential significance in other watersheds lacking a buffer capacity beyond the glacial border well below the Great Lakes watershed. An improved national effort is being made to investigate the acid precipitation problem. Harr and Coffey (1975) report that considerable progress has been made in accessing the probable causes of the acid precipitation phenomenon. The means of controlling these causes will involve national and international cooperation that will extend far beyond the Great Lakes watershed, because it is possible that activities within the watershed will contribute more to acid-rain-related problems beyond the watershed where a natural buffer capacity may be lacking than within it.

PLUARG (1978) regarded the acid rain issue as a local water quality problem at present, particularly in some of the inland lakes of upstate New York and in the Canadian Shield lakes of Ontario, and in two isolated embayments in Georgian Bay. The volume of water in the Great Lakes is so large and their buffering capacity so great that centuries would be required to reduce the pH of the lake system substantially, even if the entire buffering capacity in the inflowing waters were eliminated. The main concern of the author is the long-term stability of trace elements, metals, and so on, presently in storage in upland "disposal sites," which may be leachated in the future as soil pHs decrease, not the pH of Great Lakes water quality.

FINAL STATEMENT

Definition and control of diffuse source pollutants will require a knowledge of surficial and near-surface processes operating within watersheds, their tributaries, master streams, and receiving lakes and estuaries. These are still incompletely understood despite the U.S. commitment to resolve nonpoint pollution problems under Public Law 92-500, Section 208, and related recent legislation. Until these natural systems are better understood, attempts at applying costly remedial measures will meet with varying success. Some of the departures from the expected behavior of watersheds subjected to treatment to control nonpoint pollutants will result from our misunderstanding of the controlling physical, geochemical, and biochemical processes involved in the entrainment, transportation, and deposition of pollutants within atmospheric, fluvial, subsurface, and lacustrine systems. Thresholds are involved in these complex processes that are only poorly documented or not yet defined.

The PLUARG study approach and the environmental management strategy developed to protect water quality in the Great Lakes system should be of value to others concerned with the identification and control of nonpoint sources of pollution elsewhere in North America and in the world. The research needs were defined by PLUARG and pursued on an international scale to answer vital questions concerning the Great Lakes ecosystem and the impact of various land-use activities on this system. Much work still needs to be done, as identified in the final report. However, PLUARG's recommendations, if adopted by the International Joint Commission and, in turn, implemented by the U.S. and Canadian Governments, should go a long way toward maintaining favorable water quality in the Upper Lakes (Superior and Huron) and in Lake Michigan through the turn of the century, and in helping to return the lower lakes (Erie and Ontario) to a more beneficial state. The Lakes have been plundered and exploited with little real concern for their vital role that they will be called upon to fulfill in North America in the generations ahead or with regard to the possibility that ecological damage from past and present activities and land uses may not be reversible, or only so with a great economic burden. They are, indeed, a "beautiful lady or family" who have been called upon by two great nations to sustain fish and wildlife production, for water supplies for industries and for small and great metropolitan centers, for waste disposal, for mineral resources, recreation, transportation, and many other uses. They have helped generate an agricultural and industrial productivity and international treasure chest disproportionate to the land area they occupy in these two nations. A continued and expanded economic activity within the Great Lakes watershed will be inevitable as fresh water supplies become more fully exploited elsewhere in the United States and Canada.

At the same time, a pure water supply and favorable environmental setting will be demanded by the public, judging from the response of public involvement in the PLUARG Reference (U.S. Public Consultation Panel and Canadian Public Consultation Panel, 1978). The Commission, its boards, and reference groups are to be commended for their dedication to this effort. The IJC has adopted an ecosystem approach in their investigations of this system which is unique in the world. They have the mechanism to call upon the most qualified individuals to aid them in these investigations and to maintain a persistent and long-term investigative and surveillance effort which is a necessary step in protecting the Lake system.

ACKNOWLEDGMENTS

It is a privilege to serve on the PLUARG Reference as Pennsylvania's representative. Walter Lyon, Pennsylvania Department of Environmental Resources, is acknowledged for his support of my candidacy and appointment. Appreciation is expressed to Wayne C. Burnham, M. L. Keith, and Charles L. Hosler, my immediate academic supervisors; and to The Pennsylvania State University, for granting me permission to pursue the open-ended goals and responsibilities of this important international assignment, which has placed personal burdens on Reference Group members and their families. My other U.S. and Canadian PLUARG members and our research staff, contractors, and so on, are acknowledged for their efforts and for allowing free use of the published data included in this report. Discussions not taken directly from PLUARG reports and publications need not reflect PLUARG's final official position. These departures are the sole responsibility of the author.

REFERENCES AND BIBLIOGRAPHY

Agricultural watershed studies in the Canadian Great Lakes drainage basin; final summary report. 1978. Prepared for Task Group C (Canada), Activity 1, by Agriculture Canada, Ontario Ministry of Agriculture and Food, and Ontario Ministry of the Environment. May, 78 p.

Almer, B., Dickson, W., Ekström, C., Hörnström, E., and Miller, U., 1974, Effects of acidification on Swedish lakes, Ambio, v. 3, n. 1, p. 30–36.

American Public Works Association, Evaluation of the magnitude and significance of pollution loading from urban stormwater runoff in Ontario, Chicago.

Bahr, T. G., 1978, Felton-Herron Creek, Mill Creek pilot watershed studies; summary pilot watershed report. Submitted to PLUARG Task Group C Technical Committee—Synthesis and Extrapolation Work Group. Windsor, Ontario, January, 48 p.

Beamish, R. J., and Harvey, H. H., 1972, Acidification of the LaCloche mountain lakes and resulting fish mortalities: J. Fish Res. Bd. Can., v. 29, p. 1131–1143.

Blyth, K., and Rodda, J. C., 1973, A stream length study: Water Resources Res., v. 9, n. 5, p. 1454–1461.

Bolin, B., ed., 1971, Air pollution across national boundaries: the impact on the environment of sulfur in air and precipitation. Sweden's case study for the United Nations conference in the human environment: Kungl. Boktryckeriet. Stockholm, P. A. Norstedt et Soner, 96 p.

Calkin, P. E., 1970, Strand lines and chronology of the glacial Great Lakes in northwestern New York: Ohio J. Sci., v. 70, p. 78–96.

Castrilli, J. F., 1977, Control of water pollution from land use activities in the Canadian Great Lakes Basin: an evaluation of legislative, regulatory, and administrative programs. Submitted to PLUARG Task Group A (Canada). Windsor, Ontario, 460 p. (A separate document containing a summary of this report is also available.)

Chapra, S. C., and Dubson, H. F. H., 1978, Great Lakes trophic scales and typology. Contribution No. 113, Great Lakes Environmental Research Laboratory, National Oceanic and Atmospheric Administration, Ann Arbor, Mich., 24 p.

Chau, Y. K., Wong, P. T. S., Silverberg, B. A., Luxom, P.-L., and Bengert, G. A., 1976, Science, v. 192, p. 1130.

Cogbill, C. V., and Likens, G. F., 1974, Acid precipitation in the northeastern United States: Water Resources Res., v. 10, n. 6, p. 1133–1137.

Dunne, T., and Black, R. D., 1970, Partial area contributions to storm runoff in a small New England watershed: Water Resources Res., v. 6, 1269–1311.

——, Moore, T. R., and Taylor, C. R., 1975, Recognition and prediction of runoff-producing zones in humid regions: Int. Assoc. Hydrol. Sci. Bull., v. 20, p. 305–327.

Easton, J. S. Likens, G. E., Bormann, F. H., 1973, Throughfall and streamflow chemistry in a northern hardwood forest: J. Ecol., v. 61, p. 495–508.

Engman, W. T., 1974, Partial area hydrology and its application to water resources: Water Resources Bull., v. 10, n. 3, p. 512–521.

Flint, R. F., 1971, Glacial and Quaternary geology: New York, Wiley. 892 p.

Frank, R., and Ripley, B. D., 1977, Land use activities in eleven agricultural watersheds in Southern Ontario, Canada. 1975–1976. Prepared for PLUARG, Task C (Canada), Activity 1. (Project 80645, Education, Research and Special Services Division, Ontario Ministry of Agriculture and Food). Windsor, Ontario, March, 176 p.

Gambell, A. W., and Fisher, D. W., 1966, Chemical composition of rainfall in eastern

North Carolina and southeastern Virginia: U.S. Geol. Survey Water Supply Paper 1535-K, 41 p.

Great Lakes water quality agreement with annexes and texts and terms of references, between the United States of America and Canada, signed at Ottawa, Canada, April 15, 1972, International Joint Commission, Washington, D.C., and Ottawa, Ontario.

Gregory, K. J., and Walling, D. E., 1968, The variation of drainage density within a catchment: Int. Assoc. Hydrol. Sci. Bull., v. 13, p. 61–68.

Harr, T. E., and Coffey, P. E., 1975, Acid precipitation in New York State: New York Department of Environmental Conservation, Tech. Paper 43, Albany, N.Y., 52 p.

Hetling, L. J., Carlson, G. A., Bloomfield, J. A., Boulton, P. W. and Rafferty, H. R., 1978, Genesee river pilot watershed report; summary pilot watershed report. Submitted to Task Group C (U.S.), Activity 2. Windsor, Ontario, March, 73 p.

Hore, R. C., and Ostry, R. C., 1978a, Grand river, Ontario; summary pilot watershed report. Submitted to Task Group C (Canada), Activity 2. Windsor, Ontario, April, 63 p.

——, and Ostry, R. C., 1978b, Saugeen river basin pilot watershed study; summary pilot watershed report. Submitted to PLUARG Task Group C (Canada), Activity 2. Windsor, Ontario, April, 56 p.

International Reference Group on Great Lakes Pollution from Land Use Activities (PLUARG), 1976, Annual progress report to the International Joint Commission [Windsor, Ontario] July, 64 p.

——1974, Detailed study plan [Windsor, Ontario] February, 148 p.

——1976, Supplement to the detailed study plan to assess Great Lakes pollution from landuse activities. Submitted to the Great Lakes Quality Board [Windsor, Ontario] August, 119 p.

——1976, Inventory of land use and land use practices in the United States Great Lakes Basin (with emphasis on certain trends and projections to 1980 and where appropriate, to 2020). Prepared by the Great Lakes Basin Commission to be used as a portion of the PLUARG Task B Report (U.S.). U.S. Environmental Protection Agency Contract No. 68-01-1598. 6v. Windsor, Ontario.

——1977a, Land use and land use practices in the Great Lakes Basin, joint summary report—Task B, United States and Canada. Windsor, Ontario, September, 49 p.

——1977b, Inventory of land use and land use practices in the Canadian Great Lakes Basin (with emphasis on certain trends and projections to 1980 and where appropriate, to 2020). To be used as a portion of the PLUARG Task B Report (Canada). 5 v. Windsor, Ontario, December.

——1977c, Preliminary summary report of pilot watershed studies, May 27, 1977, International Reference Group on Great Lakes Pollution from Land Use Activities, submitted to International Joint Commission, May 27, 25 p.

——1978, Environmental management strategy for the Great Lakes system [Windsor, Ontario], July, 115 p.

Jenson, S., and Jernolov, A., 1967, Nordorek Biocidin formation, v. 10, n. 4.

Jonsson, B., and Sundberg, R., 1972, Has the acidification by atmospheric pollution caused a growth reduction in Swedish forests?: Institution for Skogspoduktion, Stockholm, Res. Notes 20, 46 p.

Junge, C. E., 1958, The distribution of ammonia and nitrate in rainwater over the United States: Amer. Geophys. Union Trans., v. 39, p. 241–248.

——, and Werby, R. T., 1958, The concentration of chloride, sodium, potassium, calcium, and sulfate in rainwater over the United States: J. Meteorol., v. 15, p. 417–425.

Klein, A. E., 1974, Acid rain in the United States: Science Teacher, v. 41, n. 5, p. 36–38.

Konrad, J. G., and Chesters, G., 1978, Menomonee river pilot watershed study; summary pilot watershed report. Submitted to PLUARG Task Group C (U.S.), Activity 2. Windsor, Ontario, May, 77 p.

Langmuir, D., 1972, Controls on the amounts of pollutants in subsurface waters: Earth and Mineral Sci. v. 42, n. 2, 4 p.

Lazrus, A. L., Gondrud, B. W., and Lodge, J. P., 1974, Atmos. Environ. (in press).

Likens, G. E., 1972, The chemistry of precipitation in the central Finger Lakes region: Cornell University Water Resources and Marine Science Center, Tech. Rept. 50, Ithaca, N.Y., 44 p.

——, and Bormann, F. H., 1974, Acid rain: a serious regional environmental problem: Science, v. 184, June 14, p. 1176–1179.

Lodge, J. P., Jr., Hill, K. C., Pate, J. B., Lorange, E., Basbergill, W., Lazrus, A. L., and Swanson, C. S., 1968, Chemistry of United States precipitation: National Center for Atmospheric Research, Boulder, Colo., 66 p.

Logan, T. J., 1978, Maumee river basin pilot watershed study; summary pilot watershed report. Submitted to PLUARG Task Group C (U.S.), Activity 2. Windsor, Ontario, April, 96 p.

Nicolson, J. A., 1977, Forested watershed studies; summary technical report. Submitted to PLUARG Task Group C, Activity 2. Windsor, Ontario, December, 23 p.

Nutter, W. L., and Hewlett, J. D., 1971, Stormflow production from permeable upland basins: *in* Monke, E. J., ed., Biological effects in the hydrological cycle: Proc. Third Int. Seminar for Hydrology Professors, p. 248–258.

Oden, S., 1968, The acidification of air and precipitation and its consequences on the natural environment: S. W. Nat. S. A. Res. Council, Ecology Committee, Bull. 1, Tr-1177 Translation Consultants Limited, Arlington, Va., p. 86.

Ogden, J. G., 1965, Pleistocene pollen records from eastern North America: Bot. Rev., v. 31, p. 481–504.

Ontario Ministry of Agriculture and Food and Ontario Ministry of the Environment, 1978, Agricultural watershed studies in the Canadian Great Lake drainage basin, final summary report, Prepared for TASK Group C (Canada), Activity 1, May, 79 p.

Ontario Ministry of the Environment, Preliminary estimates of urban stormwater unit loadings for the Canadian Great Lakes Basin, Toronto, Ontario.

Overrein, L. N., 1972, Sulfur pollution patterns observed: leaching of calcium in forest soil determined: Ambio., p. 145–147.

Proceedings of a Workshop on The fluvial transport of sediment-associated nutrients and contaminants, 1976, Sponsored by The International Joint Commission's Research Advisory Board on Behalf of the Pollution from Land Use Activities Reference Group, Kitchener, Ontario, Oct. 20–22 (edited by H. Shear and A. E. P. Watson), 309 p.

Ragan, R. M., 1967, An experimental investigation of partial area contributors: Int. Assoc. Hydrol. Sci. Publ., v. 76, p. 241–249.

Reports of the Canadian public consultation panels to the Pollution from Land Use Activities Reference Group (PLUARG) [Windsor, Ontario], March 1978.

Reports of the United States public consultation panels to the Pollution from Land Use Activities Reference Group (PLUARG) [Windsor, Ontario], March 1978, 148 p.

The legislative and institutional framework to control pollution in the United States Great Lakes Basin from land use activities, 1978, Prepared by E. Schweitzer, W. G. Stewart, and B. Roth, Linton and Co. Inc., Washington, D.C., for the Great Lakes Basin Commission, Ann Arbor, Mich. for PLUARG Task Group A (U.S.). Windsor, Ontario, April, 3 v.

Walling, D. E., 1977, Natural sheet and channel erosion of unconsolidated source material (geomorphic control, magnitude and frequency of transfer mechanisms), p. 11–33 *in*: Proceedings of a Workshop on the Fluvial Transport of Sediment-Associated Nutrients and Contaminants, Sponsored by The International Joint Commission's Research Advisory Board on Behalf of The Pollution from Land Use Activities Reference Group, Kitchener, Ontario, Oct. 20–22, 1976 (edited by H. Shear and A. E. P. Watson), 309 p.

Wolman, M. G., 1977, Changing needs and opportunities in the sediment field: Water Resources Res., v. 13, n. 1, p. 50–54.

Wong, P. T. S., Chau, Y. K., and Luxon, P. L., 1975, Nature, v. 253, p. 263.

Wood, T., and Bormann, F. H., 1975, Increases in foliar leaching caused by acidification of an artificial mist: Ambio. v. 4, n. 4, p. 169–171.

22

SOME APPLICATIONS OF THE CONCEPT OF GEOMORPHIC THRESHOLDS

S. A. Schumm

INTRODUCTION

In the past geomorphologists have concentrated their efforts on the development of an understanding of the erosional and depositional evolution of landforms through geologic time. The last generation of geomorphologists departed from this approach, when it was realized that it is the details of landscape evolution that require elaboration and explanation if traditional geomorphic problems are to be solved and if geomorphic research is to be of value to those who are managing and attempting to control various components of the landscape (river, slopes, floodplains). Therefore, an understanding of the functioning of geomorphic systems over short spans of time is mandatory. When this is attempted, it is soon apparent that average rates of erosion or deposition are misleading in the sense that they do not clearly display the complexities and the variability of the erosional development of landforms. In addition, components of the landscape are out of phase. Consideration of these problems led to a hypothesis that landforms do not necessarily evolve progressively through time, at least in areas of relatively rapid erosion and high sediment production, but instead the erosion and deposition may be episodic in nature, with periods of stability separated by relatively unstable periods of active landform adjustments. In Figure 1 a modified scheme of valley-floor evolution that involves episodic erosion is presented.

Inherent in the scheme of episodic erosion or deposition is the concept of thresholds. Periods of instability or rapid change occur when a threshold of stress or of strength of materials is exceeded. There are two types of thresholds, extrinsic and intrinsic. An extrinsic threshold is a threshold that is exceeded by the application of a force or process external to the system. For example, a climate change, although progressive, may at some stage initiate a period of rapid erosion or deposition. Similarly, slowly increasing discharge in an alluvial channel will first initiate sediment movement at a critical value of tractive force, and subsequently at increasing values of discharge and stream power the bed will develop ripples, dunes, plane bed, and antidunes. The control is external to the form that results.

The other type of threshold is intrinsic, indicating that change occurs without a change in an external variable. An example is long-term progressive weathering that reduces the strength of slope materials until eventually there is slope adjustment (Carson, 1971) and mass movement (Kirkby, 1973). Glacial surges that are not the result of climatic fluctuations or tectonics (Meier and Post, 1969) probably reflect periodic storage and release of ice as an intrinsic threshold when glacial stability is exceeded. In semi-arid regions, sediment storage progressively increases the slope of the valley floor until failure occurs by gullying. This is a special type of intrinsic threshold, the geomorphic threshold (Schumm, 1973). It is a result of landform (slope) change through time to a

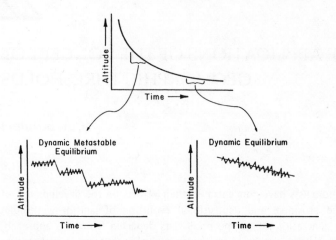

FIGURE 1 *Erosional evolution of valley floors in youth by episodic erosion (dynamic metastable equilibrium) and later by progressive erosion (dynamic equilibrium). (Schumm, 1977.)*

condition of incipient instability and then failure. Another example of a geomorphic threshold is provided by Koons (1955), who described morphologic changes resulting in the collapse of sandstone-capped shale cliffs in the Mesa Verde region in southwestern Colorado. Beneath a vertical cliff of Mesa Verde sandstone is a 32 to 38° slope of weak Mancos Shale. Through time the basal shale slope is eroded and reduced in height, thereby producing a vertical shale cliff beneath the sandstone cap. At some critical height the cliff collapses and the cycle begins again. The episodic retreat of this escarpment is the result of the change in cliff morphology under constant climatic, base level, and tectonic conditions.

Geomorphic thresholds have been defined and described as follows (Schumm, 1973):

> A geomorphic threshold is one that is inherent in the manner of landform change; it is a threshold that is developed within the geomorphic system by changes in the system itself through time. It is the change in the geomorphic system itself that is most important, because until the system has evolved to a critical situation, adjustment or failure will not occur. It may not always be clear whether the system is responding to geomorphic thresholds or to an external influence, but when a change of slope is involved, the control is geomorphic, and the change whereby the threshold is achieved is intrinsic to the system.

The geomorphic threshold, as defined, is clearly an intrinsic threshold.

The concept of geomorphic thresholds, which involve landform change without a change in external controls, challenges the well-established basis of geomorphology—that landform change is the result of some climatic, base-level, or land-use change. Therefore, the significance of the geomorphic threshold conception for the geologist is that it makes him aware that abrupt erosional and depositional changes can be inherent in the normal development of a landscape and that a change in an external variable is not always required for a geomorphic threshold to be exceeded and for a significant geomorphic event to ensue.

It is important to stress again that the geomorphic threshold as defined above is an intrinsic threshold. If, as a result of a climate change, discharge and flow velocities in a stream channel are increased, the resulting bank erosion and meander cutoffs may convert the meandering stream to a braided channel. In this case the cause of the pattern conversion is extrinsic, but if as a result of increasing channel sinuosity under unchanging discharge conditions, cutoffs occur which convert a meandering reach to a straight reach, the control is intrinsic. In the first case the change may be regional in nature, and in the second it may be local. However, explanation of locally anomalous erosional or depositional features is what is required for rational landform management.

During recent discussions with students and colleagues, it became apparent that the term "geomorphic threshold" is used in a much broader sense than was originally intended, and therefore a revision of this concept may be required. The concept of intrinsic controls was stressed because it initially provided a new approach to the understanding of the details of the landscape that could be used for predictive purposes. Nevertheless, there can also be extrinsic geomorphic thresholds. For example, in common usage "thresholds" can be the result of either cause or effect. That is, we speak of hydraulic, velocity, shear, and stream power thresholds above which sediment moves or banks fail, but we can also speak of bank, channel, and slope stability thresholds, when the forces causing the failures are not clearly identified and understood. Therefore, geomorphic thresholds can be of both types, and they can be redefined in the following way. A geomorphic threshold is a threshold of landform stability that is exceeded either by progressive change of the landform itself, that is, an intrinsic change, or by a change in an external variable. One concern is that a valuable concept that can be useful in the interpretation of the erosional and depositional details of a landscape may be so diluted by the broadening of the definition that its significance is lost. Nevertheless, if the two types of geomorphic thresholds, extrinsic and intrinsic, are identified when the term is used, confusion should be minimized.

APPLICATIONS

If the concept of intrinsic geomorphic thresholds is valid, it should have application to both geomorphic and land management problems. Such applications will be considered in the following sections.

Gully Erosion

The concept of geomorphic thresholds is useful in identifying those conditions at which a landform is incipiently unstable. Following this identification, some action can be taken either to prevent failure from occuring or to minimize the effect of the change when it does occur. For example, regime equations are an attempt to design a stable channel based on the strength of materials forming the perimeter of a canal. A critical velocity is identified above which erosion will take place, and the canal is designed so that this velocity will not be exceeded.

Erosion begins at some extrinsic threshold condition of tractive force or shear. Recent work by Graf (1978) in small valleys near Evergreen, Colorado, has demonstrated that one can locate the threshold value of tractive force at which a valley will be subjected to gullying. Figure 2 shows his relationship between the density of vegetation on the valley floor (biomass) and tractive force. It is clear that for a given biomass there is a critical tractive force at which incision of the valley floor will take place, and for a given tractive force there is a minimum value of biomass below which the valley floor becomes unstable and gullying begins.

Changes of both biomass and tractive force can be the result of land-use changes and climatic fluctuations, which weaken valley-floor vegetation and increase peak discharge and tractive force. However, with external influences unchanged, the tractive force can be increased if the valley-floor slope is steepened by aggradation, an intrinsic control.

Elsewhere in western Colorado, Patton and I (1975) have reported on a relationship between drainage area and valley-floor inclination, which in these small semiarid valleys permits identification of a threshold valley-floor slope above which incision of the valley floor is likely to take place. This situation provides an excellent example of an intrinsic geomorphic threshold. At present, research is being carried out in several field areas in Colorado and Wyoming in an attempt to substantiate this type of relationship for other areas. However, the task is difficult because in small drainage basins there are other factors that obscure this relationship. For example, unless climate, land use, and vegeta-

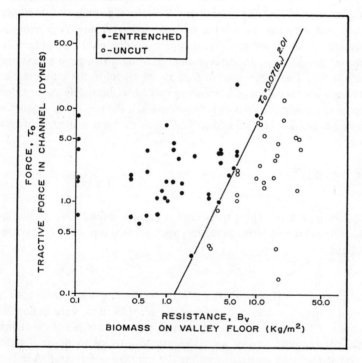

FIGURE 2 *Tractive force (calculated for flood of 10-year recurrence interval) and biomass of valley floors with discriminating function separating stable and unstable valley floors and channels. (After Graf, 1978.)*

tive cover are essentially uniform throughout an area, the valley-floor slope criterion may be masked.

Fan-head trenches are another erosional feature of variable distribution, and within a given region some fans are trenched and some are not trenched (Bull, 1968). The usual explanation for these features involves the extrinsic controls of climate, tectonics, and land use, but strong geomorphic controls may also be present. For example, during experimental studies of fan growth, precipitation was delivered to a sediment source area at a constant rate. The resulting runoff and sediment moved out of the miniature drainage basin to a piedmont area, where a miniature fluvial fan was formed. The fan is termed fluvial rather than alluvial because the streams that were carrying the water and sediment from the source area were perennial. During fan growth, the fan was trenched repeatedly, as the fan head steepened due to aggradation and then adjusted to this oversteepened condition by trenching (Fig. 3). It should be emphasized that the trenching and the reworking of the sediment and the subsequent backfilling of the channel by aggradation were not due to any change in the experimental procedure or to changes in the intensity or duration of the precipitation applied to the source area. The fan-head trenching occurred as a result of the oversteepening of the fan head, and the trenching was the result of the exceeding of an intrinsic geomorphic threshold.

It is important to realize that the threshold slope as indicated on Figure 3 will probably not be a well-defined line; rather there will be a threshold zone between trenched and untrenched fan heads that will depend on vagaries of hydrology, land use, and lithology of the source area.

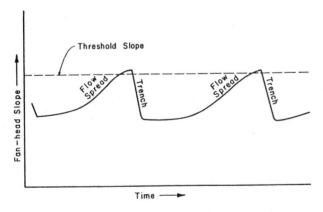

FIGURE 3 *Change in fan-head morphology through time, during an experimental study of fan growth. When flow is spread at the fan head, deposition increases the slope of the fan head to a critical slope, then the fan head is trenched and the flow is confined to the trench. Backfilling of the trench leads to a spread flow condition at the fan head. This sequence of events occurred repeatedly during the study.*

Sediment Yield

If the incision of a dendritic drainage pattern into an area of homogenous geology occurs progressively, sediment yield probably will increase exponentially, as size, length, and drainage density increase through time. However, if relief is high at some stage in the development, slope steepness and height may be such that a slope stability threshold will be exceeded, and the production of sediment may increase by an order of magnitude as extensive mass wasting begins. Over much shorter spans of time the variability in sediment yield will be related to both intrinsic and extrinsic controls.

During experimental studies of drainage pattern development (Schumm, 1977), variations in sediment yield were related to storage and flushing of sediment, as sediment was scoured in a manner analogous to the previously described fan-head trenching. Sediment was stored and then flushed as valley-floor slope thresholds were exceeded. In nature this type of erosional and depositional behavior would render any long-term predictions concerning sediment yields highly suspect. Unfortunately, sediment yield records are not sufficiently long to provide information on variations such as have been noted experimentally, but the conclusions of Trimble (1975) clearly indicate that the storage and subsequent flushing of stored sediment is a fact in high-sediment-producing areas. Certainly, the episodic erosion, as described for the small Douglas Creek drainage basin (Womack and Schumm, 1977), will produce significant short-term variations of sediment production.

Another factor that is significant in this regard is the production of salts and their introduction into the Colorado River as suspended and dissolved load. The high-sediment-producing drainage basins in the Colorado River basin are also high-salt-producing areas, which is a reflection of lithology (Laronne, 1977), but gullying provides a conduit for movement of the salts from saline bedrock outcrops into the river system.

With the concept of intrinsic thresholds in mind, it appears that sediment and salt production will increase dramatically, as valley floors are trenched, and that both sediment and salt delivery to the Colorado River will be at a maximum when stored alluvial deposits are being removed by erosion. Subsequently, deposition in the trenches will isolate the areas of high sediment and salt production, and salt and sediment delivery will decrease until the next phase of trenching. Untrenched valleys near the threshold slope can be identified, and standard erosion control practices could be used not only to reduce the sediment yields, but also salt delivery to the Colorado River.

Geomorphology—Terraces and Pediments

If as described above, channels incise as the result of the exceeding of intrinsic thresholds, it may not be possible to correlate some erosional and depositional surfaces except locally. For example, it is well known that in a piedmont region, stream capture causes dissection and abandonment of terraces (Ritter, 1972) and pediments (Rich, 1935; Hunt et al., 1953).

Figure 4 shows such a sequence of pediment formation at the base of an escarpment. This resembles the situation that exists along the base of the Book Cliffs between Grand Junction, Colorado, and Price, Utah. Streams draining from the Book Cliffs transport coarse sediments, and they require a steeper gradient than the piedmont streams that are eroding into shale (Fig. 4a). The aggressor shale streams therefore have a lower gradient, and the higher streams are vulnerable to capture by shale streams (Fig. 4 b and c).

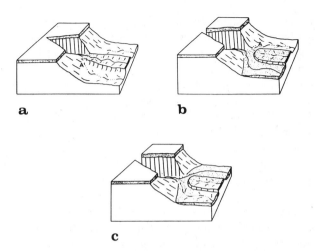

a   b

c

FIGURE 4 *Development of multiple pediments by stream capture along an escarpment. Capture of the high-level stream by the southern stream A will form a second pediment level (a, b). Continued erosion of the weaker piedmont rocks by the northern stream A will eventually cause capture and shift of the channel to the northern lowland to form a third pediment (c). (From Carter, 1978.)*

There are four erosion surfaces that have been preserved on the Mancos Shale in this area. The high-level surfaces are beautifully developed pediments with very clear contacts between the shale bedrock and the overlying gravels. In a longitudinal section the pediments sweep away from the cliffs in a smooth curve; however, where the contact can be seen in transverse sections, the bedrock surface of the pediment is very irregular (Carter, 1978), but at most locations the irregularity of the bedrock surface is masked by the alluvial or gravel deposits above it.

The shale streams at the base of the scarp (Fig. 4) open out a broad piedmont area at a level below the gravel-bearing streams, and their tributaries produce an irregular shale surface at this low level. When capture of the high-level stream takes place, the gravels from the scarp inundate this lower surface and convert it from an irregular bedrock surface to a smooth gravel-covered surface that is a new lower-level pediment surface.

The multiple pediment levels in this area can be attributed to stream capture without any external influence, although it is probable that during the erosional history of the area changes in the level of the Colorado River and climate changes have influenced pediment development. Nevertheless, the pediment surfaces can be explained primarily, as a result of the normal and intrinsic process of stream capture in this area.

The concept of geomorphic thresholds and eposodic erosion is particularly important in this respect, because the geomorphologist cannot automatically assume that each erosional or depositional discontinuity in the landscape is a result of external influence.

River Patterns

The variability of river patterns such as sinuosity variations and pattern changes from meandering to braided provide excellent examples of the effect of both intrinsic and extrinsic threshold conditions. Fisk (1944) and Winkley (1970) show that the sinuosity and length of the Mississippi River varies dramatically through time. Sinuosity decreases to a minimum when an avulsion or a series of cutoffs straighten the channel. Such changes can be related to major changes of sediment load or an increase of peak discharge, but they can also be due to a progressive increase of sinuosity with an accompanying reduction of channel gradient to the point that aggradation and cutoffs or avulsion results. Such a situation appears to exist along the sinuous parts of the Rio Puerco arroyo, New Mexico, where meander amplitude has increased to the point that in some reaches sediment is being deposited in the upstream limb of each meander, and the bends are being cut off. These changes reflect an intrinsic control by the channel pattern itself.

The work of Lane (1957) and Leopold and Wolman (1957) indicates that there is a gradient or discharge threshold above which rivers tend to be braided. The experimental work reported by Schumm and Khan (1972) showed that for a given discharge, as valley-floor slope is progressively increased, a straight river becomes sinuous and then eventually braided at high values of stream power and sediment transport. Rivers that are situated close to the meandering-braided threshold should have a history characterized by transitions in morphology from braided to meandering, and vice versa. For example, the Cimarron River in southwestern Kansas during the 1890s had a very sinuous, relatively narrow, and deep channel. It was converted to a very wide braided stream during the 1930s. It is now narrowing, and in some reaches it is resuming its sinuous pattern (Schumm and Lichty, 1963). These changes have been explained by changes in peak discharge, but the channel adjustments suggest that the Cimarron River is a river near the pattern threshold.

The identification of rivers that are near the pattern threshold would be useful, because a braided river near the threshold might be converted to a more stable single-thalweg stream. On the other hand, meandering stream near the threshold should be identified in order that steps could be taken to prevent braiding due perhaps to changes of land use.

An example of pattern instability is provided by the lower Chippewa River in Wisconsin (Schumm and Beathard, 1976). In lower 25 km of its course the Chippewa River is braided, whereas the upper reaches of the river are meandering (Fig. 5). The Chippewa River is of considerable interest because it introduces large quantities of sediment into the Mississippi River below Lake Pepin. The position of the Chippewa River with reference to the relations of Leopold, and Wolman and Lane are shown on Figures 6 and 7. The braided part of the lower Chippewa River plots below the threshold line of Leopold and Wolman and in the intermediate range of Lane, suggesting that it should be a meandering stream. Therefore, it seems that the lower Chippewa could be converted to a meandering channel and some of the sediment that is now moving into the Mississippi could be stored in point bars and in the floodplain of the lower Chippewa River. The pattern change could be accomplished by concentration of the flow into a sinuous thalweg that would be maintained by training dikes and bank stabilization work. Vegetation would stabilize the alternate bar deposits to form point bars, and the transition to a sinuous channel would be complete.

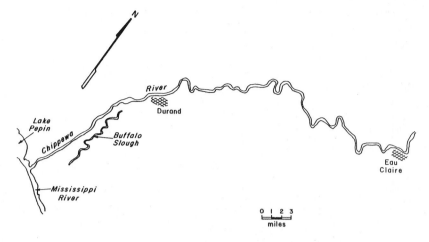

FIGURE 5 *Map of lower Chippewa River. The river is braided below Durand, Wisconsin. Buffalo Slough is an old course of the Chippewa River.*

Since the foregoing suggestions were made (Schumm and Beathard, 1976), more detailed studies of the Chippewa River basin indicate that only a 10% reduction in sediment delivery to the Mississippi would be accomplished by developing the single-thread sinuous channel. Therefore, upstream sediment productions must be controlled, especially where the upper Chippewa River is cutting into high Pleistocene outwash terraces. If the contribution of sediment from these sources were reduced, the lower Chippewa could resume its sinuous mode because historic evidence indicates that some time in the nineteenth century the Chippewa flowed through Buffalo Slough and it has a sinuosity of about 1.3 (Fig. 5).

An indication that the pattern conversion of the Chippewa could be successful if the upstream sediment sources were controlled is provided by the Rangitata River of New Zealand. The Rangitata River is the southernmost of the major rivers that traverse the Canterbury Plain of the South Island. It leaves the mountains through a bedrock gorge. Above the gorge, the valley of the Rangitata broadens, and it is choked with glacially derived sediment. Above the gorge the Rangitata is braided, and it appears that the Rangitata should be a braided stream below the gorge, as are all the other rivers that cross the Canterbury Plain. However, below the gorge, the Rangitata is meandering. A few miles farther downstream, the river cuts into high Pleistocene outwash terraces, and it abruptly converts from a meandering to a braided stream. The braided pattern persists to the sea. If the Rangitata could be isolated from the gravel terraces, it probably could be converted to a single-thalweg sinuous channel, because the Rangitata is obviously a river near the pattern threshold.

Other New Zealand rivers are near the pattern threshold, and therefore they are susceptable to pattern change. In fact, New Zealand engineers are attempting to accomplish this pattern change in order to produce a "single-thread" channel which will reduce flood damage and be less likely to acquire large sediment loads from their banks and

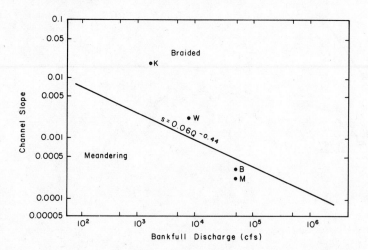

FIGURE 6 *Leopold and Wolman's (1957) relation among channel patterns, channel gradient, and bankfull discharge. The letters B and M identify the braided and meandering reaches of the Chippewa River. The letters K and W refer to the Kowhai and Wairau Rivers.*

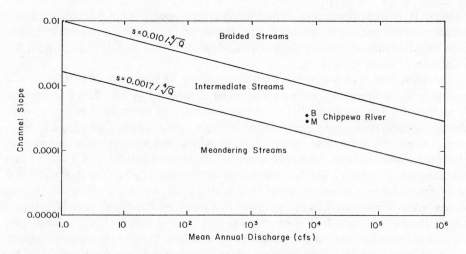

FIGURE 7 *Lane's (1957) relation among channel patterns, channel gradient, and mean discharge. The regression lines were fitted to data from streams that Lane classified as highly meandering and braided. Between the two parallel lines are located streams of intermediate character, ranging from meandering to braided. The letters B and M identify the positions of the braided and meandering reaches of the Chippewa River.*

482

terraces. For example, the engineers of the Marlborough Catchment Board in Blenheim, New Zealand, have had success in converting the Wairau River, a major braided stream, from its uncontrolled braided mode to that of a slightly sinuous, single-thalweg, relatively more stable channel (Pascoe, 1976). The increase in sinuosity was only from 1.0 to 1.05, and this was accomplished by the construction of curved training banks. On Figure 6 the Wairau River plots close to the threshold line, and with the reduction of sediment load produced by bank stabilization it appears that the pattern threshold has been crossed successfully.

Farther to the south, near Kaikoura, the Kowhai River is being modified in the same manner as was the Wairau (Thompson and MacArthur, 1969), although much of its sediment load in the Wairau River is derived from bank and terrace erosion, which can be controlled, high sediment loads are delivered to the Kowhai River directly from steep and unstable mountain slopes. On Figure 6 the Kowhai River plots well above the threshold line, and without a major reduction in upstream sediment it may be difficult to maintain a single-thalweg channel at this location.

The variability of the Rangitata River pattern indicates that braided to single-thalweg conversions should be possible for the Chippewa and Wairau Rivers. Whether all braided rivers can be so readily modified depends on their position with regard to the line defining pattern threshold on Figure 6.

SUMMARY

Geomorphic thresholds can be of two types, intrinsic or extrinsic, depending on whether the crossing of the threshold is due to influences that are internal or external to the landscape. Fan-head trenching, arroyo cutting, gully development, stream capture, pattern change, meander cutoffs can be due to both, and it is the geomorphologist's responsibility to identify each type and to advise the engineer and land manager how to work with rather than against the natural forces that modify landforms.

Erosion or deposition is not always progressive, and the crossing of thresholds in high-relief situations may lead to episodic erosion or deposition, which can significantly affect sediment yields and the correlation of pediments, terraces, and alluvial deposits.

As demonstrated, threshold and episodic erosion concepts have practical implications. They can be used to explain anomalous erosion features and to predict future erosional and deposition changes that will occur as a result of man's activities or as a normal part of the erosional and depositional evolution of a landscape.

ACKNOWLEDGMENTS

Dr. Frank Ethridge reviewed this paper, and his constructive suggestions were most useful during its revision. In addition, I thank Thomas Carter for the opportunity to use his unpublished illustration of pediment formation.

Reference is made herein to research that was supported by grants from the U.S. Army Research Office, National Science Foundation, Office of Water Research and Technology, and the Colorado Agricultural Experiment Station.

REFERENCES CITED

Bull, W. B., 1968, Alluvial fans: J. Geol. Ed., v. 16, p. 101–106.

Carson, M. A., 1971, Application of the concept of threshold slopes to the Laramie Mountains, Wyoming: Inst. Brit. Geogr. Spec. Publ. 3, p. 31–47.

Carter, T., 1978, Pediments of the Book Cliffs: M.S. thesis, Colorado State Univ., in preparation.

Fisk, H. N., 1944, Geological investigation of the alluvial valley of the lower Mississippi River: Mississippi River Commission, Vicksburg, Miss., 78 p.

Graf, W. L., 1978, The development of montane arroyos and gullies: Earth Surface Processes, in press.

Hunt, C. B., Averitt, P., and Miller, R. L., 1953, Geology and geography of the Henry Mountains Region, Utah: U.S. Geol. Survey Prof. Paper 228, 239 p.

Kirkby, M. J., 1973, Landslides and weathering rates: Geologia Applicata e Idrageologia, v. 8, p. 171–183.

Koons, D., 1955, Cliff retreat in southwestern United States: Amer. J. Sci., v. 253, p. 44–52.

Lane, E. W., 1957, A study of the shape of channels formed by natural streams flowing in erodible material: M. R. D. Sediment Series 9, U.S. Army Engineer Div., Missouri River, Corps of Engineers, Omaha, Nebr.

Laronne, J. B., 1977, Dissolution potential of surficial Mancos Shale and alluvium: unpublished Ph.D. dissertation, Colorado State Univ., 128 p.

Leopold, L. B., and Wolman, M. G., 1957, River channel patterns: braided, meandering and straight: U.S. Geol. Survey Prof. Paper 282-B, p.

Meier, M. F., and Post, A., 1969, What are glacier surges?: Can. J. Earth Sci., v. 6, p. 807–816.

Pascoe, L. N., 1976, The training of braided shingle rivers into a single thread channel with particular reference to the middle reach of the Wairau River: unpublished manuscript.

Patton, P. C., and Schumm, S. A., 1975, Gully erosion, northern Colorado: a threshold phenomenon: Geology, v. 3, p. 88–90.

Rich, J. L., 1935, Origin and evolution of rock fans and pediments: Geol. Soc. Amer. Bull., v. 46, p. 999–1024.

Ritter, D. F., 1972, The significance of stream capture in the evolution of a piedmont region: Z. Geomorphol., v. 16, p. 83–92.

Schumm, S. A., 1973, Geomorphic thresholds and complex response of drainage systems: in Morisawa, M., ed., Fluvial geomorphology: SUNY, Binghamton, Publ. Geomorphol., p. 299–310.

——, 1976, Episodic erosion: a modification of the geomorphic cycle: in Flemal, R., and Melhorn, W., ed., Theories of landform development: SUNY, Binghamton, N.Y., Publ. Geomorphol., p. 69–85.

——, 1977, The fluvial system, New York, Wiley-Interscience, 338 p.

——, and Beathard R. M., 1976, Geomorphic thresholds: an approach to river management: in Rivers 76, v. 1, 3rd Symp. Waterways, Harbors and Coastal Eng. Div., Amer. Soc. Civil Engineers, p. 707–724.

——, and Khan, H. R., 1972, Experimental study of channel patterns: Geol. Soc. Amer. Bull., v. 83, p. 1755–1770.

——, and Lichty, R. W., 1963, Channel widening and floodplain construction along Cimarron River in Southwestern Kansas: U.S. Geol. Survey Prof. Paper 352-D, p. 71–88.

Thompson, P. A., and MacArthur, R. S., 1969, Major river control, drainage and erosion control scheme for Kaikouia: Marlborough Catchment Board Report. Blenheim, New Zealand, 96 p.

Trimble, S. W., 1975, Denudation studies: can we assume stream steady state?: Science, v. 188, p. 1207–1208.

Winkley, B. R., 1970, Influence of geology on the regimen of a river: American Soc. Civil Eng., National Water Resources Meeting (Memphis), preprint 1078, 35 p.

Womack, W. R., and Schumm, S. A., 1977, Terraces of Douglas Creek, northwestern Colorado: an example of episodic erosion: Geology, v. 5, p. 72–76.

Porosity, 349, 354–357: diffuse flow aquifer, 355; secondary, 354
Power spectral analysis, 185
Precipitation: forecast, 204; La Porte Anomaly, 182; La Porte histograms, 183
Precipitation changes: chemistry, 466; man-induced, 464–465
Prigogine's rule, 210, 215
Probability density function, 197
Probability distribution, 187, 198
Process curve, 59
Public Law 92-500, Section 208, 457, 467
Puerto Rico, 349

Quasi-equilibria, 210
Quasi-equilibrium, 6, 259
Quebec, 33, 360

Radioactive waste storage, 436: time dimensions, 436
Rainfall intensity: infiltration, 397; Laguna Beach, 397; patterns, 182, 201–202
Rainfall records, Hurst phenomenon, 191
Rainsplash, 67
Ratios, 260: allometric approach, 260; denominator, 260, 262; numerator, 260, 262
Rayleigh number, 215
Regime channels, 245
Regime equations, 475
Rest and rotation, 421
Reynolds number, 15, 215, 324, 352
Richardson number, 215
Riffles, 251
Rill, 75, 227: cross section, 352; definition, 75; development, 45; gradients, 354; length, 352, 354; quadratic parabola, 352
Rio Puerco arroyo, 480
Riparian trees, ice scarred, 328
River ice, 323: field problems, 324; frazil ice, 327; jams, 324; processes, 323, 341; thresholds, 323
River ice drives, landform response, 324–325
River patterns, 479: braided, 480–481, 483; instability, 480; meandering, 480–481; sinuosity, 480–481, 483
River response, 250
Riverine: lands, 27; processes, 132
River-mouth bars, 143
Rivers: Alberta, 323–324, 328, 330–331; Athabasca, 338; Beaver, 328, 330; Bow, 331, 333, 339; braided systems, 325; Chippewa, 480, 482; Cimarron, 480; Colorado, 3, 11, 27, 231–234, 250, 253, 478–479; Des Moines, 106; Kicking Horse, 339; Kowhai, 483; Laird, 328, 335; Little Red Deer, 334; Mackenzie, 328, 337–338; Madison, 339–340; Medicine, 328; Mississippi, 9, 19, 27–28, 137, 480; Mojave, 423; Nelson, 328; Neva, 339; Niger, 190–191; Nile, 45, 179–182; North Platte, 14; North Saskatchewan, 340; Rangitata, 481; Red Deer, 333, 338; St. Lawrence, 181, 190–191, 339, 441; Salmon, 339; scientific investigation, 27; South, 104–108; South Saskatchewan, 340;

Susquehanna, 301; Swan, 330; Tarkio, 99; Wairau, 483; Western Canada, 323; Willow, 17; Yellow, 28; Yukon, 328, 338
Rock glaciers, 270, 290
Rock solution, 345: characteristics, 347; mechanical strength, 351; weakness, 397
Rockfalls, 267, 269, 291–292: acquisition of data, 274–275; causal factors, 270, 290; classification, 270; definition, 269–271; environmental association, 290–291; favorable conditions, 271; frequency, 279–284, 288, 291; high-magnitude, 278, 282, 284, 288, 290, 292; inventories, 270; landforms, 276–279; microclimate, 291; motion, 269; rates, 288, 290; scarp slopes, 271, 277; spatial distribution, 284–288, 291; thrust faulting, 290
Rockslides, 267, 269, 291–292: acquisition of data, 274–275; catastrophic, 269; causal factors, 270, 290; classification, 270; definition, 269–271; dip slopes, 271, 277; favorable conditions, 271; frequency, 279–284, 288; high-magnitude, 278, 282, 284, 288, 290, 292; inventories, 270; landforms, 276–279; motion, 269; rates, 288, 290; spatial distribution, 284–288, 291
Rogen moraines, 314–315: ice-thrust ridge, 315
Runoff, mean annual, 359
Russian shield, 44

Sahara, 47
Sahel, 16
Salt, 478: accumulation, 12
Salton Sea, 11, 27
San Juan mountains, 33, 270
Sand boil, 19
Sand deposits, stabilization, 374
Sand storms, 424
Santa Ana winds, 418
Saskatchewan, 351
Satellite photography, 415
Scablands, 45
Scandinavia, 466
Scotland, Cairngorms, 319
Sea level, 298–299, 363: budget, 383; changes, 363–364; deficit, 374; rise, 363–364, 382–383; tectono-eustatic effects, 46; Wisconsinan, 363
Sea-floor deformations: bottleneck slide, 141–142, 163; chutes, 143; collapse depressions, 139–140, 163; crown cracks, 142; retrogressive elongate slides, 142–143, 163
Sea-floor spreading, 44
Sediment, 48, 132–133, 452, 461: abrasion, 246, 251; bed-material, 232; bimodal distribution, 252–253; capacity grain size, 235, 240–241, 252, 254; chute, 162; concentration, 240; critical grain size, 235, 254; deformational features, 131, 144, 148; delivery rate, 462; delivery ratio, 439; flushing, 104, 122–123; gradients, 244; grain size, 252; grain-size range, 233; load, 251; log-normal distribution, 233; particle size, 439; pollutants, 437; production, 439; quantity supplied, 233; size deficiency, 252; specific